Long Wavelength Infrared Detectors

Optoelectronic Properties of Semiconductors and Superlattices

A series edited by *M. O. Manasreh,* Phillips Laboratory, Kirtland Air Force Base, New Mexico, USA

Volume 1
Long Wavelength Infrared Detectors
Edited by Manijeh Razeghi

In preparation

GaN and Related Materials
Edited by Stephen J. Pearton

Structural and Optical Properties of Porous Silicon Nanostructures
Edited by H. J. von Bardeleben

Characterization of Reduced Dimensional Semiconductor Microstructures
Edited by Fred H. Poliak

Antimonide-Related Strained Layer Heterostructures
Edited by M. O. Manasreh

The Optics of Semiconductor Quantum Wires and Dots: Fabrication, Characterization, Theory and Application
Edited by Garnett W. Bryant

Vertical Cavity Surface-Emitting Lasers and Their Applications
Edited by Julian Cheng and Niloy K. Dutta

Long Wavelength Infrared Detectors

Edited by

Manijeh Razeghi

Northwestern University
Evanston, Illinois, USA

CRC Press
Taylor & Francis Group
Boca Raton London New York

CRC Press is an imprint of the
Taylor & Francis Group, an **informa** business

First published 1996 by Gordon and Breach Science Publishers

Published 2020 by CRC Press
Taylor & Francis Group
6000 Broken Sound Parkway NW, Suite 300
Boca Raton, FL 33487-2742

First issued in hardback 2020

ISBN 13: 978-1-138-45587-0 (hbk)
ISBN 13: 978-2-88449-209-6 (pbk)

This book contains information obtained from authentic and highly regarded sources. Reasonable efforts have been made to publish reliable data and information, but the author and publisher cannot assume responsibility for the validity of all materials or the consequences of their use. The authors and publishers have attempted to trace the copyright holders of all material reproduced in this publication and apologize to copyright holders if permission to publish in this form has not been obtained. If any copyright material has not been acknowledged please write and let us know so we may rectify in any future reprint.

**Visit the Taylor & Francis Web site at
http://www.taylorandfrancis.com**

**and the CRC Press Web site at
http://www.crcpress.com**

British Library Cataloguing in Publication Data

Long wavelength infrared detectors. – (Optoelectronic
    properties of semiconductors and superlattices; v.1)
    1. Infrared detectors
    I. Razeghi, M. (Manijeh)
    621.3'62

CONTENTS

ABOUT THE SERIES

The series *Optoelectronic Properties of Semiconductors and Superlattices* provides a forum for the latest research in optoelectrical properties of semiconductor quantum wells, superlattices, and related materials. It features a balance between original theoretical and experimental research in basic physics, device physics, novel materials and quantum structures, processing, and systems—bearing in mind the transformation of research into products and services related to dual-use applications. The following sub-fields, as well as others at the cutting edge of research in this field, will be addressed: long wavelength infrared detectors, photodetectors (MWIR–visible–UV), infrared sources, vertical cavity surface-emitting lasers, wide-band gap materials (including blue-green lasers and LEDs), narrow-band gap materials and structures, low-dimensional systems in semiconductors, strained quantum wells and superlattices, ultrahigh-speed optoelectronics, and novel materials and devices.

The main objective of this book series is to provide readers with a basic understanding of new developments in recent research on optoelectrial properties of semiconductor quantum wells and superlattices. The volumes in this series are written for advanced graduate students majoring in solid state physics, electrical engineering, and materials science and engineering, as well as researchers involved in the field of semiconductor materials, growth, processing, and devices.

PREFACE

Infrared (IR) radiation is the most common form of electromagnetic radiation. The number of photons emitted per second (with arca unit $(10^{18}\,\mathrm{cm}^{-2})$) by a 300 K blackbody in the IR region is higher than the number of photons in the visible region. This radiation is less absorbed and scattered in the atmosphere than visible light, and provides important information about objects: their position in space, temperature, geometry, surface, distance, composition, and atmosphere.

. Recent successes in applying IR technology to remote sensing problems have been made possible by the successful development of high-performance infrared detectors over the past six decades. Infrared detectors have found military as well as civilian applications in thermal imaging, guidance, reconnaissance, surveillance, ranging, and communication systems.

The history of IR detector development follows the initial discovery of infrared radiation by Herschel in 1800, using thermometers. The thermometer was the first of a trio of detectors that were to dominate the infrared detector field until World War I. In 1821, Seebeck discovered the thermoelectric effect, and soon thereafter demonstrated the first thermocouple. In 1829, Nobili constructed the first thermopile by connecting a number of thermocouples in a series. Macedonio Melloni helped him to modify the design of series-connected thermocouples in 1833. The third member of the trio, Langley's bolometer, appeared in 1881.

The photoconductive effect was discovered by Smith in 1873, when he experimented with selenium as an insulator for submarine cables. This discovery provided a fertile field of investigation for several decades, although most of the efforts were of doubtful quality. Work on the infrared photovoltaic effect in naturally occurring lead sulfide or galena was announced by Bose in 1904; however, this effect was not used in a radiation detector for the next several decades.

The first infrared photoconductor of high responsivity was developed by Case in 1917. He discovered that a substance composed of thallium and sulfur exhibited photoconductivity. Later, he found that the addition of oxygen greatly enhanced the response. However, instability of resistance in the presence of light or polarizing voltage; loss of responsivity due to over-exposure to light; high noise; sluggish response; and lack of reproducibility seemed to be inherent weaknesses. Around 1930, the appearance of the Cs-O-Ag phototube (with more stable characteristics) to a great extent discouraged further development of photoconductive cells, until about 1940.

At that time, interest in improved detectors arose in Germany. In 1933 at the University of Berlin, Kutzscher discovered that lead sulfide (from natural galena found in Sardinia) was photoconductive and had response to about 3 μm. This work was done under great secrecy, of course, and the results were not generally known until after 1945. Lead sulfide was the first practical infrared detector deployed in a variety of applications during World War II. In 1941, Cashman improved the technology of thallium sulfide detectors, which led to successful production. Cashman, after success with thallium sulfide detectors, concentrated his efforts on lead sulfide and after the war found that other semiconductors of the lead salt family (PbSe and PbTe) showed promise as infrared detectors. Lead sulfide photoconductors were brought to the manufacturing stage of development in Germany in about 1943. They were first produced in the United States at Northwestern University in 1944 and, in 1945, at the Admiralty Research Laboratory in England.

The years during World War II saw the origins of modern infrared detector technology. The dates given in Fig. 1 show the chronology of significant development efforts on the materials mentioned. Interest has centered mainly on the wavelengths of the two atmospheric windows from 3–5 μm and 8–14 μm, though in recent years there has been increasing interest in longer wavelengths, stimulated by space applications.

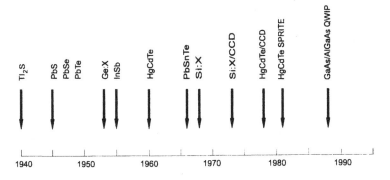

Figure 1   Chronology of infrared detector materials development.

During the 1950s, infrared detectors were built using single-element-cooled lead salt detectors, primarily for anti-air-missile seekers. Usually, lead salt detectors were polycrystalline and were produced by vacuum evaporation and chemical deposition from a solution, followed by a post-growth sensitization process. The preparation process of lead salt photoconductive detectors usually was not well understood, and reproducibility could be achieved only after following well-tried recipes. The first extrinsic photoconductive detectors were reported in the early 1950s, after the discovery of the transistor, which stimulated a considerable improvement in growth and material purification techniques. Since the techniques for controlled impurity introduction became available for germanium at an earlier date, the first high-performance extrinsic detectors were based on the use of germanium. Extrinsic photoconductors were widely used at wavelengths beyond 10 um, prior to the development of intrinsic detectors. Extrinsic detectors must be operated at lower temperatures to achieve performance similar to that of intrinsic detectors, and a sacrifice in quantum efficiency is required to avoid impractically thick detectors.

At the same time, rapid advances were being made in narrow-band gap semiconductors that would later prove useful in extending wavelength capabilities and improving sensitivity. The first such material was InSb, a member of the newly discovered III-V compound semiconductor family. The interest in InSb stemmed not only from its small energy gap, but also from the fact that it could be prepared in single-crystal form using a conventional technique. The end of the 1950s saw the introduction of semiconductor alloys in III-V, IV-VI, and II-VI material systems. These alloys allowed the band gap of the semiconductor, and hence the spectral response of the detector, to be custom tailored for specific applications. In 1959, research by Lawson and coworkers triggered development of variable band gap $Hg_{1-x}Cd_xTe$ (HgCdTe) alloys, providing an unprecedented degree of freedom in infrared detector design. HgCdTe detectors are now at the center of a major industry with a worldwide turnover of billions of dollars.

As photolithography became available in the early 1960s, it was applied to the making of infrared detector arrays. Linear array technology was first applied to PbS, PbSe, InSb, and Hg-doped germanium detectors. In the late 1960s and early 1970s, first-generation linear arrays (in which an electrical contact for each element of a multielement array is brought off the cryogenically cooled focal plane to the outside, where there is one electronic channel at ambient temperature for each detector element) of instrinsic HgCdTe photoconductive detectors were developed. These allowed LWIR forward looking infrared (FLIR) systems to operate with a single-stage cryoengine, making the system much more compact, lighter, and requiring significantly less power consumption.

In 1973, after the invention of charge-coupled devices (CCDs) by Boyle and Smith, Shepherd and Yang proposed the material-silicide/silicon Scottky-barrier detectors. For the first time, it became possible to have much more sophisticated readout schemes—both detection and readout could be implemented in one common silicon chip. These second-generation systems have at least two orders of magnitude more detector elements on the focal plane than first-generation systems, and the detector elements are in a two-dimensional array configuration rather than a linear configuration; these staring arrays are scanned electronically by circuits integrated with the arrays.

In the late 1970s, and through the 1980s, HgCdTe technology efforts focused almost exclusively on photovoltaic device development, because of the need for low power dissipation and high impedance in large arrays to interface to readout input circuits. This effort is finally paying off, with the birth of HgCdTe second-generation infrared systems that provide large two-dimensional arrays in both linear formats with time delay and integration for scanning imagers, and in square and rectangular formats for staring arrays. 640×840 hybrid HgCdTe focal plane arrays (FPAs) have been produced recently. However, HgCdTe has the most serious technological problems of any semiconductor material in mass production. The present HgCdTe FPAs are limited by the yield of arrays, which increases their cost. The difficulties with this material have made it desirable to examine other material systems to see whether performance can be improved.

Since the initial proposal in 1970 by Esaki and Tsu and the advent of two major epitaxial technologies, molecular beam epitaxy (MBE) and metal organic vapor phase epitaxy (MOVPE), interest in low dimensional structures (LDS) has increased continuously over the years, driven by technological challenges, new physical concepts and phenomena, and promising applications. LDS has formed a new branch of physics research since it has become possible to fabricate material structures whose dimensions are comparable with interatomic distances in solids. Their electronic properties are significantly different from the same material bulk form, and these properties are changed by quantum effects.

The intent of this book is to give the optical science and research community a report on the status of current efforts to investigate new ternary alloy systems and quantum well infrared photodetectors (QWIPs). A distinguishing feature of QWIPs is that they can be implemented in chemically stable wide-band gap materials, as a result of the intraband processes used. Therefore, we can use material systems such as GaAs/AlGaAs, InAs/GaInSb, GaInP/GaAs, InGaAs/InAlAs, and SiGe/Si, as well as other systems, although most experimental work has been carried out with GaAs/AlGaAs. These materials have fewer processing problems than HgCdTe, and achievement of monolithic FPAs could also be implemented. Some of the devices are sufficiently advanced that the possibility exists of their incorporation into high-performance integrated circuits. High uniformity of epitaxial growth over large areas shows promise for the production of large-area two dimensional arrays. In addition, flexibility associated with control over composition during epitaxial growth can be used to tailor the response of QWIPs to particular IR bands or multiple bands.

However, many problems concerning the performance of QWIPs must be resolved. New ideas relating to the improvement of quantum efficiency and suppression of thermal generation in QWIPs are required. This book shows efforts in these areas by physicists and materials science engineers from many leading laboratories in the world.

Manijeh Razeghi

INTRODUCTION

There has been an increasing interest in long wavelength infrared detectors fabricated from semiconductor quantum wells and superlattices. This is because modern crystal growth techniques, such as molecular beam epitaxy and metalorganic chemical vapor deposition, have made it possible to grow semiconductor thin films with precise control of composition, doping, and thickness. In addition to the growth of a single semiconductor thin film, one can deposit multiple layers with these techniques, forming what are known as quantum wells and superlattices.

There are three approaches used to fabricate infrared detectors based on epitaxially grown semiconductors. The first approach utilizes alloy combinations, such as InTlSb, in which the spectral range of interest lies between the band gaps of the two binary compounds (i.e., InSb and TlSb). The spectral range of this approach is, therefore, limited. The second approach utilizes the intersubband transitions formed in either the conduction (n-type) or valence (p-type) bands of multiple quantum wells such as GaAs/AlGaAs, InGaAs/AlGaAs, and Si/SiGe. Detectors fabricated from intersubband transitions in multiple quantum wells suffer two disadvantages:

- The incoming photons in n-type multiple quantum wells must strike the surface at the Brewster's angle of the quantum well materials to allow coupling between the incoming radiation and the electrons in the quantum wells. Part of this book is devoted to a treatment of this issue, which does not appear to be a significant problem in p-type multiple quantum wells.

- Devices fabricated from such structures are extrinsic where the materials must be doped in order to introduce the charge carriers to the quantum wells. Thus, base devices have limited optical cross sections because the dopant concentrations are limited by the epitaxial growth techniques.

The third approach is based on interband transitions in quantum wells and superlattices. The band gaps of this class of materials are engineered in a way that the electrons and holes are located in adjacent layers. An example of this quantum structure in type II superlattices is InAs/InGaSb strained-layer superlattices. Theoretically speaking, both intersubband and interband transitions in quantum wells and superlattices can be continuously adjustable from zero covering wide spectral ranges. Various aspects of the above three approaches are discussed in great detail in this book.

Intersubband transitions in multiple quantum wells, in particular GaAs/AlGaAs multiple quantum wells, have been the most studied to date. The emphasis in this book is on the basic processes in long wavelength infrared detectors based on intersubband transitions. Chapter 1 is devoted to a discussion of some of these processes. One of these processes is the dark current in the thermoionic regime. A few assumptions were considered to define such a regime: the inter-well tunneling contribution to the dark current is neglected; the electron density in each well does not vary with the applied bias; a perfect injecting contact from the contact layer is assumed to exist; and, either there exists only one bound state in the quantum wells, or the upper bound state is very close to the top of the barrier. These assumptions are relevant to quantum well infrared photodetectors having thick barriers and operating at low temperatures. The dark current in this regime is, thus, controlled by the flow of electrons above those barriers, by the emission and trapping of electrons in the wells, and by the injection of electrons from the contact layer. Generally speaking, these assumptions provide a good model which is found to be a good approximation for a wide range of device parameters, including barrier thicknesses and numbers of wells.

The basic theoretical modeling of the intersubband transitions follows the single particle effective mass approximation method. These calculations usually provide very good expressions for the matrix element of interest between the two states, the oscillator strength of the intersubband transitions, the optical absorption coefficients, and the peak position energy. However, it is found that the energy peak position of the intersubband transition observed experimentally differs from the theoretical values obtained from single particle calculations by at least 20 meV. This difference can be accounted for using many body effects such as depolarization and exchange interaction energies.

One of the most important parameters for quantum well infrared photodetectors is the photocurrent and the photoconductivity gains. These parameters can tell us how good the device is and how much of the radiation is converted into an electrical signal. Usually, the conventional theory of photoconductivity is employed to describe quantum well infrared photodetectors. In chapter 1, a model has been constructed specifically for quantum well infrared photodetectors. That model describes exactly what constitutes the mechanism of photoconductivity gain and exactly how it depends on device parameters such as the number of wells. This model also explains the observation of large photoconductivity gains.

Detector responsivity is controlled by both quantum efficiency and photoconductive gain. In general, high absorption does not necessarily result in high detector responsivity. The photoexcited electron must escape the wells efficiently to give rise to a large photocurrent. It is found that the highest peak spectral responsivity is achieved when the upper state is nearly in resonance with the top of the barrier.

An example of an intersubband transition is discussed in chapter 2. In this chapter the growth and characterization of GaInP/GaAs multiple quantum wells are discussed, bearing in mind the potential application of this system to infrared detectors. As GaInP is lattice matched to GaAs, it shows a number of unique and interesting features in comparison to the AlGaAs/GaAs system. It has been realized that the GaInP/GaAs interface is remarkably improved relative to that of AlGaAs/GaAs. In addition, the excess carrier recombination velocity at a GaInP/GaAs heterostructure is about an order of magnitude lower than the lowest reported recombination velocity at a AlGaAs/GaAs heterostructure. Another advantage of GaInP/GaAs over AlGaAs/GaAs is that GaInP is expected to have a very low DX center due to the crossover of its direct and indirect conduction band minima at $x = 0.74$, which is far from the lattice matched composition of $x = 0.51$. The strong selective etching between GaAs and GaInP makes it a very promising system for device fabrications. Other advantages of using GaInP/GaAs multiple quantum wells for long wavelength infrared detectors include the low growth temperatures of both layers that in return reduce the interdiffusion of impurities at the interfaces; excellent material characteristics when it is grown by metalorganic chemical vapor deposition, which is a good growth technique for mass production; and the fact that this system can be grown on silicon wafers, which is a desirable advantage for integrated circuits and read-out electronics.

As mentioned earlier, one major drawback of utilizing n-doped intersubband transitions in multiple quantum wells is that the incoming photon must strike the surface at Brewster's angle of the quantum well materials to allow coupling between them and the electron in the quantum wells. Therefore, it is necessary to find geometries that overcome this drawback and enhance the quantum efficiency of the detectors. The absorbance for normally incident radiation can be enhanced and become polarization-independent through the use of a crossed grating with cladding layer, and reach values close to unity for large detector mesa sizes. Theoretical and experimental investigations of grating coupled quantum well infrared detectors and the dependence of absorbance on various factors like grating type, the inclusion of cladding layer, detector mesa size, type of contact, and the geometry of the metal grating (for example, two-dimensional planar square mesh) are discussed in great detail in chapters 3 and 4. The main objective of chapter 3 is to present three types of metal grating-coupled III-V semiconductor quantum well infrared photodetectors using bound-to-miniband intersubband transition schemes for the $8-14\,\mu m$ spectral range. The miniband described in chapter 3 is formed by replacing the barrier materials in the multiple quantum wells with superlattices in which the band (miniband) is formed. This band is resonant with the excited bound state in the well. Such design is made so that the excited electrons will be transported through the miniband in order to reduce the dark current in the device itself.

The use of gratings in n-type multiple quantum wells has the undesirable effect of increasing the number of process steps in device fabrication, and adds the complication of achieving grating uniformity over a large two-dimensional array. An alternative approach to n-type multiple quantum wells is the use of p-type multiple quantum wells in which the intersubband transition is in the valence band. This is achieved by doping the structure with acceptors rather than donors such as Be-doped AlGaAs/GaAs multiple quantum wells. The selection rules in the p-type multiple quantum wells seem to allow normal incident absorption. Chapter 5 is devoted to a discussion of p-type multiple quantum wells, which have received little attention in the literature. This chapter discusses in detail the physics behind the normal incident absorption in p-type AlGaAs/GaAs multiple quantum wells that would be useful for infrared detectors.

Other structures that can be used for long wavelength infrared detectors without grating are discussed in chapter 6. This chapter provides detailed discussions of ellipsoidal-valley quantum wells and n-type multiple quantum wells exhibiting normal-incident photoresponse.

The second approach to fabricating detectors is based on the interband transitions in systems such as type II superlattices. An example of such a structure is the InGaSb/InAs strained-layer superlattice, which is discussed in great detail in chapter 7. The advantages of detectors based on this structure come from the electronic band structure of the active layer. Long wavelength photoresponse is derived from a broken gap, known as type-II band alignment, in which the valence band edge of InGaSb lies below the conduction band edge of InAs. The alloying of the antimonide layer enhances the strain in the superlattices, which increases the energy separation between the InAs conduction band and the InGaSb valence band edges. Type-II band offset tends to localize electrons in the InAs layers and holes in the InGaSb layers. The localization of electrons is weak for layers this thin. Thus, electron-hole overlap is significant in the antimonide layers despite the type-II band alignment, resulting in appreciable optical matrix elements for the superlattice valence-to-conduction band transitions.

The last approach used to fabricate infrared detectors that is considered in this book is based on the alloy combinations in which the spectral range lies between the band gaps of the binary compounds. An example of this approach is discussed in chapter 8. This chapter presents the physical properties of InTlSb alloy and the advancements made toward realizing InTlSb infrared photodetectors. This chapter also describes the first successful growth of InTlSb alloys exhibiting extended infrared response. Detailed discussions of the various structural, electrical, and optical characterization results are presented, providing insight into the material properties of InTlSb. In addition, the first InTlSb photodetectors are demonstrated and their detecting capabilities are evaluated to assess the merits of this alloy as a novel III-V material for long wavelength infrared detectors.

M. O. Manasreh

CHAPTER 1

The Basic Physics of Photoconductive Quantum Well Infrared Detectors

H. C. LIU

Institute for Microstructural Sciences, National Research Council, Ottawa, Ontario K1A 0R6, Canada

1. INTRODUCTION

Suggestions to use optical transitions between quantum confined states for infrared (IR) applications were made in the early days of

quantum well research [1, 2]. A series of theoretical papers considering various schemes of long wavelength infrared (LWIR) detection
were published by Coon et al. [3–5], and Goossen and Lyon [6, 7].
The first observation of intersubband transitions (ITs) in GaAs-AlGaAs
multiple quantum wells (MQWs) was made by West and Eglash [8].
Stark shifts were then observed by Harwit and Harris [9]. The first
unambiguous demonstration of quantum well infrared photodetectors
(QWIPs) was made by Levine et al. [10]; since then tremendous progress has been made by the AT&T group [11–14]. Other groups
around the world have also investigated detector performance
[15–20] and various aspects of the physics involved [21–26]. More
recently, large focal plane arrays have been demonstrated with excellent uniformities [27–29]. Other material systems [30–33] have also
been investigated. Research activities in this area are expanding very
rapidly, with more and more groups starting to investigate various
aspects from both basic physics and application points of view. Several review articles have been published [34–38]. Two collections of
papers, solely on ITs in quantum wells, have been published as a
result of two NATO Advanced Research Workshops [39, 40].

The device operation of photoconductive QWIPs is similar to that
of conventional semiconductor photoconductors. The distinct feature
of QWIPs, in contrast with the conventional intrinsic or extrinsic
photoconductors, is the discreteness, i.e., incident photons are only
absorbed in discrete quantum wells which are normally much narrow
than the inactive barrier regions. Other quantitative differences, resulting from the thin structural nature of QWIPs, are high operating
electric fields, long carrier mean free paths in comparison with the
device thickness, and high intrinsic device speeds. Furthermore, unlike
most conventional photoconductors, current-voltage characteristics of
photoconductive QWIPs are highly nonlinear.

QWIP materials are normally grown by molecular beam epitaxy
(MBE). The GaAs-AlGaAs MBE technology is mature enough for the
fabrication of QWIPs [41]. Devices are made by standard GaAs
microfabrication techniques. The operation of a photoconductive
QWIP is easily summarized using the bandedge profile shown in
Fig. 1. Upon application of a finite bias, incident photons excite

---

Note added in proof: Although interpreted in terms of free carrier absorption, the
idea of using quantum wells for infrared detection was first tested experimentally by
J.S. Smith, L.C. Chui, S. Margalit, A. Yariv, and A.Y. Cho [J. Vac. Sci. Technol. B1, 376
(1983). I thank Prof. Amnon Yariv for pointing out this reference.

electrons out of the quantum wells creating a photocurrent. The wells are doped to provide a finite population of electrons in the ground state subbands. The most widely studied materials system is GaAs-$Al_xGa_{1-x}As$, where x is the Al alloy fraction. The conduction band is commonly used. For this system, the conduction band offset (barrier height) as a function of x is given approximately by [22]

$$\Delta E_C \approx (0.7\text{--}0.8) \times x(\text{eV}), \quad \text{for} \quad x < 0.5. \tag{1}$$

The range 0.7–0.8 reflects the spread of values reported in the literature. The detection wavelength is determined by the difference between the first excited state and the ground state of the quantum well. Keeping the first excited state close to the top of the barrier (to within a few meV), the shortest wavelength that GaAs-AlGaAs system with $x < 0.5$ can achieve is about 5 μm [22]. Employing thin AlAs barriers, a detector with cutoff wavelength of about 3 μm has been demonstrated [23]. Shorter wavelength detectors down to about 3 μm should be achievable using strained InGaAs wells. Shallow wells can be made with low x, consequently there are no limitations for designing detectors with cutoff wavelengths much longer than 10 μm [42]. Properties of GaAs and AlGaAs can be found in review articles by Blakemore [43] and Adachi [44]. The advantages of using GaAs-and Si-based materials in large focal plane array applications, especially monolithically integrated detector arrays and processing circuits, are obvious. Because of the current interest in producing electronic and optoelectronic circuits with InP-based materials, this material system has potential as well [45].

The goal of this chapter is to present the current physical understanding of QWIPs. We concentrate on the simplest isotropic one band electron case and on MQWs made of simple (square) wells. Other more complicated structures, such as coupled wells [46, 47], are interesting, but are not considered here. We assume that the reader has a basic knowledge of quantum wells [48]. The organization of the chapter is as follows.

- Section 2 considers the detector dark current which is of fundamental importance because the dark current related noise commonly limits detector performance. We only discuss the "thermionic" regime where inter-well tunneling can be neglected.
- Section 3 treats the optical intersubband transition and gives a simple calculation example of a symmetric quantum well.
- Section 4 considers the photoconductivity of MQW structures and emphasizes the differences from the conventional theory of photoconductivity.

- Section 5 discusses the primary noise sources associated with dark current and background photon fluctuation, and the gain associated with the dark current noise.
- Section 6 addresses performance and points out possible improvements.
- Section 7 concludes the chapter.

2. DARK CURRENT IN THE THERMIONIC REGIME

2.1. Physical Model and Assumptions

Understanding the dark current in QWIPs is one of the most important aspects in optimizing detector performance. Several published papers have addressed and modeled the dark current [13, 16, 18, 49, 50, 51]. In this section, we consider the regime where inter-well tunneling can be neglected, which is normally accomplished by employing thick barriers. We call this regime the "thermionic regime" for categorization (and, in fact, tunneling occurring near the top of the barriers under a large bias dominates the current as seen later). We first evaluate the physical validity of the assumptions behind different models [51].

Figure 1 shows schematic potential profiles for a QWIP under zero and non-zero bias. To define precisely the physical regime that we are considering we assume that:

(a) the inter-well tunneling contribution to the dark current can be neglected;
(b) the electron density in each well does not vary with the applied bias;
(c) a perfectly injecting contact from the heavily doped contact layer to the MQW region exists;
(d) either there exists only one bound state in the quantum wells or the upper bound state is very close to the top of the barrier.

This assumed regime is most relevant to practical QWIPs having thick barriers (e.g., $> 250\text{Å}$) and operating at low temperatures (e.g., 77 K). The dark current in this regime is controlled (1) by the flow of electrons above the barriers, (2) by the emission and trapping of electrons in the wells, and (3) by the injection of electrons from the emitter contact. Assumption (a) is then satisfied by requiring a thick enough barrier (note that the devices studied, e.g., by Choi et al. [52] and by Liu et al. [53] do not satisfy this condition). Assumption (b) is

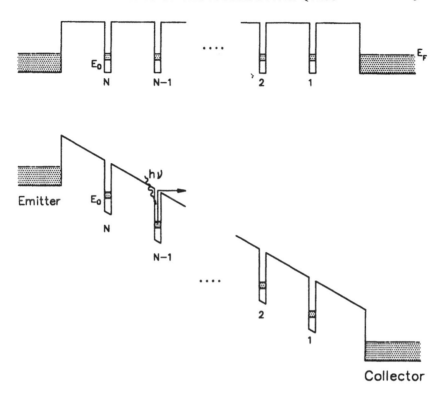

Fig. 1 Bandedge profile of a quantum well infrared photodetector under zero bias (above) and a finite bias (below). The lowest subband (E_0) in the wells are populated with electrons by doping. Because of the unipolar nature of the device, we label the two contacts as emitter and collector. The contacts are degenerately doped, and resulting Fermi energy is E_F.

not strictly valid at large bias voltages [24], but is a good approximation. Assumption (c) is approximately valid for QWIPs, but a substantial deviation occurs when an intense IR source, e.g., from a CO_2 laser, is used [26]. To produce good detectors, condition (d) is required (see subsection 4.3) [19, 54].

Fig. 2 presents diagrammatically the electron distribution (top) and the processes controlling the dark current (bottom). The top part of the figure indicates that, at finite temperature, electrons are not only bound in the well but are also distributed outside of the well and on top of the barriers. The energy region for electrons contributing to the dark current is indicated by the large brace bracket. The lower part of

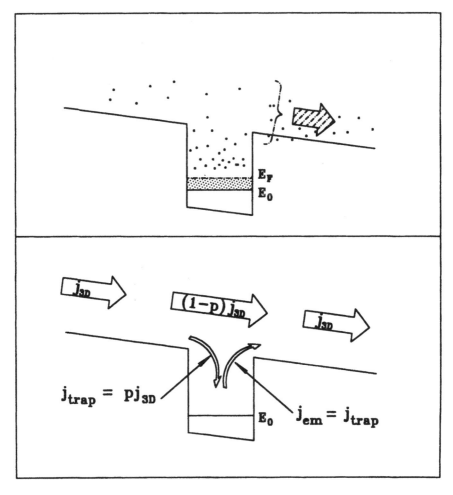

Fig. 2 Schematic representation of the electron distribution (top) and the processes controlling the dark current (bottom). Symbols are defined in the text.

the figure shows the dark current paths. In the barrier regions (on top of the barriers), the current flows in a three-dimensional (3D) fashion, and the current density is labeled as j_{3D} which *equals* the dark current J_{dark}. In the vicinity of each well, the emission of electrons from the well contributes a current component j_{em} to the dark current. This current, which tends to lower the electron density in the well, *must* be balanced by the trapping into the well under steady state ($j_{trap} = j_{em}$).

Since the dark current is the same throughout the structure, j_{3D} and j_{em} (or j_{trap}) are related. If we define a trapping probability p for an electron traversing a well with energy larger than the barrier height, we must have $j_{trap} = p j_{3D}$, and the sum of the trapped (p) and untrapped $(1-p)$ fractions must equal the current in the barrier region: $j_{3D} = j_{trap} + (1-p)j_{3D} = j_{em} + (1-p)j_{3D}$. One can model the dark current J_{dark} by calculating either j_{3D} directly [18] or j_{em} [13, 49, 50], and in the latter case $J_{dark} = j_{em}/p$.

At this point we comment on the physical validity of different calculation schemes. The model of Petrov and Shik [49] estimates j_{em} and takes this as the total dark current. This estimate was done by integrating the product of the "velocity" in z-direction, the transmission coefficient, and the Fermi distribution, where the z-coordinate is in the current flow direction (the epitaxial growth direction). This approach neglects (1) the trapping process for refilling the quantum well, and (2) the scattering-assisted escape of electrons associated with the confined ground state in the well, distributed on the two dimensional in-plane dispersion curve. The model of Pelvé et al. [50] estimates the emission current by the standard 3D thermionic emission formalism, but neglects the trapping process for balancing the loss of electrons. The model of Kane et al. [18] estimates j_{3D} by calculating the 3D electron density on top of the barriers n_{3D}. The dark current is then given by evn_{3D}, where v is the electron drift velocity which depends on electric field E. We have compared the result of this model with our experiments on several samples. Acceptable agreement was obtained *only* for very low electric fields. The reason for the failure at finite fields is probably due to the deviation of n_{3D} from its zero field equilibrium value. The model that we will follow [51], which is proposed by Levine et al. [13] and also used by Andrews and Miller [16], produces a good fit to experiment. The process of trapping, however, has not been discussed explicitly by these authors. Luc et al. [55] have recently reported a study of trapping by impedance measurements on a single well structure. This technique provides a direct way to infer trapping parameters relevant to the operation of QWIPs.

2.2. Derivation of the Dark Current Expression

We now derive the dark current expression and discuss its physical basis. We use the important theoretical result of Meshkov [56] which can be stated as follows: including scattering processes, an electron tunneling rate in a one dimensional potential is controlled by the *total* energy rather than the energy associated with the tunneling direction. For thick barriers, scattering assisted processes determine the tunneling transmission probability. We stress that the inclusion of this scattering effect is necessary because otherwise all electrons on the ground state subband would have an equal escape probability since they have the same energy associated with the one-dimensional confinement, and hence without scattering the dark current would be only weakly dependant on the electron density in the well, contrary to experiments. We calculate j_{em} associated with the escape of electrons from the ground state subband (see the lower part of Fig. 2). We model the emission of electrons from a two-dimensional (2D) subband into the continuum as a scattering process, and use the following expression:

$$j_{em} = e \int_{E_0}^{x} dn_{2D} \, D(E, \varepsilon)/\tau_{scatt} , \qquad (2)$$

$$dn_{2D} \equiv \frac{m}{\pi \hbar^2} \left[1 + \exp\left(\frac{E - E_0 - \varepsilon_F}{k_B T} \right) \right]^{-1} dE, \qquad (3)$$

where m is the carrier effective mass which equals the product of the reduced effective mass and the electron rest mass ($m^* \times m_0$), E_0 is the ground state eigenenergy, E is the total energy of an electron, $D(E, \varepsilon)$ is the transmission coefficient which is taken to be unity for E higher than the barrier, τ_{scatt} is the scattering time for electron transfer from the 2D ground state subband to a non-confined 3D transport state on top of the barrier, $m/\pi \hbar^2$ is the 2D density of states, and ε_F is the Fermi energy (referenced to E_0). Strictly speaking, the scattering time τ_{scatt} is energy dependent, but here as an approximation we will take τ_{scatt} out of the integral in Eq. (2). One can view this as taking an "averaged" τ_{scatt}. This step is necessary to obtain a simple final result. At low and moderate electric fields, τ_{scatt} corresponds physically to the lifetime τ_{life} associated with the carrier trapping process because the two processes (trapping into the well and scattering out of the well) are the inverse of each other. We then have (using $\tau_{scatt} = \tau_{life}$):

$$j_{em} = e N_{2D}/\tau_{life}, \qquad (4)$$

$$N_{2D} \equiv \int_{E_0}^{x} D(E, \varepsilon) \, dn_{2D}. \qquad (5)$$

The emission current j_{em} constitutes only part of the total dark current. Taking trapping and current injection into account (see the lower part of Fig. 2), the total dark current is

$$J_{dark} = j_{em}/p, \tag{6}$$

where the trapping or capture probability is $p = \tau_{trans}/(\tau_{life} + \tau_{trans})$ and τ_{trans} is the transit time for an electron across *one* quantum region, partly including the surrounding barriers. In the limit of $p \ll 1$, i.e., $\tau_{life} \gg \tau_{trans}$, as is true for actual devices at operating electric fields [57], the dark current expression becomes

$$\begin{aligned} J_{dark} &= \frac{eN_{2D}}{\tau_{life}} \times \frac{\tau_{life} + \tau_{trans}}{\tau_{trans}} \\ &\approx eN_{2D}/\tau_{trans} \\ &\equiv evN_{2D}/\mathscr{L}, \end{aligned} \tag{7}$$

where the electron "drift" velocity on top of the barriers is $v \equiv \mathscr{L}/\tau_{trans}$ and $\mathscr{L}$ is a relevant length scale.

There is an ambiguity concerning this relevant length scale $\mathscr{L}$ involved in both τ_{life} and τ_{trans}. The uncertainty is related to the quantum mechanical nature of the electron and to the question of exactly what length scale an electron is distributed over within its coherent lifetime. The definition of p and its physical meaning is, however, rigorous and clear since the same uncertainty exists in both τ_{life} and τ_{trans}. As an approximation, the length $\mathscr{L}$ is taken to be the period of the multiple quantum well structure $L_p = L_w + L_b$, i.e., the sum of the well width L_w and the barrier width L_b. Physically, taking $\mathscr{L} = L_p$ should give a good approximation as long as the barrier width is large enough, but is not so large that L_b is longer than the electron coherence length. One should therefore expect a reduction in dark current when L_p is increased [13], but increasing L_b beyond a certain point will not reduce the dark current further. Another factor neglected in the development of our expression for the dark current is the possible field dependence of the relevant length scale.

Using Eqs. (3) and (7), and setting $\mathscr{L} = L_p$, the final expression for dark current is:

$$\begin{aligned} J_{dark} &= evN_{2D}/L_p \\ &= ev \int_{E_0}^{\infty} \frac{m}{\pi \hbar^2 L_p} D(E, \varepsilon) \left[1 + \exp\left(\frac{E - E_0 - \varepsilon_F}{k_B T} \right) \right]^{-1} dE. \end{aligned} \tag{8}$$

Physically, we could arrive at the above expression only when the trapping process was taken into account. Early derivations were done simply by "converting" the 2D electron density N_{2D} into an "average" 3D density by multiplying by $1/L_p$ [13, 16].

In our derivation, we have not explicitly included the effects of an upper bound state. Because the upper bound state (if it exists) is very close to the top of the barrier (by assumption) and the broadening is substantial, we believe that the contribution to the dark current from the upper subband is part j_{3D} and, therefore, has been effectively taken into account.

Following Levine et al. [13], the final ingredient in the dark current expression, i.e., the v vs. ε relationship is given by

$$v = \frac{\mu \varepsilon}{[1 + (\mu \varepsilon / v_{sat})^2]^{1/2}}, \tag{9}$$

where μ is the low field mobility and v_{sat} is the saturated drift velocity. Different v vs. ε expressions can also be used to produce a good fit to experiments.

We stress that the dark current result Eq. (8) is only a model and not a "first principles" calculation. This is apparent from the many physically plausible assumptions and approximations used in deriving the final result. To go beyond the present model, rigorous calculations are needed to predict the relative strengths of various scattering processes, such as electron-phonon, electron-electron, interface roughness, and doping ion scattering. Because of the relatively high electron density in the wells, a many particle calculation may be unavoidable. A quantitative understanding of scattering and trapping should also give us clues to improve the carrier lifetime and thereby the photoconductive gain, resulting in detectors with better temperature performance (see Sec. 6).

2.3. Comparison with Experiments

We now compare the dark current expression with the experimental results obtained from several samples to show that the model provides a good approximation for a wide range of device parameters, including barrier thicknesses from 250 Å to 700 Å and numbers of wells from 4 to 32 [51].

Figs. 3 and 4 compare Eq. (8) with measured dark currents obtained for seven samples. The sample parameters, which were refined by

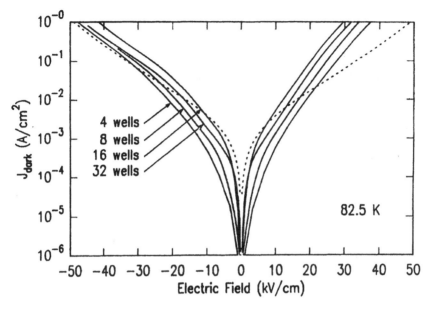

Fig. 3 Measured dark current for four samples with different number of wells (solid lines) at 82.5 K and calculated (dashed line) dark current vs. applied electric field for this potential profile.

X-ray and secondary ion mass spectroscopy (SIMS) measurements, are given in Table 1. Notice the difference of 1 Å in the well widths among the last three samples 346 Å B–709 Å B. This is real and results in a measurable difference in the peak positions associated with inter-subband absorption. The parameters used in the model calculations are listed in Table 2. The measured current-voltage characteristics on 120 µm square mesa devices were converted into current density vs. electric field characteristics, taking into account the depletion on the collector side (which was a small effect for these samples). The 77 K results were obtained by immersing samples in liquid nitrogen (LN$_2$), and those at 82.5 K were for samples mounted on test packages affixed to a 77 K cold finger. The difference of 5.5 K was determined accurately by lowering the cold finger temperature until the measured current coincided with its value obtained in a 77 K LN$_2$ immersion measurement. It is important to note that there is a non-negligible difference between the actual device temperature and the cold finger temperature in a typical measurement geometry. This

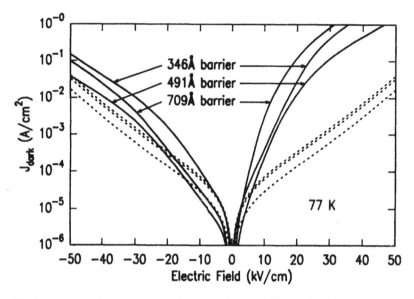

Fig. 4 Measured dark current for three samples with different barrier widths (solid lines) at 77 K and calculated (dashed lines) dark current vs. applieed electric field for each potential profile. The model predicts a decrease in dark current as the barrier thickness is increased.

TABLE 1

Sample parameters. The accuracies for the Al fraction x, thicknesses L_b and L_w, and doping concentration, are estimated to be ± 0.005, ± 1 Å, and 10%, respectively. The Si doping density in the two contact layers was 1.5×10^{18} cm^{-3}, and the center Si δ-doping density in the quantum wells was 9×10^{11} cm^{-2}, for all samples.

Sample	x	L_w(Å)	L_b(Å)	Repeats
32W	0.275	60	246	32
16W	0.275	59	244	16
8W	0.270	61	254	8
4W	0.266	62	254	4
346ÅB	0.297	58	346	32
491ÅB	0.300	57	491	32
709ÅB	0.292	59	709	32

TABLE 2

Parameters used in calculations. The Fermi energy was 32 meV corresponding to the two-dimensional doping density in the wells, the mobility was 1000 $cm^2V^{-1}s^{-1}$, and the saturated velocity was 10^7 $cm s^{-1}$. The symbol m_b^* is for the barrier reduced effective mass.

Sample	$V_0(eV)$	$E_0(meV)$	m_b^*	$L_w(Å)$	$L_b(Å)$	$T(K)$
32W	0.22	59	0.090	60	250	82.5
16W	0.22	59	0.090	60	250	82.5
8W	0.22	59	0.090	60	250	82.5
4W	0.22	59	0.090	60	250	82.5
346ÅB	0.24	63	0.092	58	346	77.0
491ÅB	0.24	64	0.092	57	491	77.0
709ÅB	0.24	61	0.091	59	709	77.0

increase in temperture resulted in a 50% increase in measured dark currents.

The electric field dependence in Eq. (8) is explicitly through (9) and implicitly through $D(E, \varepsilon)$. Using the WKB approximation, $D(E, \varepsilon)$ is given by

$$D(E, \varepsilon) = \exp\left[-2 \int_0^{z_c} dz \sqrt{2m_b(V - E - e\varepsilon z)}/\hbar \right] \qquad (10)$$

for erergy less than the barrier height $E < V$, where m_b is the barrier mass, $V = V_0 - e\varepsilon L_w/2$, and V_0 is the barrier height at zero field. This value for V takes into account the effective barrier lowering because E_0 (referenced to the center of the well) is approximately independent of ε, and $z_c = (V - E)/e\varepsilon$ defines the classical turning point. Figure 5 shows schematically the relevant quantities. The WKB approximation compares well with more exact calculations using the transfer matrix approach [16]. We neglect non-parabolic effects which would result in a small correction in the values of the calculated transmission coefficients. Furthermore, the effect of the image potential is negligible [58] in GaAs-AlGaAs which is overestimated in Ref. [16] using an approximation appropriate for metaldielectric interfaces.

To show the relative contributions from pure thermionic emission $(E > V)$ and tunneling $(E < V)$ (sometimes termed thermionic-field emission or thermionic-assisted tunneling), we plot in Fig. 6 the ratio $j_{tunnel}/j_{thermionic}$, where Eq. (8) integrated from E_0 to V is $j_{tunnel} = J_{dark}(E < V)$, and that integrated from V to ∞ is $j_{thermionic} = J_{dark}(E > V)$. At low fields, the thermionic contribution dominates,

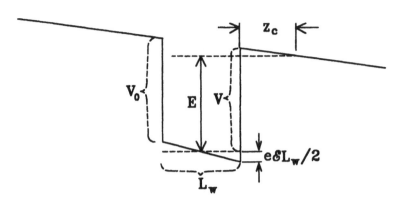

Fig. 5 Schematic illustration of relevant energies and the effective barrier lowering. The electron energy E is referenced to the center of the well.

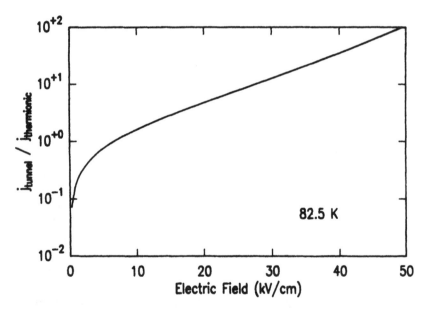

Fig. 6 Ratio of tunneling $(E < V)$ and pure thermionic $(E > V)$ contributions to the dark current vs. electric field as determined by integrating the model expression in the appropriate energy region.

and the transmission coefficient increases exponentially with increasing field. At high fields, the tunneling current constitutes the major part of the current.

Figure 7 shows a comparison between simulations and measured dark currents at different temperatures. The model compares well with

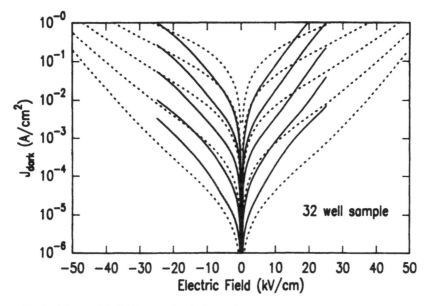

Fig. 7   Measured (solid lines) and calculated (dashed lines) current vs. electric field at different temperatures: 65, 75, 85, 95, and 105 K. The dark current increases monotonically with increasing temperature.

experiments over four orders of magnitude. Below about 70 K the model deviates from the experiment; thereafter the agreement becomes worse with decreasing temperature. In the low temperature limit, the assumptions (a) and (c) made in Sec. 2.1 no longer hold.

2.4. Discussions

Since we have compared the model systematically both with samples having different numbers of wells while keeping other device parameters fixed (4W–32W), and with samples having different barrier widths while keeping other parameters fixed (346 Å B–709 Å B), we can comment on several aspects of the model and the experiment.

The measured dark currents from the 4, 8, 16, and 32 well samples (4W–32W), shown in Fig. 3 as a function of electric field, agree with each other to within a discrepancy consistent with the small variation of device parameters. This means that a sample with as few as

4 periods is sufficiently large enough that the assumptions made for multiple quantum well samples are valid. This is in distinct contrast to experiments performed on single quantum well samples, where the electron density in the well is determined almost solely by dynamic charge storage effects [59–62].

The overall agreement (over several orders of magnitude in current density) between the model and experiments as shown in Figs. 3 and 4 is acceptable. A better fit to the sample of Levine et al. [13] was reported. Of course, one can vary the parameters in Table 2 to obtain a fit to experiments for each sample, but non-systematic variations between samples make such an exercise almost pointless. We can observe in each of the model curves a qualitative feature that indicates inadequacies of the model: the dark current is overestimated at low fields probably due to the breakdown of our assumption that $p \ll 1$ (i.e., $\tau_{life} \gg \tau_{trans}$); whereas at high fields, the equality $\tau_{life} = \tau_{scatt}$ (used explicity in Eq. (4) and thereafter) no longer holds resulting in an underestimate of the dark current.

The barrier height V_0 used in our simulations corresponds to 65% of the band gap difference between GaAs and AlGaAs [$\approx 0.8 \times x$ in Eq. (1)]. This is an empirical rule consistent with other experiments. All higher order effects (such as many body effects [24, 63]) are included in this *effective* barrier height value.

We now discuss possible reasons for the observed asymmetry between currents for positive and negative bias voltages. Positive polarity is defined as the top of the mesa biased positively relative the bottom contact. The same asymmetry has been observed in all the similar samples from the AT&T group [64]. Intuitively, many factors may give rise to the asymmetry including (1) a difference in roughness of the two heterointerfaces confining a quantum well (it is commonly stated that the interfaces resulting from AlGaAs grown on GaAs are smoother than those of GaAs on AlGaAs), (2) an asymmetry in the height of the two barriers, from unintentional asymmetries of the Al fractions during growth, and (3) an asymmetry caused by segregation of dopants during growth, which results in a spreading of the dopants in the well regions and into the barriers. Realistically, factor (1) is difficult to model quantitatively because the details of interface roughness are not known. Factor (2) relates to molecular beam flux transients commonly occurring on the opening of the cell shutters. This effect has been carefully evaluated for our MBE system, in which both Ga and Al cell transients are less than 1.5%. Taking into account the growth recipies used for our samples, we estimate that the resulting

TABLE 3
Sample parameters. The Al fraction is 0.26, the substrate temperature was 605°C, the well electron density is 5×10^{11} cm^{-2},
the device area is 120×120 μm^2, and the number of wells is 25,
for all samples.

Sample	δ shift(Å)	L_w(Å)	L_b(Å)
δ0Å	0	59	350
δ6Å	6	59	350
δ11Å	11	55	326
δ12Å	12	59	350
δ22Å	22	55	326

maximum possible asymmetry of the barriers is about 1 meV which
would produce negligible amounts of dark current asymmetry. Focus-
ing on factor (3), we have carried out a systematic study of the effect of
intentional shifts of the Si δ-doping position away from the center of
the quantum well on the current-voltage (I-V) asymmetry. We have
also modeled the effect of dopant segregation [65].

The δ-doping is used to create the ground state electron population
needed for the operation of the QWIP. Should segregation of the Si
atoms occur during growth, an asymmetry in the doping profile, and
hence in the quantum well potential, will result. Consequently, the
effective barrier heights, as seen by the electrons in the well under
forward and reverse bias conditions, will be different. Since the current
vs. barrier height relation is exponential [13, 51], the I-V measurements
are very sensitive to the amount of asymmetry in the barrier heights.

Five samples with similar structures except for the position of the
δ-doping in the well were grown by MBE at a substrate temperature
of $T_{\text{sub}} = 605$ °C. We shifted the position of the δ-doping towards the
substrate to compensate for the effect of segregation. Sample parameters are given in Table 3. I-V measurements for all samples were
made for devices immersed in liquid nitrogen, with the bottom contact
as the ground.

Measured I-V characteristics for all samples are shown in Fig. 8(a).
The ratios of the forward to reverse current are plotted in Fig. 8(b).
We see a consistent trend in the change of the I-V asymmetry, which
diminishes with the shift in the position of the δ-doping away from the
center. For a 22 Å shift (Sample δ22 Å), the resulting I-V is remarkably symmetric over several orders of magnitude of the measured
current.

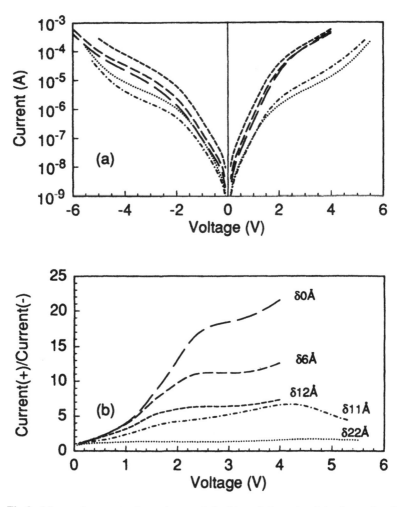

Fig. 8 Measured current-voltage characteristics (a), and the ratio of the forward and reverse currents (b) for all six samples at 77 K and with the bottom contacts chosen as the ground.

From our measurements and modeling [65], the most plausible cause of the observed asymmetry is the segregation of Si atoms. This type of segregation effect has been demonstrated for Si doping above $10^{13}\,cm^{-2}$ and growth temperatures over 550 °C [66]. Further studies, on samples grown at different substrate temperatures and at different As overpressure during growth, have also been carried out [67]. We find that as the growth temperature approaches 550 °C the segregation becomes negligible, and that an increased As overpressure suppresses Si segregation for both As_2 and As_4 forms of As.

3. INTERSUBBAND TRANSITION

In this section we present the basic physics of ITs in a single particle picture. We use the effective mass approximation and deal with a single spherical band which is a good approximation for the conduction band of GaAs-AlGaAs system. Higher order effects such as scattering assisted ITs [21] and many particle effects [46, 24] are not discussed.

3.1. Basic Formulas

Following Coon and Karunasiri [3], the interaction potential appropriate for calculating the radiative transition absorption rate using Fermi's golden rule is given by

$$H_{rad} = \frac{e}{m} \left(\frac{\phi \hbar}{2 \varepsilon_0 n_r wc} \right)^{1/2} \hat{\varepsilon} \cdot \vec{p} \, e^{i \vec{q} \cdot \vec{x}}, \tag{11}$$

where ϕ is the incident photon flux (number of photons per unit area per unit time), ω is the photon angular frequency, $\hat{\varepsilon}$ is the polarization unit vector, $\vec{p}$ is the electron momentum operator, n_r is the material refractive index, and $\vec{q}$ is the photon momentum which is perpendicular to $\hat{\varepsilon}$. We assume that the photon flux is sufficiently small so that the interaction in Eq. (11) can be treated as a perturbation. For high IR intensities, the perturbation approach fails and the higher order term proportional to ϕ [in addition to the $\phi^{1/2}$ term in (11)] must be included in the analysis. The derivation of Eq. (11) is straightforward starting from the dipole interaction Hamiltonian $(e/m)\vec{A} \cdot \vec{p}$ and keeping only the contribution for photon absorption, where $\vec{A}$ is the vector potential.

The envelope wavefunction in a conduction band quantum well structure can be written as

$$\psi(\vec{x}) = \psi_{xy} \psi_z, \tag{12}$$

where ψ_{xy} and ψ_z are two separable components of the envelope function ψ in the x-y plane and in the z direction, respectively. We have chosen the quantum well direction (the epitaxial layer growth direction) as the z coordinate. The in-plane envelope function is simply a planewave

$$\psi_{xy} = A^{-1/2} e^{i\vec{k}_{xy} \cdot \vec{x}}, \tag{13}$$

where A is the in-plane normalization area, and the wavevector $\vec{k}_{xy}$ is two-dimensional $\vec{k}_{xy} = (k_x, k_y)$. The energy associated with the x-y motion is $E_{xy} = \hbar^2 k_{xy}^2/(2m)$.

The matrix element of interest between two the states ψ and ψ' is

$$M = \langle \psi' | H_{rad} | \psi \rangle. \tag{14}$$

Some general properties can be obtained using (14). We first express (14) in terms of the z-component and the in-plane envelope function in the quantum well:

$$
\begin{aligned}
M &= \langle \psi' | H_{rad} | \psi \rangle \\
&= \frac{e}{m} \left(\frac{\phi\hbar}{2\varepsilon_0 n_r \omega c} \right)^{1/2} \left[\left(\frac{1}{A} \int dx\, dy\, e^{-i\vec{k}'_{xy} \cdot \vec{x}} \hat{\varepsilon} \cdot \vec{p}_{xy} e^{i\vec{k}_{xy} \cdot \vec{x}} \right) \left(\int dz \psi^*_{z,n'} e^{iq_z z} \psi_{z,n} \right) \right. \\
&\quad \left. + \left(\frac{1}{A} \int dx\, dy\, e^{i(\vec{k}_{xy} - \vec{k}'_{xy}) \cdot \vec{x}} e^{i\vec{q}_{xy} \cdot \vec{x}} \right) \left(\int dz \psi^*_{z,n'} \hat{\varepsilon} \cdot \hat{z} p_z \psi_{z,n} \right) \right] \\
&\approx \frac{e}{m} \left(\frac{\phi\hbar}{2\varepsilon_0 n_r \omega c} \right)^{1/2} [\hbar(\hat{\varepsilon} \cdot \vec{k}_{xy}) \delta_{\vec{k}'_{xy}, \vec{k}_{xy}} \langle \psi_{z,n'} | e^{iq_z z} | \psi_{z,n} \rangle \\
&\quad + \delta_{\vec{k}_{xy}, -\vec{k}'_{xy}, \vec{q}_{xy}} (\hat{\varepsilon} \cdot \hat{z}) \langle \psi_{z,n'} | p_z | \psi_{z,n} \rangle], \tag{15}
\end{aligned}
$$

where $\hat{z}$ is a unit vector in the z direction. The factor $e^{iq_z z}$ in the first term of the above can normally be neglected by setting $e^{iq_z z} = 1$. The photon wavelength (λ) of interest is greater than about $3/n_r \approx 1\,\mu m$ and the domain of the z direction integration is about $w \sim 100$ Å (the bound state wavefunction extent), giving $q_z z \sim 2\pi w/\lambda \ll 1$, where $n_r \approx 3$ for GaAs. The first term in (15) is proportional to $\delta_{\vec{k}'_{xy}, \vec{k}_{xy}} \delta_{nn'}$ which vanishes because $\vec{k}'_{xy} = \vec{k}_{xy}$ and $n = n'$ correspond a transition between the same state (forbidden by energy conservation). The first term can be shown to vanish for a superlattice, where $e^{iq_z z} \approx 1$ is no longer true in general. The physical reason for a vanishing contribution from the first term in (15) is that photons can not cause direct transitions between free carrier states. Note that the in-plane motion of the carriers is free. The commonly referred "free carrier absorption" is in fact a higher order process assisted by scattering events (see, e.g., Ref. [43]). In general, if the incident light is polarized in the x-y plane, the matrix element in (15) vanishes. By a similar argument that the photon momentum can be neglected, we can neglect the $\vec{q}_{xy}$ in the δ-function of the second term in (15). The matrix element then becomes

$$M = \frac{e}{m} \left(\frac{\phi\hbar}{2\varepsilon_0 n_r \omega c} \right)^{1/2} (\delta_{\vec{k}_{xy}, \vec{k}'_{xy}}) \sin\theta \langle \psi_{z,n'} | p_z | \psi_{z,n} \rangle, \qquad (16)$$

where $(\hat{\varepsilon} \cdot \hat{z}) = \sin\theta$ and θ is the internal angle of incidence. The polarization selection rule [3] can now be stated as: *optical intersubband transitions associated with a single spherical band are induced by light polarized in the quantum well direction.* Using Fermi's golden rule

$$W = \frac{2\pi}{\hbar} \sum_{f,i} |M|^2 f_i (1 - f_f) \delta(E_f - E_i - \hbar\omega), \qquad (17)$$

the total transition rate (W) is easily calculated by summing over initial and final states, where f_i and f_f are Fermi factors for the initial and final states. Here, for a pure bound-to-bound transition, we sum over only the in-plane two-dimensional states for the initial and final state summations, which is appropriate for transitions between two two-dimensional subbands. At zero temperature, Fermi factors become $f_i = 1$ and $f_f = 0$, assuming that only the ground state subband is occupied. The momentum conserving $\delta_{\vec{k}_{xy}, \vec{k}'_{xy}}$-function in Eq. (16) takes care of the final state summation in (17), and the initial state summation gives simply the two-dimensional density of electrons in the ground state subband (n_{2D}) multiplied by the area (A). Equation (17) is then trivially evaluated (defining the absorption quantum efficiency η):

$$\eta \equiv W/(\phi A \cos\theta) = \frac{e^2 h}{4\varepsilon_0 n_r mc} \frac{\sin^2\theta}{\cos\theta} n_{2D} f \delta(E_1 - E_0 - \hbar\omega), \qquad (18)$$

where the oscillator strength [8, 22] is defined by

$$f \equiv \frac{2}{mh\omega} |\langle \psi_1 | p_z | \psi_0 \rangle|^2, \qquad (19)$$

where ψ_1 and ψ_0 are the first excited and the ground state z-direction envelope functions, and E_1 and E_0 are the corresponding eigenenergies. An equivalent expression for the oscillator strength is given by

$$f = \frac{2m\omega}{\hbar} |\langle \psi_1 | z | \psi_0 \rangle|^2. \qquad (20)$$

The equivalence between (19) and (20) is shown by considering the matrix element of commutator $\langle \psi_1 | [H_z, z] | \psi_0 \rangle$ and noting $E_1 - E_0 = \hbar\omega$, where H_z is the quantum well effective mass Hamiltonian in z-direction. For finite temperatures, as long as $E_{10} - \varepsilon_F \gg k_B T$, Eq. (18) gives an excellent approximation, where $E_{10} \equiv E_1 - E_0$ and ε_F is the Fermi energy related to the two-dimensional electron density n_{2D} in the ground state subband by

$$\varepsilon_F = (\pi\hbar^2/m)n_{2D}. \tag{21}$$

The expression $m/(\pi\hbar^2)$ is the two-dimensional density of states. Relevant energy scales of interest are $E_{10} - \varepsilon_F \sim 120$ meV and $k_B T < 10$ meV. (Note that LWIR detectors must be cooled to $T < 100$ K). Taking into account the finite lifetime of the excited state (limited mainly by phonon emission), the δ-function in (18) becomes a Lorentzian:

$$\eta = \frac{e^2 h}{4\varepsilon_0 n_r mc} \frac{\sin^2\theta}{\cos\theta} n_{2D} f \frac{1}{\pi} \frac{\Delta E}{(\hbar\omega - E_{10})^2 + \Delta E^2}, \tag{22}$$

where ΔE is the half width. Ideally, in the absence of other elastic broadening mechanisms (e.g., by interface roughness and well width fluctuations), ΔE is related to the lifetime by $\tau_{\text{life}} = \hbar/(2\,\Delta E)$. For convenience, we will use (22) as a model, even when elastic broadening is not negligible as in the following discussions. For GaAs at $10\,\mu m$ wavelength, the constant $e^2 h/4\varepsilon_0 n_r mc$ in Eq. (22) equals 5.2×10^{-16} eV-cm^2. Taking $n_{2D} = 10^{12}$ cm^{-2} and $\Delta E \sim 0.01$ eV, peak absorption (at $\hbar\omega = E_{01}$) of a fraction of a percent is expected, which is consistent with measured results for absorption per quantum well [20].

3.2. Calculations for a Symmetric Well

The proposal by Coon and Karunasiri [3] to use ITs for IR detection predicted large absorption quantum efficiencies (up to 50% for only one well) when the upper state (the IT absorption final state) is nearly in resonance with the top of the barrier. The calculations of Coon and Karunasiri did not include lifetime broadening. When a 10 meV Lorentzian broadening factor is included in their expressions, we still obtain a large absorption quantum efficiency (more than 20%). This large absorption has never been observed experimentally: an absorption of a fraction of a percent per well per IR path at internal angles less than 45° is commonly observed [8, 10, 68, 20]. Asai and Kawamura studied the well width dependence of intersubband absorption using a series of samples which include the situation where the upper state is nearly in resonance with the top of the barrier, but no absorption enhancement was observed [68]. Some of our recent detector samples had the upper state very close to the top of the barrier, but we only observed absorptions less than one percent per well per IR path [20]. The question of whether one can obtain an extremely large absorption by positioning the upper state in resonance with the top of the barrier, motivated the work reported in Ref. [54]. This regime of having the upper state close to the top of the barrier is of practical importance, since it corresponds to a near-optimum detector design [19].

We reproduce the analytical results of Ref. [54] for IT absorption including both bound-to-bound and bound-to-continuum contributions and including Lorentzian broadening for the finite final state lifetime. Using these analytical results, the physical reason for the large predicted absorption in the limit of zero broadening and for the case when the upper (final) state is in resonance with the top of the barrier is discussed. Calculated absorption spectra are compared with the experiments of Asai and Kawamura [68].

Analytical results for the case of ITs between bound states have been given by West and Eglash for infinite barriers [8], and by Liu et al. for a general symmetric quantum well [22]. The result [22] for the bound-to-bound IT is first given below. The symmetric quantum well potential profile is shown schematically in Fig. 9. The ground (ψ_0) and the first excited (ψ_1) state wavefunctions are

$$
\psi_0 = C_0 \begin{cases} e^{\kappa_0(z + L_{w/2})} \cos k_0 L_{w/2} & \text{if } z < -L_{w/2} \\ \cos k_0 z & \text{if } -L_{w/2} \leqslant z \leqslant L_{w/2} \\ e^{-\kappa_0(z - L_{w/2})} \cos k_0 L_{w/2} & \text{if } z > L_{w/2} \end{cases} \tag{23}
$$

$$
C_0 = \frac{1}{\sqrt{L_{w/2} + (V/\kappa_0 E_0)\cos^2 k_0 L_{w/2}}}, \tag{24}
$$

with

$$
\cos k_0 L_{w/2} - \frac{m_b k_0}{m \kappa_0} \sin k_0 L_{w/2} = 0, \tag{25}
$$

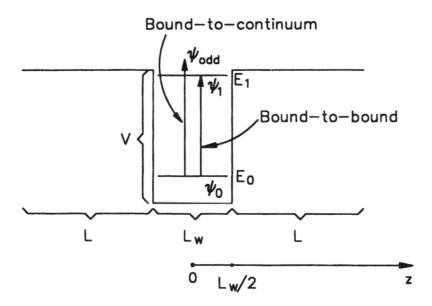

Fig. 9 The quantum well potential profile. The ground and the first excited states are labeled by E_0 and E_1, respectively. For small well widths (L_w), the first excited state is "pushed out" of the well and only the ground state exits. Both bound-to-bound and bound-to-continuum intersubband transitions are considered.

and

$$\psi_1 = C_1 \begin{cases} -e^{\kappa_1(z+L_{w/2})} \sin k_1 L_{w/2} & \text{if } z < -L_{w/2} \\ \sin k_1 z & \text{if } -L_{w/2} \leqslant z \leqslant L_{w/2} \\ e^{-\kappa_1(z-L_{w/2})} \sin k_1 L_{w/2} & \text{if } z > L_{w/2} \end{cases} \quad (26)$$

$$C_1 = \frac{1}{\sqrt{L_{w/2} + (V/\kappa_1 E_1)\sin^2 k_1 L_{w/2}}}, \quad (27)$$

with

$$\cos k_1 L_{w/2} + \frac{m\kappa_1}{m_b k_1} \sin k_1 L_{w/2} = 0, \quad (28)$$

where $k_{0,1} = \sqrt{2mE_{0,1}}/\hbar$, $\kappa_{0,1} = \sqrt{2m_b(V-E_{0,1})}/\hbar$, $E_0(E_1)$ is the ground (first excited) state eigenenergy, $m(m_b)$ is the well (barrier) effective mass, V is the barrier height, and $L_{w/2}$ is the half well width (i.e., the well width is $L_w = 2L_{w/2}$). Eqs. (25) and (28) determine the eigenenergies E_0 and E_1. In fact, all even parity bound states satisfy Eqs. (23), (24) and (25), and all odd parity bound states satisfy Eqs. (26), (27) and (28). We have taken the origin ($z = 0$) as the center of the well.

The above-barrier continuum eigenstates can be chosen to have even or odd parity. For ITs from the (even parity) ground state, only the odd parity continuum states are allowed as final states:

$$\psi_{odd} = \frac{1}{\sqrt{L}} \begin{cases} \sin[k'(z + L_{w/2}) - \beta] & \text{if } z < -L_{w/2} \\ (\sin^2 kL_{w/2} + (m_b k/mk')\cos^2 kL_{w/2})^{-1/2}\sin kz & \text{if } -L_{w/2} \leqslant z \leqslant L_{w/2} \\ \sin[k'(z - L_{w/2}) + \beta] & \text{if } z > L_{w/2} \end{cases} \quad (29)$$

where $k = \sqrt{2mE_z}/\hbar$, $k' = \sqrt{2m_b(E_z - V)}/\hbar$, E_z is the energy associated with only the z-direction motion, β is given by $\tan\beta = (mk'/m_b k)\tan kL_{w/2}$, and L is a normalization length on either side of the well (see Fig. 9). We have used a box-normalization scheme ($L \gg L_w$), and the above-barrier continuum states are normalized in a length $2L$.

The IT oscillator strength [8] mentioned in subsection 3.1 is given by

$$f \equiv \frac{2m\omega}{\hbar}|\langle z\rangle|^2 = \frac{2\hbar}{m\omega}|\langle \partial/\partial z\rangle|^2, \quad (30)$$

where $\langle \cdots \rangle$ represents a matrix element wavefunctions.

After some algebraic manipulation, the oscillator strength for the bound-to-bound transition [using ψ_0 and ψ_1 in (30)] is found to be

$$f_{B-B} = \frac{8\hbar C_0^2 V^2}{m\omega(E_1 - E_0)^2}\cos^2 k_0 L_{w/2}\frac{E_1\kappa_1\sin^2 k_1 L_{w/2}}{E_1\kappa_1 L_{w/2} + V\sin^2 k_1 L_{w/2}}, \quad (31)$$

and that for the bound-to-continuum (using ψ_0 and ψ_{odd}) is

$$f_{B-C} = \frac{8\hbar C_0^2 V^2}{m\omega L(E_z - E_0)^2}\cos^2 k_0 L_{w/2}\frac{(E_z - V)\tan^2 kL_{w/2}}{E_z + (E_z - V)\tan^2 kL_{w/2}}. \quad (32)$$

For analytical clarity, we have set $m = m_b$ in deriving the results given in Eqs. (31) and (32). This is a good approximation since ψ_0 is localized mainly in the well, with a very small exponential tail into the barrier. In determining the eigenenergies, however, we do include the difference between m and m_b, which is important especially for the excited state E_1.

We consider the two example measurement geometries shown in Fig. 10(a) and (b). The first geometry is relevant to the 45° facet detectors [13] and to the 45° zigzag waveguides. The second is

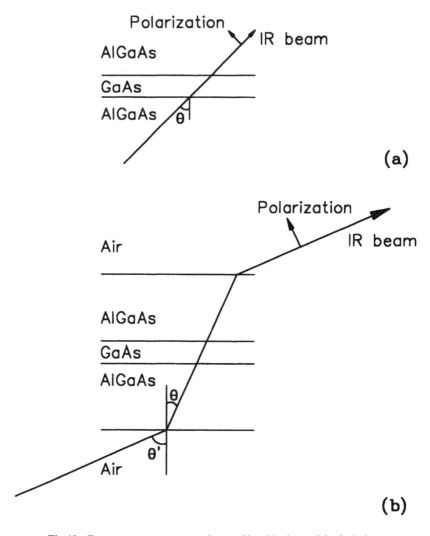

Fig. 10 Two measurement geometries considered in the model calculation.

commonly used for the Brewster angle transmission measurements [68]. From Eq. (22), the absorption quantum efficiency explicitly for Fig. 10(a) is given by

$$\eta = \frac{e^2 h}{4\varepsilon_0 n_r mc} \frac{\sin^2\theta}{\cos\theta} n_{2D} \sum_{\text{final}} \frac{1}{\pi\Delta E} \frac{1}{1 + [(E_{\text{final}} - E_0 - \hbar\omega)/\Delta E]^2} f, \quad (33)$$

where E_{final} is the final state energy associated with the z-direction motion, the polarization of the light is in the plane of incidence (p-polarized), and the summation is over the final states. The contribution from the bound-to-bound transition is trivially evaluated, and the contribution from bound-to-continuum transitions involves converting the sum into a integral by $\sum \to \int_0^x L\,dk'/\pi$. We then have

$$\eta = \frac{e^2 h}{4\varepsilon_0 n_r mc} \frac{\sin^2\theta}{\cos\theta} n_{2D} \frac{1}{\pi\Delta E} \left[\frac{1}{1 + [(E_1 - E_0 - \hbar\omega)/\Delta E]^2} f_{B-B} \right.$$

$$\left. + \frac{L\sqrt{2m}}{2\pi\hbar} \int_V^x \frac{dE_z}{\sqrt{E_z - V}} \frac{1}{1 + [(E_z - E_0 - \hbar\omega)/\Delta E]^2} f_{B-C} \right]. \quad (34)$$

Note that the above is independent of L, as it should be, because $f_{B-C} \propto L^{-1}$ [see Eq. (32)].

Because of the divergence in the one-dimensional density of states at $E_z = V$ [the $1/\sqrt{E_z - V}$ factor in the second term in Eq. (34)], the bound-to-continuum contribution can be large when the broadening is small. In fact, in the limit of $\Delta E \to 0$, one obtains a divergent result. We will discuss this divergence explicitly to show that it is an effect of the density of states together with a special arrangement of the quantum well. In the $\Delta E \to 0$ limit, the Lorentzian lineshape becomes a δ-function, and the second term in in Eq. (34), i.e., $\sum f_{B-C}$, becomes

$$\sum_{final} f_{B-C} = \frac{8 C_0^2 V^2 \sqrt{2m}}{2\pi^2 \Delta E m\omega (E_z - E_0)^2} \cos^2 k_0 L_{w/2}$$

$$\times \frac{1}{\sqrt{E_z - V}} \frac{(E_z - V)\tan^2 kL_{w/2}}{E_z + (E_z - V)\tan^2 kL_{w/2}}. \quad (35)$$

The conservation of energy gives $E_z = E_0 + \hbar\omega$. Let us concentrate on the following factor from Eq. (35):

$$\frac{1}{\sqrt{E_z - V}} \frac{(E_z - V)\tan^2 kL_{w/2}}{E_z + (E_z - V)\tan^2 kL_{w/2}}$$

$$= \frac{1}{\sqrt{\hbar\omega - (V - E_0)}} \frac{[\hbar\omega - (V - E_0)]\tan^2 kL_{w/2}}{(E_0 + \hbar\omega) + [\hbar\omega - (V - E_0)]\tan^2 kL_{w/2}}. \quad (36)$$

One can see immediately that the divergence in the density of states at $\hbar\omega = V - E_0$ is normally canceled by the factor $\hbar\omega - (V - E_0)$ in the numerator from the oscillator strength f_{B-C} except when $\tan^2 kL_{w/2} = \infty$, i.e., $kL_{w/2} = \pi/2 + P\pi$, where $P = 0, 1, 2, \ldots$.

In order to clarify the physical correspondence of this "selection rule" for $kL_{w/2}$, we rewrite the upper state eigenenergy condition [Eq. (28)] in the following form:

$$\tan k_1 L_{w/2} = -\frac{m_b k_1}{m \kappa_1}. \tag{37}$$

When $\kappa_1 \to 0$, we have $\tan^2 k_1 L_{w/2} \to \infty$ for which $k_1 L_{w/2} = \pi/2 + P\pi$. This implies that the condition $k_1 L_{w/2} = \pi/2 + P\pi$ corresponds to an odd parity "bound" state which is exactly in resonance with the top of the barrier. The physical reason for the divergent absorption is clear: it arises from the combination of the (1) one-dimensional density of states and (2) having the final state in resonance with the top of the barrier. One therefore might expect that a very large absorption could be obtained in a quantum well specially designed to this situation [3]. However, this does not occur in reality, because of the broadening factor. This, is shown below, where calculated examples with realistic values for the broadening are given and are compared with experiments.

Figure 11 shows calculated absorption spectra for a single path through one quantum well. The measurement geometry corresponds to Fig. 10(a). The internal angle of incidence is 45°, and the IR light is p-polarized. Parameters appropriate for an $Al_{0.33}Ga_{0.67}As$-GaAs quantum well were used. The GaAs well width was varied from 35 Å to 65 Å in 3 Å steps. The well reduced effective mass is 0.067 and that for the barrier is 0.094. The barrier height used was 0.25 eV which corresponds to 61% of the bandgap difference [$0.76 \times x$ eV in Eq. (1)]. The broadening full width $2\Delta E$ was taken to be 10 meV. The two-dimensional electron density in the well n_{2D} is 9×10^{11} cm^{-2}. The crossover from one bound state to two bound states occurs in the range of 48–49 Å. From Fig. 11, the absorption spectra become narrower in lineshape and higher in peak strength when the well width is increased. There is no abrupt change in the spectra when crossing from the pure bound-to-continuum case ($L_w \leqslant 47$ Å) to the case where both bound-to-bound and bound-to-continuum ITs contribute to the absorption ($L_w \geqslant 50$ Å).

To compare with experiments, including the interesting situation of crossing from one bound state to two bound states, one needs to measure a series of samples having either different well widths or different barrier heights. The only systematic study of the well width dependence of ITs in this crossover regime found in the literature was by Asai and Kawamura [68]. Some of their results are reproduced in

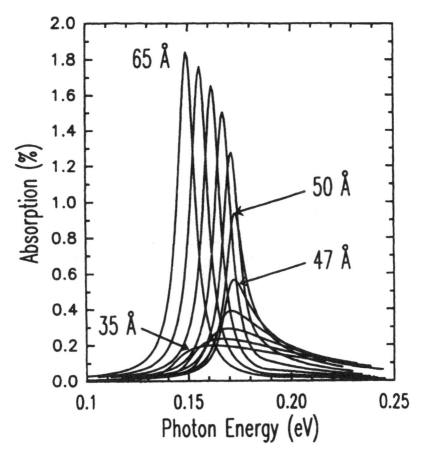

Fig. 11 Calculated absorption quantum efficiency vs. photon energy for one well for different well well widths from 35 Å to 65 Å. Parameters appropriate to $Al_{33}Ga_{67}As$-GaAs were used. The electron density in the well is $9 \times 10^{11}\,cm^{-2}$, and the internal angle of incidence is 45° (p-polarized).

Fig. 12 (dashed line). Unfortunately, only the relative transmittance was reported [68] and it is not possible to convert their data into absolute absorption. The calculated absorption spectra, using parameters appropriate for their 70-period InGaAs-InAlAs-quantum well structure grown on InP, are shown in Fig. 12. The InGaAs well width is varied from 35 Å to 100 Å. The well reduced effective mass is 0.042 and that for the barrier is 0.075. The barrier height used is 0.47 eV. The broadening 2 ΔE is taken to be 20 meV. The two-dimensional electron density in the well n_{2D} is $L_w \times (1.5 \cdot 10^{18}\,cm^{-3})$.

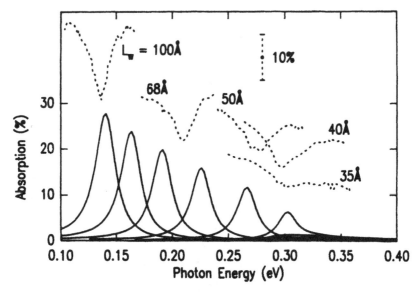

Fig. 12 Experimental transmittance data (dashed lines) of Asai and Kawamura [Appl. Phys. Lett. **56**, 1146 (1990)], and calculated absorption quantum efficiency vs. photon energy (solid lines) for 70 wells for different well well widths of 35, 40, 50, 60, 70, 80, 90, and 100 Å. Parameters appropriate to InGaAs-InAlAs quantum wells grown in InP were used. The electron density in the well is $L_w \times (1.5 \cdot 10^{18} \text{cm}^{-3})$, and the internal angle of incidence is 17° (p-polarized), where L_w is the well width.

The Brewster angle measurement geometry is shown in Fig. 10(b). The external (internal) angle of incidence is 73° (17)° and the IR light is p-polarized. Note that for the Brewster geometry, the factor $\sin^2\theta/(n_r\cos\theta)$ reduces to $1/(n_r^2\sqrt{n_r^2+1})$ [8]. Since the bandgap of the well material (InGaAs) is much narrower than that of GaAs, it is necessary to include the effect of band nonparabolicity to obtain an agreement of the calculated absorption peak positions with experiments. The calculated curves in Fig. 12 were obtained using the same formalism for the calculation of Fig. 11, but included an energy dependent effective mass [69] with parameter $\alpha = 1.24 \text{ eV}^{-1}$. The general agreement between our calculations and the measurements of Asai and Kawamura is good.

The major conclusions of this subsection are that (1) the expected divergence [3] of the absorption coefficient is an artifact arising from neglecting the final-state broadening, and (2) in a more realistic calculation the absorption spectra change smoothly in the region of the crossover between a one bound state and a two bound state configurations. For a pure bound-to-bound transition, the broadening determines the width of the absorption spectrum and the height of the absorption changes slowly. When approaching the crossover, the width of the absorption spectrum broadens further, due to contributions of the continuum states, and the height of the absorption peak decreases. The absorption spectrum for a pure bound-to-continuum case becomes even broader and asymmetrical in lineshape, while the peak absorption decreases even more. To obtain a high detector responsivity, however, one must trade-off between the high absorption and the fast electron escape (see Sec. 4.3). The highest responsivity is achieved approximately at the crossover point.

4. PHOTOCURRENT

4.1. Photocurrent and Photoconductive Gain

In this subsection, we discuss the resulting photocurrent due to ITs and introduce the concept of photoconductive gain. We consider only the case of positive photoconductivity, i.e., the effect of the incident IR light is to make the device resistance smaller. Negative photoconductivity is achievable, e.g., if one has a device having a negative differential resistance region [59].

The conventional theory of photoconductivity [70] is normally employed to describe QWIPs [13, 16, 18, 71, 72]. A model specifically for QWIPs has been constructed [57], which answers exactly what constitutes the mechanism of photoconductive gain and exactly how it depends on device parameters such as the number of wells. The model also explains observations of large ($\gg 1$) photoconductive gains [20, 71].

For a photoconductive device, the photocurrent given by the conventional theory is [70]

$$I_{photo} = e\Phi\eta\frac{\tau_{life}}{\tau_{trans,tot}} \equiv e\Phi\eta g, \tag{38}$$

where Φ is the incident number of photons per second, η is the absorption quantum efficiency (i.e., number of photons absorbed divided by the incident number of photons), τ_{life} is the lifetime of the carrier, and $\tau_{trans,tot}$ is the total transit time across the detector active region. The photoconductive gain is defined by $g \equiv \tau_{life}/\tau_{trans,tot}$. For a bulk photoconductor, derivation of Eq. (38) is straightforward using a simple rate equation consideration. Let n_{ex} be the number of the 3D photoexcited electrons, we have $n_{ex} = \Phi \eta \tau_{life}$ by balancing the photo-excitation and recombination. Then, the photocurrent is $I_{photo} = e(n_{ex}/L_{tot}A_{dev})vA_{dev} = e\Phi\eta\tau_{life}v/L_{tot}$, where A_{dev} is the device area, L_{tot} is the device length, and v is the drift velocity of the excited electrons travelling towards the collector. The total transit time is, by definition, $\tau_{trans,tot} = L_{tot}/v$. Eq. (38) is then obtained. A simple consequence is that g is inversely proportional to the length (L_{tot}) of the photoconductor because $\tau_{trans,tot}$ is proportional to the length. Moreover, if η is proportional to the length of the active region, the magnitude of the photocurrent is *independent* of the device length. (This becomes that the photocurrent is independent of the number of wells for QWIPs [20] as shown later). The conventional theory is constructed for a uniform and homogeneous photoconductor. For a QWIP (see Fig. 1) on the other hand, photons are only absorbed in the wells. The width of the barriers is normally much thicker than that of the wells. Some photoexcited electrons are recaptured in traversing subsequent wells before reaching the collector contact. It is not obvious that the conventional theory should accurately describe a QWIP with discrete active regions (wells) separated by wide inactive regions (barriers). Here we will show that the conventional theory of photoconductivity can be used for QWIPs under appropriate assumptions.

We have constructed a model [57] for evaluating the gain under the same assumptions (a)–(d) in Sec. 2. In addition, we assume that (e) The incident light intensity is low, so that the photocurrent is at most a fraction of the dark current, and the photoexcited electrons follow the same path as dark electrons.

This assumption is expected to be a good approximation, especially for an optimized QWIP where the barrier is thick and the first excited state is in resonance with the top of the barrier [19, 54]; whereas in an unoptimized device with narrow wells, so that the photoexcited electrons have a somewhat higher initial energy than the barrier height, the assumption (e) is not strictly valid. Moreover, in most applications the incident IR signal intensity is low. A high intensity tends to drive the device away from the assumed conditions (b) and (c) [26].

Based on the analysis specifically for QWIPs [57] (see next subsection) we can arrive at an expression if we re-express the quantities in Eq. (38) and define the trapping or capture probability for an electron traversing a well:

$$p = \frac{\tau_{esc}}{\tau_{life} + \tau_{esc}}, \tag{39}$$

where τ_{esc} is the escape time for a photoexcited electron from the well. Making the following key approximation: $\tau_{esc} \approx \tau_{trans}$, where τ_{trans}, is the transit time for an electron across *one* quantum region (as in Sec. 2). We point out that, physically, the escape time (τ_{esc}) for an excited electron is identical to the time that an electron spends traversing a well region, if we view the "size" of the electron is approximately equal to the length of one quantum well region. (This, of course, cannot be valid if the MQW period is much larger than the excited electron coherence length.) The photoconductive gain becomes

$$g = \frac{1-p}{Np},$$ (40)

so that

$$I_{photo} = e\Phi\eta g,$$ (41)

where $\tau_{trans,tot} = N\tau_{trans}$ and N is the number of wells or periods in the MQW structure. For a QWIP, the lifetime τ_{life} is associated *only* with those processes that scatter an electron into the bound state in the well (trapping). The lifetime therefore equals the intersubband relaxation time. We stress that the definitions of various times here are not rigorous: since the electron de Broglie wavelength is comparable to the length scales (for wells and barriers) involved, a direct identification of an electron located in the well or the barrier is not valid. The escape time for an excited electron in the well is therefore somewhat ill-defined, and could include a fraction of the electron transit across a barrier region. In addition, the intersubband relaxation time introduced here is somewhat "loose" because one could calculate the intersubband relaxation rate from two very different viewpoints, simply by choosing different initial wavefunctions: an extended state or a wave packet located in the well region. We therefore have expressed the gain in terms of a more precisely defined quantity: the capture or trapping probability p. A similar photoconductive gain expression for QWIPs, valid only for $p \ll 1$, was given by Serzhenko and Shadrin [73].

According to Eq. (40) the numerical value of g is arbitrary: $g \geqslant 0$ because $0 < p \leqslant 1$. Gain values much larger than unity have been observed [20, 71]. This seems unreasonable at first because one can have more than one electron in I_{photo} for one photon absorbed. Here we present a simple physical picture [57] of the gain mechanism using an example of a single well structure. The top part of Fig. 13 shows the dark current paths (same as in Fig. 2 but now in dark arrows). All

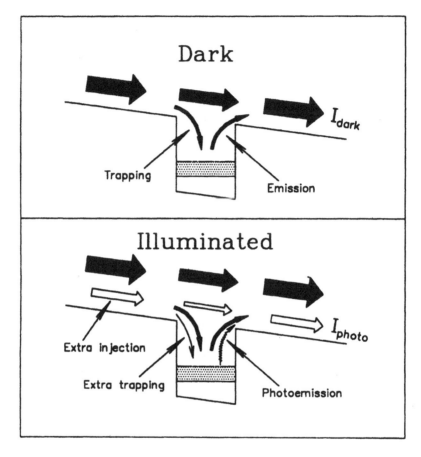

Fig. 13 The photoconductive gain mechanism. The top part shows the dark current paths, while the bottom indicates the direct photoemission and the extra current injection from the contact to balance the loss of electrons from the well. The dark current paths remain the same under illumination. The collected total photocurrent is the sum of the direct photoexcited and the extra injection contributions.

these dark current paths remain *unchanged* when the infrared illuminates the detector (by assumption). The additional processes as a result of the infrared are shown in the bottom part of Fig. 13. There is a *direct* photoemission of electrons from the well which, of course, contributes to the observed photocurrent. This direct photoemission tends to lower the electron density in the well, and therefore must be balanced in the steady state. The photoconductive gain is a result of the extra current injection from the perfect contact. The injection is necessary to balance the loss of electrons from the well due to photoemission. The amount of the extra injection must be sufficiently large so that the fraction trapped in the well equals the direct photoemission current. The fraction of the extra injected current that reaches the collector contact is in fact indistinguishable from the direct photoemitted current, and therefore contributes to the observed photocurrent. The total photocurrent consists of contributions from the direct photoemission and the extra current injection. Note that the physical mechanism given here is the same as for a conventional photoconductor [70], although this simple physical picture was presented only recently [57]. The common physical picture used to explain larger than unity gain states that photoexcited electrons circulate around the circuit several times. Although this seems plausible and it appears in text books (see, e.g., page 97 of Ref. [74]), this picture is completely wrong since a collector "absorbs" all electrons and the excess energy of the electrons give rise to the Ohmic heating. The basic picture of an "absorbing" collector contact is widely used in treating ballistic transport.

Using Eq. (40) for different values of p, the calculated photoconductive gain vs. the number of wells in is plotted Fig. 14, along with some existing experimental data [13, 18, 20, 71, 72]. Our experiments [20] were performed on samples with comparable parameters (grown one after the other) except the number of wells. A capture probability of about 0.08 is inferred from our data. Most of the reported detector samples have 50 quantum wells. It is seen from Fig. 14 that a range of gain values from about 0.27–0.80 for 50 well samples have been observed, and hence the gain is very sample dependent. The difference in τ_{trans} between samples is a possible reason for the spread of the observed gain values. Another possibility is the variation in τ_{scatt}. Processes that result in trapping are due to scattering by impurities and electrons in the well region, phonons, and interface roughness. Experiments [72] suggest that the impurity and electron-electron scattering may not be the dominant mechanism because the observed gain values did not decrease systematically with increasing well doping density. Phonon scattering would probably result in comparable values of capture probability for similar structures, and may not explain the strong sample dependence. Interfaces between AlGaGs and GaAs could be very different from sample to sample and from one crystal growth facility to another. Further investigations on minimizing trapping effects are needed, including theoretical work to calculate capture probability for all possible mechanisms and experimental

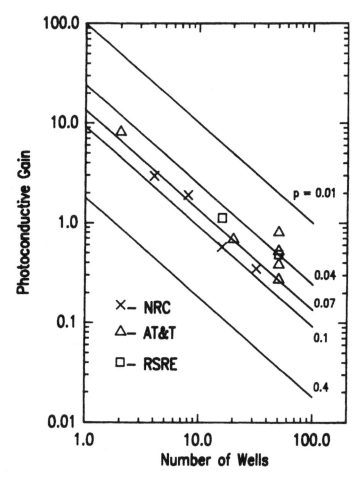

Fig. 14 Calculated photoconductive gain vs. the number of wells for capture probability $p = 0.01, 0.04, 0.07, 0.1$, and 0.4, respectively. Experimental data are taken from Liu,
et al. (cross), Levine, et al. (triangle), and Kane et al. (square).

work to obtain the optimum growth conditions for minimizing scattering processes. Hot electron and above-barrier resonance effects
[75] can be studied by analyzing the energy dependence of the capture
probability.

We can make some estimates of the time scales involved. From the
measured intersubband absorption line width (normally about a tenth
of the absorption energy), a lower bound on τ_{life} of about 50 fs is

obtained for 9–10 μm peak wavelength QWIPs. From time resolved experiments [76], an upper bound on τ_{life} of about 500 fs is expected. Therefore, we have 50 fs $\leqslant \tau_{life} \leqslant$ 500 fs for normal samples. The electron escape time can be *estimated* by $\tau_{esc} \sim (L_w/2v)/D$, where L_w is the well width, v is the "velocity" defined by the kinetic energy of an excited electron in the well region, and D is the transmission coefficient across a potential step (of which a well is formed). This estimate of τ_{esc} is a lower bound since we do not include any portion of the barrier region. For a typical 10 μm QWIP, τ_{esc} is estimated to be about 5 fs assuming a kinetic energy in the *barrier* region of about a half of the longitudinal optical phonon energy. One therefore expects a capture probability $[p = \tau_{esc}/(\tau_{life} + \tau_{esc})]$ in the range of $0.01 \leqslant p \leqslant 0.1$ consistent with existing experiments (see Fig. 14).

Quantitatively calculating τ_{life} is a highly complex problem [77, 78]. This problem is also of key importance to the operation of (interband) quantum well lasers [79]. Many experiments have been carried out in an attempt to measure the fundamental carrier intersubband relaxation time [76, 80, 81, 82, 83]. There is a need for both theoretical and experimental investigations of intersubband relaxation and lifetime specifically for QWIP structures, i.e., heavily doped quantum wells with at most two bound states. A detailed understanding may lead to better QWIP designs in terms of enhancing the lifetime. Large variations of carrier relaxation time from about 1 ps to 20 ps have been predicted. Experimental evidence obtained using a series samples with varying well width supports the predictions [79, 84, 85]. One may therefore design a quantum well with a special shape or with a certain range of parameters in order to enhance the lifetime, and hence the detector performance.

4.2. Derivation of the Photoconductive Gain Expression

In this subsection, we present a detailed derivation of the photoconductive gain expression Eq. (40).

Further clarification of assumption (c) is required since this is an important point both for the dark current model and for the theory of photoconductivity (see previous subsection). To provide a good injecting contact, the barrier between the emitter contact and the MQW region must not be substantially higher than the barrier of MQW. In most cases, this barrier approximately equals the barrier separating wells in the MQW region. If needed, an extra injection of electrons is achieved by increasing the electric field at the emitter-MQW junction.

This is a self-consistent process: for example, if one adds an extra emission channel of electrons in the wells (e.g., by photoemission), the wells will tend to become slightly positive, which increases the electric field at the emitter-MQW junction and hence increases the injection to balance the loss of electrons isn the wells [26]. The mechanism is shown schematically in Fig. 15.

We first calculate the photocurrent resulting from the *direct* excitation of electrons into the continuum or the first excited state subband which is close to the top of the barrier (the zigzag arrow shown in the bottom part of Fig. 13). The photocurrent directly ejected from *one* well is

$$i^{(1)}_{\text{photo}} = e\Phi\eta^{(1)}\frac{\tau_{\text{life}}}{\tau_{\text{life}} + \tau_{\text{esc}}} \equiv e\Phi\eta\frac{1-p}{N}, \qquad (42)$$

where the superscript (1) indicates quantities for one well, $p \equiv \tau_{\text{esc}}/(\tau_{\text{life}} + \tau_{\text{esc}})$ is the capture or trapping probability for an excited electron traversing a well, $\eta \equiv N\eta^{(1)}$ is the total absorption quantum efficiency, and N is the number of wells. The derivation of Eq. (42) is straightforward from a rate equation consideration. Letting n_{ex} be the number of the excited electrons, we have

$$\frac{dn_{\text{ex}}}{dt} = \Phi\eta^{(1)} - \frac{n_{\text{ex}}}{\tau_{\text{esc}}} - \frac{n_{\text{ex}}}{\tau_{\text{life}}}. \qquad (43)$$

Under steady state $dn_{\text{ex}}/dt = 0$, then we may solve for n_{ex} from Eq. (43). The photocurrent emitted from one well is $en_{\text{ex}}/\tau_{\text{esc}}$, which gives

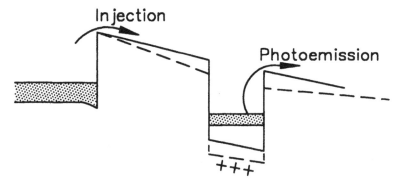

Fig. 15 The mechanism of injection. The bandedge profile with and without the photoemission is shown in dashed and solid lines, respectively.

Eq. (42). In the second half of Eq. (42), we have assumed that the absorption is weak ($\eta^{(1)} \ll 1$). The photon flux could depend on the location of the well, but this is very specific to the detector light coupling geometry (e.g., 45° facet coupling [10] or grating coupling [17]). Taking into account the capture of photoexcited electrons traversing the subsequent wells, the direct photocurrent collected in the collector contact from all the wells is then

$$i_{\text{photo}} = i_{\text{photo}}^{(1)} \sum_{n=1}^{N} (1-p)^{n-1} = i_{\text{photo}}^{(1)} \frac{1-(1-p)^N}{p}, \tag{44}$$

where each term in the summation represents the contribution from the nth well.

We then calculate the contribution from the extra injection (the empty arrows in the bottom part of Fig. 13). The extra dark current injected from the emitter contact must be such that none of the quantum wells gain or lose electrons, i.e., the well electron densities remain constant as function of time. An injection of

$$\delta I_{\text{inject}} = i_{\text{photo}}^{(1)}/p \tag{45}$$

would balance the *net* loss of electrons due to photoemission and due to recapture of the photoexcited electrons for *all* the wells. The demonstration is trivial for the well next to the emitter (the Nth well) and can be shown easily for all wells. For the jth well, the loss of electrons (same for any well) comes from the photoemission $i_{\text{photo}}^{(1)}$, and the refilling comes from (1) trapping of the dark current injection δI_{inject} $(1-p)^{N-j}p = i_{\text{photo}}^{(1)} (1-p)^{N-j}$ and (2) trapping of the photoemitted electrons from all the up-stream wells (N to $j+1$) $\sum_{k=j+1}^{N} i_{\text{photo}}^{(1)}$ $(1-p)^{k-j-1}p = i_{\text{photo}}^{(1)} [1-(1-p)^{N-j}]$. Adding the two refilling contributions balances the loss, and hence the electron density remains constant as a function of time. Using Eq. (45), the fraction of extra current reaching the collector contact is

$$\delta i = (1-p)^N \delta I_{\text{inject}} = i_{\text{photo}}^{(1)} \frac{(1-p)^N}{p}. \tag{46}$$

This extra current (though due to dark current injection) would be observed as photocurrent; hence, using Eqs. (44), (46), and (42), the total photocurrent is

$$I_{\text{photo}} = i_{\text{photo}} + \delta i = i_{\text{photo}}^{(1)}/p = e\Phi\eta \frac{1-p}{Np}. \tag{47}$$

From the above, the photoconductive gain for a QWIP is given by $g \equiv (1 - p)/(Np)$, which is Eq. (40). Note that the total photocurrent I_{photo} equals the extra dark current injection $i^{(1)}_{\text{photo}}/p$ because the photocurrent must be the same at the emitter and the collector. From the above [Eq. (47)], the current responsivity $[I_{\text{photo}}/(h\nu\Phi)]$ is *independent* of the number of wells if $\eta \propto N$, which is approximately true in the limit of weak absorption per well or for the grating-waveguide light coupling geometry [17]. This does not mean that the detector performance is independent of the number of wells because of noise consideration [18] (see Secs. 5 and 6). This independence on N can be easily understood by examining a two well case as shown in Fig. 16. The well shown on the left follows the same argument as in the one well case (Fig. 13). By simply realizing that the extra injection (photocurrent) is the same in the barrier separating the two wells, the well shown on the right would be equivalent to the left well because the photocurrent is the "extra injection" for this well. The same argument can be made for any subsequent wells. This means that the magnitude of the photocurrent is unaffected by adding more wells as long as the magnitude of absorption and hence photoemission from all the wells remains the same. To compare with experiments [20], Fig. 17 shows measured absorption and responsivity data for samples with different well numbers. The absorption experiment is done at room temperature under P-polarized IR light at Brewster angle incidence. The absorption increases nearly proportional to the number of wells as expected. The responsivity is measured in the 45°-facet geometry at 80 K detector temperature and at about 30 kV/cm applied field. It is clearly seen that the responsivity is independent of the number of wells [20].

Since the photoconductive gain is given by $g = (1 - p)/(Np)$, another direct consequence is that for the same drift velocity the gain should

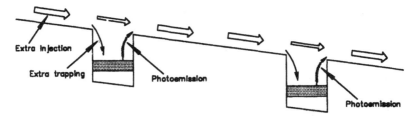

Fig. 16 A two well case which illustrates the independence of photocurrent on number of wells. The left well is next to the emitter. The extra injection is the observed photocurrent, and the same injection balances the photoemission from both wells.

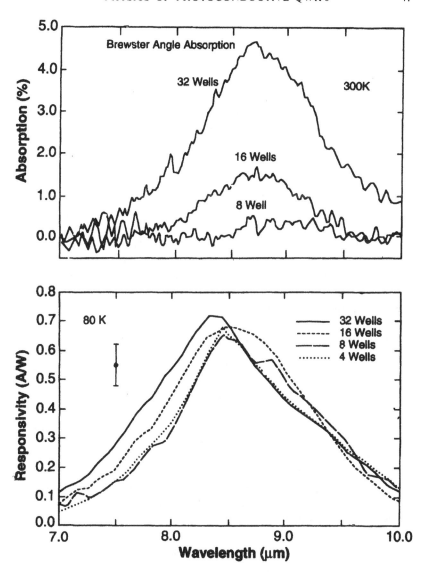

Fig. 17 Room temperature Brewster angle incidence absorption (top) and 45 -facet detector responsivity at 80 K (bottom) for samples having 4, 8, 16, and 32 wells.

be insensitive to the thickness of the barriers. Experimental testing proves this to be true [86].

The model discussed in this subsection can also be easily extended to treat the inter-well tunneling regime with relatively thin barriers where, in the absence of avalanche gain, the photoconductive gain does not exceed unity. The responsivity for a QWIP with tunneling barriers would depend on the number of wells.

4.3. Detector Responsivity

In this subsection, we estimate QWIP's responsivities using practical values for the photoconductive gain. The estimated responsivity as a function of well width and consequently as a function of the upper state position, is compared with experiments [19].

The responsivity magnitude is controlled by both quantum efficiency and photoconductive gain. A high absorption does not necessarily result in a high detector responsivity. There must not only be high absorption but the photoexcited electrons must also escape the wells efficiently to give rise to a large photocurrent. From Eq. (41), the spectral current responsivity of a photoconductive detector is given by

$$\mathcal{R}_i = I_{photo}/(h\nu\Phi) = eg\eta/(h\nu). \qquad (48)$$

For QWIPs, g is given by Eq. (40). which is simplified to be

$$g = \tau_{life}/(N\tau_{esc}). \qquad (49)$$

For quantum wells having only one bound state (the bound-to-continuum case) this escape time τ_{esc} is simply estimated by the transit time across the quantum well region; whereas for two bound states (the bound-to-bound case) photoexcited electrons must tunnel out of the well from the upper state. Generally, we write

$$\tau_{esc} = \tau_{tunnel} + \tau_{QW}, \qquad (50)$$

where τ_{tunnel} is the time for an electron tunneling out of the upper state. There is an uncertainty in defining τ_{QW}, because the time spent by an electron in the "vicinity" of a quantum well region includes the well and *some portion* of the barriers on both sides of the well (assuming that the barriers are wide, e.g., wider than about 250 Å). As a result, we can only make estimates based on existing experiments.

For 50-period QWIPs ($N = 50$), experimental gain values are in the range of $0.2 \leqslant g \leqslant 1$ [57] at the typical operating field of about 25 kV/cm. Consequently, we have [using Eq. (49)] $10 \leqslant (\tau_{life}/\tau_{esc}) \leqslant 50$. The intersubband lifetime is in the range of $50 \leqslant \tau_{life} \leqslant 500$ fs as discussed in Sec. 4.1. For concreteness, let us take $\tau_{life} = 500$ fs and $g = 0.4$ (case 1), and $\tau_{life} = 500$ fs and $g = 0.7$ (case 2). The tunneling contribution to τ_{esc} in Eq. (50) for a bound-to-continuum QWIP is non-existent. We then have (using $N = 50$ and $\tau_{life} = 500$ fs) $\tau_{QW} = 25.0$ fs for case 1 and $\tau_{QW} = 14.3$ fs for case 2. For a quantum well with two bound states, the upper state tunneling time is estimated easily by considering an "attempt frequency" $v_1/2L_w$ and a transmission coefficient D: $\tau_{tunnel} \approx (2L_w/v_1)D^{-1}$ where $v_1 = \sqrt{2}_1/m$. This estimate is semiclassical but does produce excellent results in comparison with rigorous calculations. The electric field dependent transmission coefficient is easily estimated using a WKB method.

Fig. 18(a) shows the calculated peak spectral responsivity vs. well width using the same parameters as those for Fig. 11. The solid line is for case 1($g = 0.4$), and the dashed line is for case 2 ($g = 0.7$). We assume that the spectral peaks correspond to the maxima in Fig. 11 for different well widths. This assumption is not rigorously valid [87]. The electric field value is 25 kV/cm. The detector geometry considered is that of a 45° edge-facet detector [13]. The IR beam passes through the quantum well structure twice (equivalent to having 100 wells) due to reflection from the top of the detector. We divide the absorption by two to convert into an absorption for *un-polarized* IR light, i.e., Fig. 18(a) is the responsivity for un-polarized light. The crosses in Fig. 18(a) are our experimental results [19]. An additional experimental data point (the square) is included, but the detector parameters are somewhat different in both the Al fraction and well width [20]. Because the most sensitive parameter is the energy separation between the upper state and the top of the barrier, this point is plotted at the well width corresponding to a separation value of about 10 meV. The estimated results in Fig. 18(a) have large uncertainties (up to almost a factor of two) as shown by the two different gain values, but the agreement with experiments is clearly seen.

For direct comparison, Fig. 18(b) shows the peak absorption vs. well width taken from Fig. 11. Comparing Fig. 18(a) and (b) clearly indicates that the highest peak spectral responsivity is achieved when the upper state is nearly in resonance with the top of the barrier [19, 20]. The slight shift of the maximum peak responsivity to a larger well width than that for the crossover (48–49 Å) is due to the applied

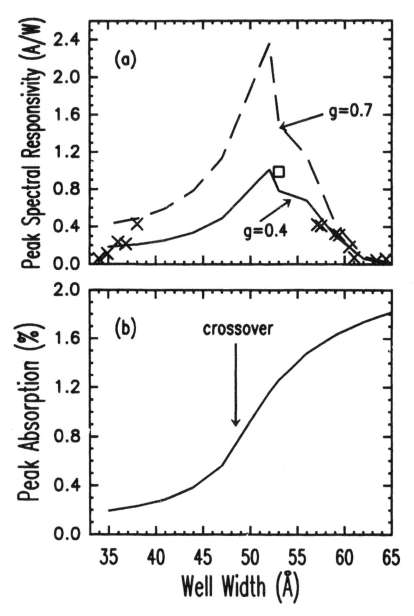

Fig. 18 (a) Estimated peak spectral responsivity vs. well width for photoconductive gain values of 0.4 (solid line) and 0.7 (dashed line), and experimental results (crosses and square), and (b) calculated peak absorption quantum efficiency vs. well width for one well and one infrared path at an internal angle of incidence of 45°. The arrow indicates the crossover from the bound-to-continuum to the bond-to-bound case.

field which "pulls" down the effective barrier for a photoexcited electron to escape the quantum well.

5. DARK CURRENT NOISE AND BACKGROUND NOISE

In general, a photoconductive detector has several sources of noise: the $1/f$ noise, Johnson noise, noise caused by the dark current, and noise results from the fluctuation of the number of photons (photon noise). The physical mechanism of the $1/f$ is still not well understood. The $1/f$ noise is important at low frequencies and it increases as the contact quality decreases. For GaAs QWIPs, good Ohmic contacts are achievable and the $1/f$ noise seldom limits the detector performance. We therefore neglect the contribution of the $1/f$ noise. Johnson noise is inherent to all resistive devices; the noise mean square current is

$$i_{n,J}^2 = 4k_B T \, \Delta f / R_d, \qquad (51)$$

where Δf is the measurement bandwidth and R_d is the device differential resistance. Johnson noise is easily calculated once the device current-voltage (I-V) curve is known. The contribution is usually small in a QWIP at operating bias voltage. Contributions from the dark current noise and the photon noise often limit the detector's ultimate performance. Here we concentrate on these two mechanisms.

In deriving noise expressions, the author has found the simple physical picture given by Rose (see p. 97–99 of Ref. [70]) to be extremely convenient. One identifies the noise source α_n and the magnification factor F in observation. Then the noise (square-average) is

$$I_n^2 = 2F^2 \alpha_n \, \Delta f. \qquad (52)$$

For the dark current noise, we treat (as in Sec. 3) the case where inter-well tunneling can be neglected. Because of the dark current transport mechanism (see the top part in Fig. 13), the dark current noise is generation-recombination (g-r) in nature.

The noise current should be given by the standard g-r noise expression [70]

$$i_{n,\text{dark}}^2 = 4eg_{\text{noise}}I_{\text{dark}} \, \Delta f, \qquad (53)$$

where g_{noise} is the noise gain, and I_{dark} is the device dark current. In a conventional photoconductor the noise gain is the photoconductive gain $g_{\text{noise}} = g$ (at least as an very good approximation for all pratical purposes). We show in the following that the two gains are, strictly speaking, not identical and display measurable difference for QWIPs. In fact, if we label the emission current (see the top part in Fig. 13) from one well as $i_{em}^{(1)}$ then the dark current is

$$I_{dark} = i_{em}^{(1)}/p = i_{em}/(Np),$$ (54)

where $i_{em} \equiv N \times i_{em}^{(1)}$ is the total emission current from all N wells. Equivalently, one can express the dark current in the alternative form $I_{dark} = i_{trap}^{(1)}/p$, where $i_{trap}^{(1)}$ is the trapping current per well and $i_{trap}^{(1)} = i_{em}^{(1)}$. The "$g$-$r$" (emission and trapping here) noise therefore consists of two contributions: the fluctuations in i_{em} and in i_{trap}. The magnification factor is $1/(Np)$ according to Eq. (54). Then we have

$$
\begin{aligned}
i_{n,dark}^2 &= 2e\left(\frac{1}{Np}\right)^2 (i_{em} + i_{trap})\,\Delta f \\
&= 4e\left(\frac{1}{Np}\right)^2 i_{em}\,\Delta f \\
&= 4e\frac{1}{Np}I_{dark}\,\Delta f \\
&\equiv 4eg_{noise}I_{dark}\,\Delta f,
\end{aligned}
$$ (55)

where the noise gain is defined by $g_{noise} \equiv 1/(Np)$. We can immediately see that g_{noise} is different from the photoconductive [Eq. (40)] $g = (1-p)/(Np)$. We recall that p is the capture or trapping probability, and $1-p$ therefore represents the probability that a dark current transport electron escapes from a well. Note that because we neglect inter-well tunneling, the capture probability for transport electrons associated with the dark current and that for photoelectrons are the same, i.e., the dark current and the photocurrent follow the same path with only the difference in emission mechanism from the wells. Experiments of Levine et al. [14, 42] reported a gain derived from the ratio of the measured current responsivity and the obsorption (i.e., the photoconductive gain), and a gain derived from the direct noise measurements (the noise gain). The ratio of the two measured gains is the escape probability $1-p$, and experimentally $1-p$ approached unity as the bias voltage was increased [14, 42].

We now give an empirical expression for the escape probability p_e vs. electronic field. Physically, the trapping process is controlled by events that scatter electrons from a barrier transport state onto the confined two-dimensional subband. Therefore, one expects the functional dependence of the trapping probability on the field to be similar

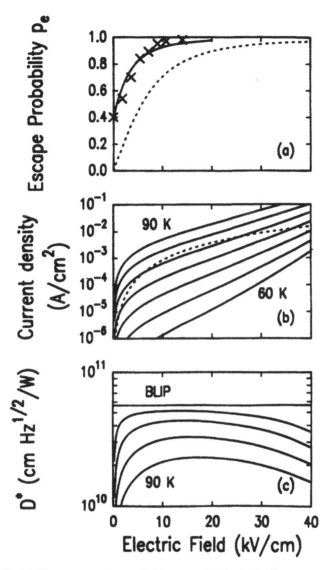

Fig. 19 (a) Electron trapping probability vs. electric field. The crosses represent measured results, and the solid curve is the fit to these data using our empirical expression. A model (dashed) curve for the detector used in part (b) and (c) is also shown. (b) Calculated dark current density vs. electric field for an 8.7 μm detector at different temperatures of 60, 65, 70, 75, 80, 85, and 90 K. the intersecting points of the dark current curves with the dashed curve determine the background limited infrared performance (BLIP) temperatures. The dashed curve is plotted for evaluating the BLIP temperature. (c) Calculated detectivity (*D**) vs. electric field for the 8.7 μm detector at different temperatures of 90, 85, 80, and 75 K. The horizontal line is the BLIP *D**.

to that of drift mobility. We write

$$p_e = 1 - p = p_0 \frac{(\varepsilon + \varepsilon_0)/\varepsilon_{sat}}{\sqrt{1 + [(\varepsilon + \varepsilon_0)/\varepsilon_{sat}]^2}} \tag{56}$$

where the fitting constants are p_0, ε_0, and ε_{sat}. When $\varepsilon_0 > 0$ and $\varepsilon \to 0$, a nonvanishing escape probability is obtained. This corresponds to the case where the quantum well confines only one bound state, i.e., a bound-to-continuum transition QWIP. The asymptotic high field value of p_e is p_0 which sould be nearly 100%. Fitting to the experimental data is Levine et al. [14] for the bound-to-continuum case, we found $p_0 = 1$, $\varepsilon_0 = 2.2 \, \text{kV/cm}$, and $\varepsilon_{sat} = 5 \, \text{kV/cm}$. The fitted p_e vs. ε curve is shown in Fig. 19(a). The case of $\varepsilon_0 < 0$ corresponds to a bound-to-bound transition QWIP (i.e., two bound states exist in the well), and one must take $p_0 = 0$ for $\varepsilon < -\varepsilon_0$ instead of Eq. (56). Physically, a finite field is needed to obtain $p_0 > 0$ for the bound-to-bound case. The optimum detector design (as discussed in subsection 4.3) corresponds to the upper state nearly degenerate with the barrier height for which $\varepsilon_0 \approx 0$. We stress that Eq. (56) should be viewed as an empirical equation used for curve fitting. Levine et al. [14] used an exponential function which also produced good fit to the experimental data of p_e vs. ε.

The detector noise associated with the background radiation is caused by fluctuations in the number of background photons absorbed by a detector ($\eta \Phi_B$). For a photoconductor, the noise current due to this mechanism is given by

$$i_{n,B}^2 = 4e^2 g^2 \eta \Phi_B \, \Delta f. \tag{57}$$

The above includes both emission and trapping processes. Unlike the dark current noise $i_{n,dark}$ which is proportional to $g_{noise}^{1/2}$, the background noise $i_{n,B}$ is directly proportional to the photoconductive gain, i.e., $i_{n,B} \propto g$.

6. DETECTOR PERFORMANCE

In a given system and application the background photon flux is fixed. One would therefore like to reduce dark current to reach the regime where the noise contribution from the dark current is negligible. The background limited infrared performance (BLIP) is defined as the regime where the dominant noise source is due to the background photon fluctuations. BLIP can be achieved at higher detector operating temperatures for devices with larger gain values. The BLIP temperature (T_{BLIP}) is defined as the temperature at which $i_{n,dark} = i_{n,B}$. Equating (55) and (57), we obtain an equation to solve for T_{BLIP}:

$$e\frac{(1 - \bar{p})^2}{Np}\eta\Phi_B = I_{dark}|_{T = T_{BLIP}}, \tag{58}$$

where we have used $g = (1 - p)/Np$ and $g_{noise} = 1/Np$. In the weak absorption limit, Eq. (58) is independent of N because $\eta\Phi_B \propto N$ and because I_{dark} vs. electric field is independent of N [51]. To improve T_{BLIP}, one needs to maximize the quantity on the left-hand-side of Eq. (58). In designing a QWIP, N must be large enough to obtain a high absorption quantum efficiency, but increasing N indefinitely will result in a lower T_{BLIP}. Minimizing p would result in an improvement of T_{BLIP}. Current GaAs-AlGaAs QWIPs have values of p in the range of 0.02–0.1 at high electric fields [57].

For completeness, we give an expression for the most important detector figure of merit. The detectivity (D^*) is given by

$$D^* = \frac{eg\eta/hv}{\sqrt{i_{n,dark}^2/\Delta f + i_{n,B}^2/\Delta f}}$$

$$= \frac{g\eta/hv}{\sqrt{4g_{noise}J_{dark}/e + 4g^2\eta\Phi_B/A_{dev}}}, \tag{59}$$

where η is the absorption quantum efficiency, hv is the photon energy, and J_{dark} is the dark current density. Higher T_{BLIP} and D^* can be achieved with higher gain values. At high enough electric field (e.g., $> 10 \, kV/cm$), the two gains are approximately equal: $g_{noise} \approx g$ because $p \ll 1$. Once the gain value is large enough so that the second term in the denominator of Eq. (59) dominates, the detectivity D^* is independent of both the dark current and the gain.

Some results calculated based on our model are shown in Figs. 19 and 20. The dashed curve in Fig. 19(a) is the escape probability vs. electric field, with parameters $p_0 = 1$, $\varepsilon_0 = 0$, and $\varepsilon_{sat} = 10 \, kV/cm$. These parameters are appropriate for our detectors with the upper state nearly in resonance with the top of the barrier [20]. The dark current density for an 8.7 μm peak response detector at different temperatures is shown in Fig. 19(b). The dashed curve is the left-hand-side of Eq. (58), and therefore the intersecting points correspond to T_{BLIP}. Fig. 19(c) shows the calculated peak detection wavelength D^* including dark current and background noise contributions. The horizontal line is for the BLIP D^*, i.e., D^* calculated without the dark current

H.C. LIU

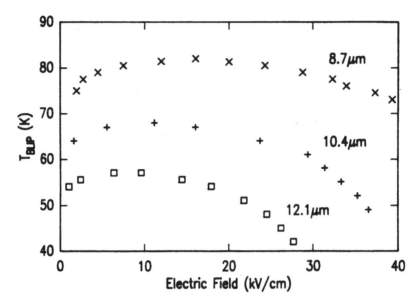

Fig. 20 Dector operating temperature for background limited infrared performance. The three curves correspond to an 8.7 μm peak response detector with half points at 7.8 and 9,6 μm (crosses), a 10.4 μm peaked detector covering 9.4–11.4 μm (pluses), and a 12.1 μm peaked detector covering 10.9–13.3 μm (squares).

TABLE 4

Parameters used in calculations. The Fermi energy was 18 meV corresponding to a 5×10^{11} cm^{-2} two-dimensional doping density in the wells, the mobility was 1000 cm^{2}V^{-1}s^{-1}, the saturated velocity was 10^7 cms^{-1}, and detectors all have 50 wells and 50.0 nm wide barriers. The symbols V_0, E_0, m_b^*, and L_w are for the barrier height, the ground state eigenenergy, the barrier reduced effective mass, and the well width, respectively.

Sample	V_0(eV)	E_0(meV)	m_b^*	L_w(nm)
8.7 μm	0.211	59.4	0.089	5.9
10.4 μm	0.178	49.4	0.085	6.5
12.1 μm	0.154	42.3	0.083	7.1

noise. The detector spectral bandwidth is 1.8 μm, and this detector therefore covers the 7.8–9.6 μm range. The parameters used in these calculations are listed in Table 4, and the dark current model is that given in Sec. 2 [13, 51]. In calculating $\eta\Phi_B$, we have used a 90° (full cone angle) FOV for a 300 K blackbody background, and a

Lorentzian lineshape for the absorption vs. photon energy with a full width at the half maximum corresponding to 20% of the photon energy at the peak absorption. The peak absorption is taken to be 100%, which is achievable with 50 wells and a grating-waveguide light coupling scheme [17, 88]. Figure 20 shows the calculated T_{BLIP} vs. applied field for three detectors corresponding to an 8.7 μm peak response detector with half points 7.8 and 9.6 μm peaked detector covering 9.4–11.4 μm, and a 12.1 μm peaked detector covering 10.9–13.3 μm. Our estimated T_{BLIP} values are higher than those of Kinch and Yariv [89], who used device parameters appropriate for lower absorption quantum efficiencies.

Both T_{BLIP} and D^* depend on the FOV angle. Because of fewer background photons on the detector, a smaller FOV angle results in a higher D^* once the BLIP condition is reached, however a lower tempertaure is required to reach BLIP. For the 10.4 μm peaked detector and for 30, 60, and 90° FOV, the BLIP D^* is 1.53×10^{11}, 7.91×10^{10}, and 5.60×10^{10} cm Hz$^{1/2}$/W, and T_{BLIP} at the optimum electric field of above 10 kV/cm is 61, 66, and 68 K, respectively. The dependence of T_{BLIP} on FOV for two examples is shown in Fig. 21. The recent work of Andersson and Lundqvist [88] indicates that the quantum efficiency remains constant at least up to 30° FOV.

We now comment on the possible improvements of T_{BLIP}. Looking at the factors involved in Eq. (58) an obvious approach to try is to reduce the dark current. For the 8.7 μm detector, if we change the mobility or the saturated velocity in Table 4 by a factor less than 10, the predicted T_{BLIP} values hardly change. The most sensitive parameter is the well electron density which affects the dark current exponentially. It is therefore highly desirable to decrease the well doping density, but one must insure a high absorption while decreasing the density. Recent experiments using random scattering optical couplers [90] have shown a large improvement in the device responsivity. This should allow a five-fold reduction in electron density while maintaining a nearly 100% absorption. Changing the density from 5×10^{11} to 1×10^{11} cm^{-2} for the 8.7 μm detector results in a maximum T_{BLIP} of 92 K for a 90° FOV, which is a 10 K improvement in comparison with Fig. 20.

7. CONCLUDING REMARKS

In this chapter, we have given a comprehensive summary of the physics of QWIP in its simplest embodiment. We have emphasized the

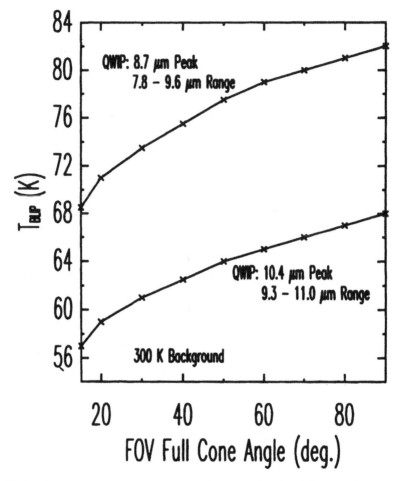

Fig. 21 Detector operating temperature for background limited infrared performance vs. field of view (FOV) full cone angle. The two curves correspond to an 8.7 μm peak response detector with half points at 7.8 and 9.6 μm, and a 10.4 μm peaked detector covering 9.4–11.4 μm.

physical picture and have presented the basic formulas for modeling. We like to stress that almost all the discussions are based on phenomenological models, which reflects the present stage of our understanding. Clearly, further work on many different aspects is required. A full calculation of various scattering processes is of critical importance. This will likely result in ways to enhance the carrier lifetime and photoconductive gain. Schemes to reduce the dark current should also be investigated, e.g., by thin AlAs barriers [91], to enhance the T_{BLIP}. More detailed modeling of dark current, e.g., using the Monte Carlo technique [92], would be helpful.

By making use of the MBE capability, many new approaches which are both interesting scientifically and practically will certainly evolve. The natural potentials of QWIPs are high speed [93–95], multiband and multicolor [96–99], and monolithic integration with electronic circuits.

In conclusion, the continued research and development will bring this technology out of laboratories into practical applications.

Acknowledgments

I thank B.F. Levine of AT&T Bell Laboratories for many discussions. I have also benefited from communications with E.R. Brown of MIT Lincoln Laboratory, E. Rosencher of Thomson CSF, J.G. Simmons of McMaster University, and J.M. Xu and R.Q. Yang of University of Toronto. The work at NRC has been carried out in collaboration with M. Buchanan, H. Chu, E. Dupont, G.E. Jenkins, M. Lamm, J. Li, A.G. Steele, J.R. Thompson, Z.R. Wasilewski, and P.H. Wilson. I also thank R. Borris, P. Chow-Chong, M. Davies, P. Marshall, and J. Stapledon for sample preparation, and E.V. Kornelsen for his strong support in starting up the IT activity at NRC. This work was supported in part by DND DREV.

8. LIST OF VARIABLES

A — normalization area
A_{dev} — device area
$\vec{A}$ — vector potential
C_0 — normalization constant for ψ_0
C_1 — normalization constant for ψ_1
c — speed of light
D — transmission coefficient
D^* — detectivity
E — total energy
E_z — energy associated with z motion
E_{xy} — in-plane motion energy
E_0 — ground state eigenenergy
E_1 — first excited state eigenenergy

$\Delta E-$ broadening half width

ΔE_C – conduction band offset

ε – electric field

ε_{sat} – field parameter in empirical p expression

e – elementary charge

F – magnification factor in observation

f – oscillator strength

f_{B-B} – bound-to-bound oscillator strength

f_{B-C} – bound-to-continuum oscillator strength

$f_{i/f}$ – Fermi factor for initial/final state

Δf – measurement bandwidth

g – photoconductive gain

g_{noise} – noise gain

H_{rad} – Hamiltonian for radiative absorption

H_z – z direction effective mass Hamiltonian

h – Planck constant

$\hbar$ – reduced Planck constant

I_{dark} – dark current

I_n – noise spectral density

I_{photo} – photocurrent

δI_{inject} – injection current

$i_{n,B}$ – background noise current spectral density

$i_{n,dark}$ – dark current noise spectral density

$i_{n,J}$ – Johnson noise spectral density

i_{photo} – direct photoemission current reaching collector

$i_{photo}^{(1)}$ – direct photoemission current from one well

i_{em} – thermionic emission current

i_{trap} – trapping current

δi – extra current reaching collector

J_{dark} – dark current density

j_{em} – emission current density from well

$j_{thermionic}$ – pure thermionic current density from well

j_{trap} – trapping current density into well

j_{tunnel} – tunneling current density from well

j_{3D} – three-dimensional current density in barrier

k – electron wavevector in well

k' – electron wavevector in barrier

k_B – Boltzmann constant

k_0 – ground state wavevectorr in well

k_1 – first excited state wavevectorr in well

$\vec{k}_{xy}, \vec{k}'_{xy}$ – in-plane wavevectors

L – wavefunction normalization length
L_b – barrier thickness
L_p – length of one period
L_{tot} – total length of multiple quantum well structure
L_w – well width
$L_{w/2}$ – half well width
$\mathscr{L}$ – a relevant length scale
M – matrix element
m – effective mass in well
m_0 – electron rest mass
m_b – effective mass in barrier
m^* – reduced effective mass in well
m_b^* – reduced effective mass in barrier
N – number of wells
n, n' – wavefunction indices
n_r – refractive index
n_{ex} – number of photoexcited electron
n_{2D} – two-dimensional electron density in well
n_{3D} – three-dimensional electron density in barrier
P – non-negative integer
p – trapping or capture probability
$\vec{p}$ – momentum operator
p_e – escape probability
p_z – momentum operator in z direction
p_0 – zero field trapping probability
$\vec{q}$ – photon momentum
q_{xy} – in plane photon momentum
q_z – photon momentum in z direction
R_d – device dynamic resistance
$\mathscr{R}_i$ – current responsivity
T – temperature
T_{BLIP} – temperature for background limited infrared performance
V – barrier height
V_0 – barrier height at zero bias
v – drift velocity
v_{ac} – ac voltage
v_{sat} – saturated velocity
v_1 – velocity in first excited state
x – Aluminum mole fraction
W – radiative transition rate
w – wavefunction extend

z_c – classical turning point

$\hat{z}$ – unit vector in z direction

α – nonparabolicity parameter

α_n – source of noise

β – phase parameter in continuum wavefunction

ε_0 – vacuum permittivity

ε_F – Fermi energy in well

$\hat{\varepsilon}$ – polarization unit vector

η – total absorption quantum efficiency

$\eta^{(1)}$ – absorption quantum efficiency for one well

κ_0 – imaginary wavevector for first excited state in barrier

κ_1 imaginary wavevector for first excited state in barrier

μ – mobility

ν – photon frequency

ω – photon angular frequency

Φ – photon number per unit time incident on detector

Φ_B – background photon number per unit time incident on detector

ϕ – photon flux

ψ, ψ' – envelope wavefunctions

ψ_{odd} – odd parity continuum state

ψ_{xy} – in plane envelope function

ψ_z – envelope functin in z direction

ψ_0 – ground state eigenfunction

ψ_1 – first excited state eigenfunction

θ – internal angle of incidence

θ' – external angle of incidence

τ_{esc} – escape time

τ_{life} – lifetime

τ_{QW} – time spent in vicinity of a quantum well

τ_{scatt} – scattering time

τ_{trans} – transit time across one well

$\tau_{trans,tot}$ – total transit time

τ_{tunnel} – tunneling time

References

1. L.L. Chang, L. Esaki, and G.A. Sai-Halaz, IBM Technical Disclosure Bulletin, **20**, 2019 (1977).
2. L. Esaki and H. Sakaki, IBM Technical Disclosure Bulletin, **20**, 2456 (1977).
3. D.D. Coon and R.P.G. Karunasiri, *Appl. Phys. Lett.*, **45**, 649 (1984).
4. D.D. Coon, R.P.G. Karunasiri, and L.Z. Liu, *Appl. Phys. Lett.*, **47**, 289 (1985).
5. D.D. Coon, R.P.G. Karunasiri, and H.C. Liu, *J. Appl. Phys.*, **60**, 2636 (1986).

6. K.W. Goossen and S.A. Lyon, *Appl. Phys. Lett.*, **47**, 1257 (1985).
7. K.W. Goossen and S.A. Lyon, *J. Appl. Phys.*, **63**, 5149 (1988).
8. L.C. West and S.J. Eglash, *Appl. Phys. Lett.*, **46**, 1156 (1985).
9. A. Harwit and J.S. Harris, Jr., *Appl. Phys. Lett.*, **50**, 685 (1987).
10. B.F. Levine, K.K. Choi, C.G. Bethea, J. Walker, and R.J. Malik, *Appl. Phys. Lett.*, **50**, 1092 (1987).
11. B.F. Levine, C.G. Bethea, K.K. Choi, J. Walker, and R.J. Malik, *Appl. Phys. Lett.*, **53**, 231 (1988).
12. B.F. Levine, C.G. Bethea, G. Hasnain, J. Walker, and R.J. Malik, *Appl. Phys. Lett.*, **53**, 296 (1988).
13. B.F. Levine, C.G. Bethea, G. Hasnain, V.O. Shen, E. Pelve, R.R. Abbott, and S.J. Hsieh, *Appl. Phys. Lett.*, **56**, 851 (1990).
14. B.F. Levine, A. Zussman, S.D. Gunapala, M.T. Asom, J.M. Kuo, and W.S. Hobson, *J. Appl. Phys.*, **72**, 4429 (1992).
15. B.K. Janousek, M.K. Daugherty, W. L. Bloss, M.L. Rosenbluth, M.J. O'Loughlin, H. Kanter, F.J. De Luccia, and L.E. Perry, *J. Appl. Phys.*, **67**, 7608 (1990), a comment by W.A. Beck and the reply appeared in **69**, 4129 (1991).
16. S.R. Andrews and B.A. Miller, *J. Appl. Phys.*, **70**, 993 (1991).
17. J.Y. Andersson and L. Lundqvist, *Appl. Phys. Lett.*, **59**, 857 (1991).
18. M.J. Kane, S. Millidge, M.T. Emeny, D. Lee, D.R.P. Guy, and C.R. Whitehouse, in Ref. [39], pp. 31–42.
19. A.G. Steele, H.C. Liu, M. Buchanan, and Z.R. Wasilewski, *Appl. Phys. Lett.*, **59**, 3625 (1991).
20. A.G. Steele, H.C. Liu, M. Buchanan, and Z.R. Wasilewski, *J. Appl. Phys.*, **73**, 1062 (1992).
21. H.C. Liu and D.D. Coon, *Superlatt. Microstruct.*, **3**, 357 (1987).
22. H.C. Liu, D.D. Coon, B.O, Y.F. Lin, and M.H. Francombe, *Superlatt. Microstruct.*, **4**, 343 (1988).
23. H. Schneider, F. Fuchs, B. Dischler, J.D. Ralston, and P. Koidl, *Appl. Phys. Lett.*, **58**, 2234 (1991).
24. H.C. Liu, M. Buchanan, Z.R. Wasilewski, and H. Chu, *Appl. Phys. Lett.*, **58**, 1059 (1991).
25. J.D. Ralston, M. Ramsterner, B. Dischler, M. Maier, P. Koidl, and S.J. As, *J. Appl. Phys.*, **70**, 2195 (1991).
26. E. Rosencher, F. Luc, Ph. Bois, and S. Delaitre, *Appl. Phys. Lett.*, **61**, 468 (1992).
27. C.G. Bethea, B.F. Levine, V.O. Shen, R.R. Abbott, and S.J. Hsieh, IEEE Trans. Electron Devices **38**, 1118 (1991).
28. L.J. Kozlowski, G.M. Williams, G.J. Sullivan, C.W. Farley, R.J. Anderson, J. Chen, D.T. Cheung, W.E. Tennant, and R.E. DeWames, IEEE Trans. Electron Devices **38**, 1124 (1991).
29. B.F. Levine, C.G. Bethea, K.G. Glogovsky, J.W. Stay, and R.E. Leibenguth, *Semicond. Sci. Technol.*, **6**, C114 (1991).
30. R.P.G. Karunasiri, J.S. Park, and K.L. Wang, *Appl. Phys. Lett.*, **59**, 2588 (1991).
31. H. Hertle, G. Schuberth, E. Gornik, G. Abstreiter, and F. Schäffler, *Appl. Phys. Lett.*, **59**, 2977 (1991).
32. J.S. Park, R.P.G. Karunasiri, and K.L. Wang, *Appl. Phys. Lett.*, **60**, 103 (1992).
33. H.C. Liu, J.-P. Noël, L.Li, M. Buchanan, and J.G. Simmons, *Appl. Phys. Lett.*, **60**, 3298 (1992).
34. B.F. Levine, *J. Appl. Phys.*, **74**, R1 (1993).
35. J.P. Loehr and M.O. Manasreh, in **Semiconductor Quantum Wells and Superlattices for Long-Wavelength Infrared Detectors**, Artech House, Boston, 1993, ch. 2, pp. 19–54, edited by M.O. Manasreh.
36. R.L. Whitney, K.F. Cuff, and F.W. Adams, in **Semiconductor Quantum Wells and Superlattices for Long-Wavelength Infrared Detectors**, Artech House, Boston, 1993, ch. 3, pp. 55–108, edited by M.O. Manasreh.

37. K.L. Wang and R.P.G. Karunasiri, in **Semiconductor Quantum Wells and Superlatti-ces for Long-Wavelength Infrared Detectors**, Artech House, Boston, 1993, ch. 5, pp. 139–205, edited by M.O. Manasreh.
38. F.F. Sizov and A. Rogalski, *Progress in Quantum Electronics*, **17**, 93 (1993).
39. E. Rosencher, B. Vinter, and B.F. Levine (editors), *Intersubband Transitions in Quantum Wells*, Plenum, New York, 1992.
40. H.C. Liu, B.F. Levine, and J.Y. Andersson (editors), *Quantum Well Intersubband Transition Physics and Devices*, Kluwer, Dordrecht, Netherlands, 1994.
41. P.H. Wilson, M. Lamm, H.C. Liu, J. Li, M. Buchanan, Z.R. Wasilewski, and J.G. Simmons, *Semicond. Sci. Technol.*, **8**, 2010 (1993).
42. B.F. Levine, A. Zussman, J.M. Kuo, and J. de Jong, *J. Appl. Phys.*, **71**, 5130 (1992).
43. J.S. Blakemore, *J. Appl. Phys.*, **53**, R123 (1982).
44. S. Adachi, *J. Appl. Phys.*, **58**, R1 (1985).
45. S.D. Gunapala, B.F. Levine, D. Ritter, R. Hamm, and M.B. Panish, SPIE Proc. **1541**, 11 (1991).
46. H.C. Liu, M. Buchanan, and Z.R. Wasilewski, *J. Appl. Phys.*, **68**, 3780 (1990).
47. H.C. Liu, A.G. Steele, M. Buchanan, and Z.R. Wasilewski, *J. Appl. Phys.*, **70**, 7560 (1991).
48. C. Weisbuch and B. Vinter, *Quantum Semiconductor Structures*, AP, Boston, 1991.
49. A.G. Petrov and A. Ya Shik, *Semicond. Sci. Technol.*, **6**, 1163 (1991).
50. E. Pelve, F. Beltram, C.G. Bethea, B.F. Levine, V.O. Shen, S.J. Hsieh, and R.R. Abbott, *J. Appl. Phys.*, **66**, 5656 (1989).
51. H.C. Liu, A.G. Steele, M. Buchanan, and Z.R. Wasilewski, *J. Appl. Phys.*, **73**, 2029 (1993).
52. K.K. Choi, B.F. Levine, R.J. Malik, J. Walker, and C.G. Bethea, *Phys. Rev.*, *B*, **35**, 4172 (1987).
53. H.C. Liu, J. Li, M. Buchanan, Z.R. Wasilewski, and J.G. Simmons, *Phys. Rev.*, *B*, **48**, 1951 (1993).
54. H.C. Liu, *J. Appl. Phys.*, **73**, 3062 (1992).
55. F. Luc, E. Rosencher, and B. Vinter, *Appl. Phys. Lett.*, **62**, 1143 (1993).
56. S.V. Meshkov, *Sov. Phys.*, JETP **64**, 1337 (1986).
57. H.C. Liu, *Appl. Phys. Lett.*, **60**, 1507 (1992).
58. D.D. Coon and H.C. Liu, *Superlatt. Microstruct.*, **3**, 95 (1987).
59. H.C. Liu, M. Buchanan, and Z.R. Wasilewski, *Phys. Rev.*, *B*, **44**, 1411 (1991).
60. H.C. Liu, G.C. Aers, M. Buchanan, Z.R. Wasilewski, and D. Landheer, *J. Appl. Phys.*, **70**, 935 (1991).
61. H.C. Liu, M. Buchanan, G.C. Aers, and Z.R. Wasilewski, *Semicond. Sci. Technol.*, **6**, C124 (1991).
62. K.M.S.V. Bandara, B.F. Levine, G. Sarusi, R.E. Leibenguth, M.T. Asom, and J.M. Kuo, in Ref. [40], pp. 111–122.
63. M.O. Manasreh, F. Szmulowicz, T. Vaughan, K.R. Evans, C.E. Stutz, and D.W. Fischer, in Ref. [39], pp. 287–297.
64. B.F. Levine and C.G. Bethea, private communication.
65. H.C. Liu, Z.R. Wasilewski, M. Buchanan, and H. Chu, *Appl. Phys. Lett.*, **63**, 761 (1993).
66. J.J. Harris, J.B. Clegg, R.B. Beall, J. Castagné, K. Woodbridge and C. Roberts, *J. Crystal Growth.*, **111**, 239 (1991), and references therein.
67. Z.R. Wasilewski, H.C. Liu, and M. Buchanan, *J. Vac. Sci. Technol.*, *B*, **12**, 1273 (1994).
68. H. Asai and Y. Kawamura, *Appl. Phys. Lett.*, **56**, 1149 (1990).
69. W. Chen and T.G. Andersson, *Phys. Rev.*, *B*, **44**, 9068 (1991).
70. A. Rose, *Concepts in Photoconductivity and Allied Problems*, Interscience Publishers, John Wiley & Sons, New York, 1963.
71. G. Hasnain, B.F. Levine, S. Gunapala, and N. Chand, *Appl. Phys. Lett.*, **57**, 608 (1990).

72. S.D. Gunapala, B.F. Levine, L. Pfeiffer, and W. West, *J. Appl. Phys.*, **69**, 6517 (1991).
73. F.L. Serzhenko and V.D. Shadrin, *Sov. Phys. Semicond.*, **25**, 953 (1991).
74. E.L. Dereniak and D.G. Crowe, *Optical Radiaiton Detectors*, John Wiley & Sons, New York, 1984.
75. B.F. Levine, C.G. Bethea, G. Hasnain, J. Walker, R.J. Malik, and J.M. Vandenberg, *Phys. Rev. Lett.*, **63**, 899 (1989).
76. M.C. Tatham, J.F. Ryan, and C.T. Foxon, *Phys. Rev. Lett.*, **63**, 1637 (1989).
77. S.M. Goodnick and P. Lugli, *Phys. Rev.*, *B*, **37**, 2578 (1988).
78. P. Sotirelis, P. von Allmen, and K. Hess, *Phys. Rev.*, *B*, **47**, 12744 (1993).
79. P.W.M. Blom, J.E.M. Haverkort, P.J. van Hall, and J.H. Wolter, *Appl. Phys. Lett.*, **62**, 1490 (1993).
80. A. Seilmeier, H.-J. Hübner, G. Abstreiter, G. Weimann, and W. Schlapp, *Phys. Rev. Lett.*, **59**, 1345 (1987).
81. J.A. Levenson, G. Dolique, J.L. Oudar, and I. Abram, *Phys. Rev. B*, **41**, 3688 (1990).
82. B. Deveaud, A. Chomette, D. Morris, and A. Regreny, *Solid State Commun.*, **85**, 367 (1993).
83. A. Seilmeier, U. Plödereder, J. Baier, and G. Weimann, in Ref.[40], pp. 421–432.
84. P.W.M. Blom, C. Smit, J.E.M. Haverkort, and J.H. Wolter, *Phys. Rev.*, *B*, **47**, 2072 (1993).
85. D. Morris, B. Deveaud, A. Regreny, and A. Auvray, in Ref. [40], pp. 433–442.
86. E. Rosencher, F. Luc, P. Bois, and Y. Cordier, *Appl. Phys. Lett.*, **63**, 3312 (1993).
87. E. Rosencher, E. Martinet, F. Luc, P. Bois, and E. Bockenhoff, *Appl. Phys. Lett.*, **59**, 3255 (1992).
88. J.Y. Andersson and L. Lundqvist, *J. Appl. Phys.*, **71**, 3600 (1992).
89. M.A. Kinch and A. Yariv, *Appl. Phys. Lett.*, **55**, 2093 (1989), see also a comment by B.F. Levine and the reply in **56**, 2354–2356 (1990).
90. B.F. Levine, G. Sarusi, S.J. Pearton, K.M.S.V. Bandara, and R.E. Leibenguth, in Ref. [40], pp. 1–12.
91. H.C. Liu, P.H. Wilson, M. Lamm, A.G. Steele, Z.R. Wasilewski, J. Li, M. Buchanan, and J.G. Simmons, *Appl. Phys. Lett.*, **64**, 475 (1994).
92. M. Artaki and I.C. Kizilyalli, *Appl. Phys. Lett.*, **58**, 2467 (1991).
93. C.G. Bethea, B.F. Levine, G. Hasnain, J. Walker, and R.J. Malik, *J. Appl. Phys.*, **66**, 963 (1993).
94. E.R. Brown, K.A. McIntosh, F.W. Smith, and K.M. Molvar, *Appl. Phys. Lett.*, **62**, 1513 (1993).
95. H.C. Liu, G.E. Jenkins, E.R. Brown, K.A. McIntosh, K.B. Nichols, and M.J. Manfra, *IEEE Elec. Dev. Lett.*, **16**, 253 (1995).
96. I. Gravé, A. Shakouri, N. Kruze, and A. Yariv, *Appl. Phys. Lett.*, **60**, 2362 (1992).
97. A. Köck, E. Gornik, G. Abstreiter, G. Böhm, M. Walther, and G. Weimann, *Appl. Phys. Lett.*, **60**, 2011 (1992).
98. H.C. Liu, J. Li, J.R. Thompson, Z.R. Wasilewski, M. Buchanan, and J.G. Simmons, *IEEE Elect. Dev. Lett.*, **14**, 566 (1993).
99. H.C. Liu, J. Li, Z.R. Wasilewski, M. Buchanan, P.H. Wilson, M. Lamm, and J.G. Simmons, in Ref. [40], pp. 123–134.

CHAPTER 2

Growth and Characterization
of GaInP/GaAs System for Quantum
Well Infrared Photodetector Applications

M. RAZEGHI

*Center for Quantum Devices, Department of Electrical Engineering
and Computer Science, Northwestern University, Evanston,
Illinois 60208 USA*

1. INTRODUCTION

The developement of new infrared (IR) detectors and imaging systems with high sensitivity and ease of manufacture is necessary for numerous military and commercial applications. The current technology, $Hg_x Cd_{1-x} Te$, has received the most attention to date, owing to the relative ease of changing the band gap over a wide range by small variations in composition x. However, the same versatility in band gap adjustment leads to one of the chief disadvantages of the HgCdTe system: small variations in composition produce large variations in the band gap and therefore lead to non-uniform optical absorption. For example, a 3% change in Hg content leads to a shift in the absorption edge from 12 μm to 14 μm. Compositional uniformity of better than 3% is quite difficult to obtain, especially in the growth of HgCdTe. The resultant non-uniformity is not well suited to large focal plane arrays. A second disadvantage of HgCdTe system is its lack of compatibility with monolithic formation on the Si integrated circuits that are used to derive information from the array. Poor expansion coefficient match makes the monolithic heteroepitaxial growth of HgCdTe on Si substrate quite difficult. In addition, HgCdTe system is fragile which restrict device processing yield and the application circumstances.

   GaAs-AlGaAs quantum well infrared photodetectors (QWIP) have demonstrated high detectivity and long wavelength photoresponse with narrow spectral linewidth in photoconductive devices [1]. It is certainly a viable approach that can replace HgCdTe.

However, AlGaAs has serious problems owing to the strong reaction between aluminum and oxygen. Even trace quantities of oxygen have a dramatically deteriorate effect on the quality of GaAlAs layers due to the easily introduction of deep level defects. In addition, AlGaAs alloys present a high surface recombination velocity. As a result, photonic devices based on GaAs-AlGaAs suffer from the catastrophic dark line defect formation, which result in accelerated degradation of the device. Oxidization of AlGaAs layers makes further regrowth and device fabrication difficult. High growth temperature, which is a common practice to decrease the oxygen incorporation in the layers, is also not compatible with monolithic integration. High density of donor-related deep traps, also known as DX center, in AlGaAs barrier is another drawback that affects the electrical properties of AlGaAs.

GaInP/GaAs system has, recently, received a large amount of attention as the potential replacement of AlGaAs/GaAs system for opto-electronic applications. In comparison with AlGaAs, GaInP lattice matched to GaAs shows a number of unique and interesting features in comparison to AlGaAs. It has been realized that the GaInP/GaAs interface is remarkable improved relative to AlGaAs/GaAs. The excess carrier recombination velocity at a GaInP/GaAs heterointerface [2] is more than a factor of ten lower than the lowest reported recombination velocity at a AlGaAs/GaAs heterointerface [3]. GaInP is also expected to have very low DX center due to the crossover of its direct (Γ) and indirect (X) conduction points lies at $x = 0.74$ for GaInP [4] which is far from the lattice matched composition ($x = 0.51$). It is logical to conclude from the above two features that the use of GaInP as the wide bandgap layer in a superlattice structure could result in an improvement in the performance of a superlattice detector.

Strong selective etching between GaAs and GaInP also make GaInP/GaAs system very promising on device fabrication.

In the following sessions, we are going to discuss the growth conditions of GaInP and the doping behavior in the GaInP. We will discuss the characterization result of GaInP and GaInP/GaAs heterostructure. Then, the characterization of GaInP/GaAs quantum confinement structure will be present as a glimpse to the state of art of the GaInP/GaAs quantum well and superlattice.

2. MOCVD GROWTH AND DOPING OF GaInP

2.1. Growth Condition of GaInP

The GaInP layers can be grown by MOCVD at the growth temperature between 500 to 600°C at either atmospheric pressure or low
pressure. Alkyls of Ga and In, and hydrides or alkyls of P can be used
as sources. Chemical reactions occuring among these sources are as
follows:

$$0.51 R_3 Ga + 0.49 R'_3 In + EH_3 \rightarrow Ga_{0.51} In_{0.49} P + nC_n H_{2n} \quad (1)$$

where R, R' and E can be methyl, ethyl, alkyl or hydride. Reaction
happens in a hydrogen ambient.

Follows are some examples:

$$0.49(C_2 H_5)_3 In + 0.51(C_2 H_5)_3 Ga + PH_3 \rightarrow InGaP + nC_2 H_6 \quad (2)$$

or

$$0.49(CH_3)_3 In + 0.51(CH_3)_3 Ga + PH_3 \rightarrow InGaP + nCH_4 \quad (3)$$

or

$$0.49(CH_3)_3 In + 0.51(C_2 H_5)_3 Ga + PH_3 \rightarrow InGaP + nC_2 H_6 \quad (4)$$

or

$$0.49(C_2 H_5)_3 In + 0.51(CH_3)_3 Ga + PH_3 \rightarrow InGaP + nC_2 H_6 \quad (5)$$

The optimum growth conditions by using triethylgallium (TEGa),
trimethylindium (TMIn), and pure phosphine (PH_3) as sources of Ga,
In and P, respectively, and H_2 as the carrier gas are listed in Table 1.

The growth rate of GaInP depends on the flow rates of TMIn and
TEGa (group III elements) only and is independent of PH_3 flow rate
(group V elements) and growth temperature under the growth

TABLE 1
Optimum growth parameters for GaAs and GaInP by MOCVD.

	GaAs	GaInP
Growth pressure	76 Torr	76 Torr
Growth temperature	510°C	510°C
Total H_2 flow rate	3 liter/min	3 liter/min
AsH_3 Flow rate	30 cc/min	–
H_2 through TMIn bubbler at 18°C	–	200 cc/min
H_2 through TEGa bubbler at 0°C	120 cc/min	120 cc/min
PH_3 flow rate	–	300 cc/min
Growth rate (dx/dt)	150 Å/min	200 Å/min

conditions listed in Table 1. The distribution coefficient of indium and gallium defined as

$$K = \frac{X_{Ga}^{S}}{X_{Ga}^{V}} \tag{6}$$

or

$$K = \frac{X_{In}^{S}}{X_{In}^{V}} \tag{7}$$

are nearly equal to unity. Where X_{α}^{β} is the flux of element α in the vapor phase $(\beta = V)$ or solid phase $(\beta = S)$. Figure 1 shows variation of growth rate of GaInP lattice matched to GaAs at growth temperature of $T_G = 510°C$ and growth pressure of 76 Torr.

Similar results have been reported [5] at growth temperatures from 600 to 650°C. They showed that there was no gas-phase reaction in their reactor leading to premature depletion of In or Ga.

Undoped GaInP layer grown under the conditions in Table 1 has a free electron carrier concentration of 5×10^{14} cm^{-3} with a mobility of

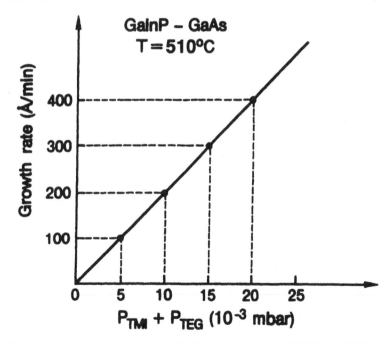

Fig. 1 Variation of growth rate of GaInP layers grown by MOCVD at 510°C as a function of flow rate of group III elements (Ref. 8).

$6000 \text{ cm}^2 \text{V}^{-1}\text{s}^{-1}$ at 300 K and $40,000 \text{ cm}^2 \text{V}^{-1} \text{ s}^{-1}$ at 77 K. No GaAs buffer layer is grown in this case [6].

2.2. Doping Behavior in GaInP

2.2.1. n-type Doping

GaInP layers grown by MOCVD can be doped n-type using group VI or group IV elements such as: S, Se, Te, Si or Ge. The hydride or organometallic sources of these elements are:

- Hydrogen sulphide H_2S
- Diethyl selenium $(C_2H_5)_2Se$
- Dimethyl selenium $(CH_3)_2Se$
- Hydrogen selenide H_2Se
- Diethyltelurium $(C_2H_5)_2Te$
- Dimethyltelurium $(CH_3)_2Te$
- Silane SiH_4
- Disilane Si_2H_6
- Germane GeH_4

Sulfur (S):

GaInP layer grown by MOCVD can be doped n-type using H_2S. Figure 2 shows the electrochemical profile of a GaInP-GaAs layer grown by MOCVD using the growth conditions in Table 1. 1000 ppm H_2S diluted in H_2 is used as the source of S. It is clear from this experimental result that under the same growth conditions with the same H_2S flow rate, the donor concentration in GaInP is much higher than in GaAs layer. The different curves in the figure correspond to the different H_2S flow rate. Figure 2 shows that the doping concentration in $Ga_{0.51}In_{0.49}P$ saturates at $4 \times 10^{18} \text{cm}^{-3}$. It was found that doping is still in the saturation region even with H_2S flow rate as low as 1 cc/min. It means that much more diluted H_2S is necessary to reach the linear zone where the doping concentration is proportional to the H_2S flow. This saturation effect is in good agreement with work done in a Chloride VPE system [7].

The effect of growth temperature on the dopant incorporation in the layers is shown in Fig. 3. The concentration of donor decreases as growth temperature increases. It can be fitted by the relation:

$$N_d - N_a = A \exp(-E_i/kT) \qquad (8)$$

with $E_i = -2.16 \text{ eV}$. Where A is a constant.

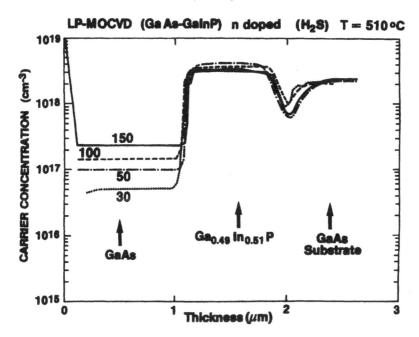

Fig. 2 Electrochemical polaron profile of a GaInP layer grown by MOCVD using H_2S as doping source. Different curves are related to different H_2S flow rate into the reactor (Ref. 8).

The phosphine flow rate D_{PH3} has been varied between 300 and 750 cc/min while keeping the growth temperture at 510°C. A relation between the doping concentration and the phosphine flow is found to be the form: $N_d - N_a \propto (D_{PH3})^{-0.7}$

In conclusion, the doping concentration of 1000 ppm H_2S in $Ga_{0.51}In_{0.49}P$ grown under a fixed group-III element flow (i.e. a growth rate of 200 Å/min) can be expressed as:

$$N_d - N_a = 7.0 \times 10^5 \exp(2.16/kT) \cdot (D_{PH3})^{-0.7} \qquad (9)$$

where kT is in eV and D_{PH3} is in cc/min.

Silicon (Si):

GaInP layer grown by MOCVD can be doped n-type using SiH_4 or Si_2H_6. In GaInP, Si substitutes Ga or In and plays the role of a donor. The doping concentration of Si in GaInP depends on different

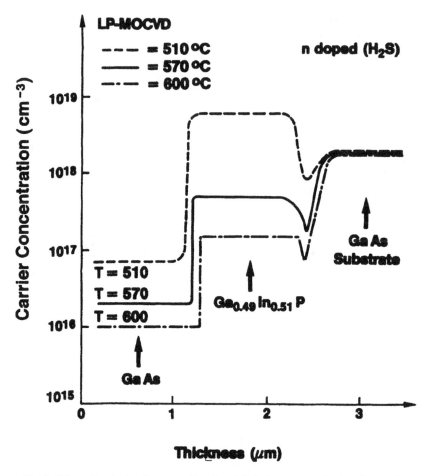

Fig. 3 Electrochemical polaron profile of a GaInP layer grown by MOCVD using H_2S as doping source. Different curves are related to different growth temperature, keeping other growth parameters constant (Ref. 8).

growth parameters such as growth temperature, growth pressure, growth rate, the ratio of III/V, and the flow rate of SiH_4 or Si_2H_6.

In contrast with H_2S doping of $Ga_{0.51}In_{0.49}P$, the doping concentration using SiH_4 varies linearly with SiH_4 flow (Fig. 4). The growth temperature is at 540°C and phosphine flow rate is 450 cc/min. By changing the growth temperature, an exponential dependence is found

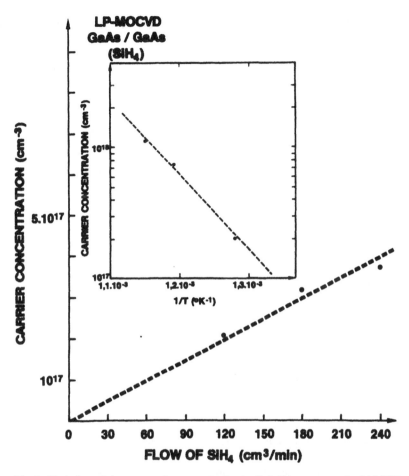

Fig. 4   Variation of electron carrier concentration in GaInP layer grown by MOCVD at 540 °C, using SiH$_4$ as a·doping source, as a function SiH$_4$ flow rate: inset is variation of carrier concentration, as a function of $1/T$ (Ref. 8).

(Fig. 5). The doping concentration can be written in the form:

$$N_d - N_a = A \exp(-E_i/kT) \qquad (10)$$

with $E_i = 1.26$ eV.

The influence of PH$_3$ flow rate on doping concentration in GaInP layers grown by MOCVD using SiH$_4$ as a n-type dopants has been studied [8]. By keeping the growth temperature equal to 540°C and

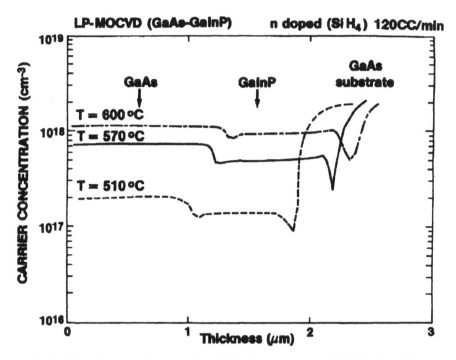

Fig. 5 Electrochemical polaron profile of GaInP layers grown by MOCVD using SiH$_4$ as a doping source for different growth temperature (Ref. 8).

the SiH$_4$ flow at 120 cc/min, PH$_3$ flow has been varied from 300 up to 750 cc/min.

Figure 6 shows that the doping concentration varies very slowly with PH$_3$ flow rate as $(D_{PH_3})^{-0.2}$. Similar results have been reported by Hsu et al. [5] and Kawamura et al. [9]. Due to the low diffusion coefficient and low vapor pressure, a sharp doping profile can be obtained for Si. Very high electroluminescence quality has been reported for Si-doped GaAs layer [10].

Gomyo et al. [11] studied the effects of Si doping on the E_g behaviour and on the $\langle \Omega, \Omega, \Omega \rangle$ sublattice ordering in GaInP grown at 700°C by MOCVD. The bandgap increased from a low value (1.836 eV) for an undoped crystal to a normal E_g value of 1.92 eV when the Si doping concentration increased to $n = 3.8 \times 10^{17} \text{cm}^{-3}$. At this doping condition, the transmission electron microscope study revealed substantial disordering of the $\langle \Omega, \Omega, \Omega \rangle$ sublattice. In a range from

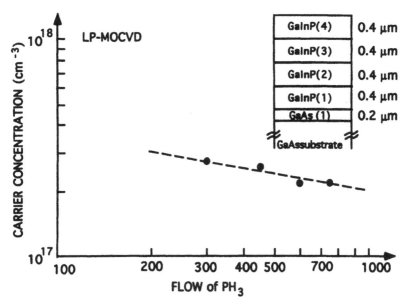

Fig. 6 Variation of elctron carrier concentration in GaInP layers grown by MOCVD, using SiH$_4$ as a doping source, as a function of PH$_3$ flow rate. Inset is the structure of epilayer (Ref. 8).

$n = 3.8 \times 10^{17}$cm^{-3} to $n = 2 \times 10^{18}$cm^{-3}, the normal E_g value remained virtually unchanged. They observed a Burstein shift (i.e., the energy shift towards a higher value due to a band-filling effect by free electrons) for $n > 2 \times 10^{18}$cm^{-3}.

The above behaviour was not observed in Si doped GaInP layer up to 5×10^{18}cm^{-3} at low growth temperature between 510 up to 550°C [6]. The E_g variation appears to depend more on the alloy composition other than on the doping concentration at low growth temperature.

Selenium (Se):

Gomyo et al. [11] also studied GaInP layer doped with Se under the same condition as those of Si doping. They found that the threshold for the sublattice disordering by Si doping ($n = 3.8 \times 10^{17}$cm^{-3}) is higher than that for the Se doping ($n = 2 \times 10^{18}$cm^{-3}). Two reasons were attributed for this difference. The first one is the difference in the sublattice occupation between Si and Se. Silicon atoms substitute the group-III sublattice and Se atoms group-V sublattice. This sublattice occupation affinity difference between Si and Se could be the cause for the difference in sublattice disordering. The second reason was the difference between the growth conditions. The V/III ratios of the growth for Si and Se doping were different.

However, further studies are required to understand the reason of ordering and disordering mechanisms in this system.

Telurium (Te):

GaInP layer grown by MOCVD can be doped n-type using Te atom as a doping element. In GaInP usually Te occupies the P site. Hsu et al. [12] used Diethyltelluride (DETe) diluted to 5.45 ppm in H_2 as the source for the n-type dopant. They found a linear relation between the dopant mole fraction in the gas phase and carrier concentrations in the range $10^{17} cm^{-3}$ up to $10^{19} cm^{-3}$ at growth temperature of 625°C. They reported a high distribution coefficient for Te with

$$K_{Te} = \frac{X_{Te}^S}{X_{Te}^V} = 54 \tag{11}$$

Hsu et al. [12] reported that the PL intensity of Te doped GaInP increases with carrier concentration until $n = 2 \times 10^{18} cm^{-3}$ and the PL half-width increases with increasing carrier concentration. They found that the electron mobility decreases from 1020 to 500 $cm^2 V^{-1} s^{-1}$ as carrier concentration increases from 10^{16} to $10^{19} cm^{-3}$.

2.2.2. p-type Doping of GaInP

GaInP layer grown by MOCVD can be doped p-type using Mg, Zn, and Cd. The sources for these elements are mainly organometallics.

Zinc (Zn):

$Ga_{0.51}In_{0.49}P$ can be doped p-type using diethylzinc (DEZn). The effect of growth temperature has been investigated between 500 and 600°C by keeping the DEZn flow constant and the phosphine flow equal to 450 cc/min. The doping concentration is found to vary linearly with DEZn flow (Fig. 7).

Figure 8 shows the evolution of the doping concentration as a function of $1/T$ under the growth temperature of 510°C and the phosphine

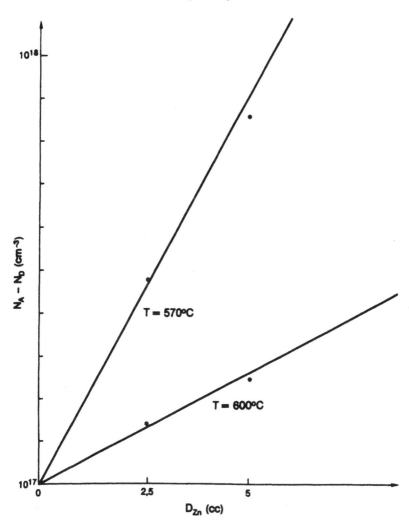

Fig. 7 Variation of hole concentration in GaInP as a function of flow rate of DEZn for different growth temperatures (Ref. 8).

flow rate of 450 cc/min. It varies exponentially and decreases as the growth temperature increases.

$$N_a - N_d = 1.6 \times 10^2 \, \text{cm}^{-3} \exp(-E_i/kT) \tag{12}$$

where $E_i = -2.5$ eV.

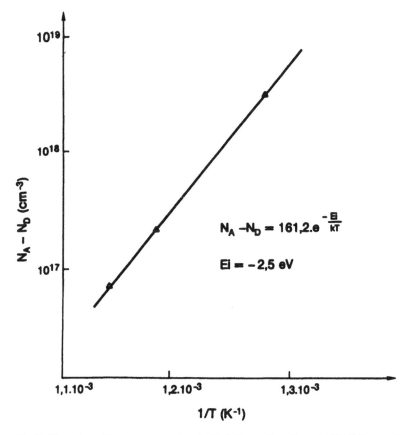

Fig. 8 Variation of hole concentration in GaInP as a function of $1/T$ (T is growth temperature) (Ref. 8).

No clear influence of group-V element flow has been found on p-type doping of GaInP with DEZn [6]. The acceptor concentration in GaInP doped by DEZn can be written in the form:

$$N_a - N_d = 10^2 D_{DEZn} \exp(2.5/kT) \tag{13}$$

where DEZn is in cc/min and kT in eV.

Magnesium (Mg):

Suzuki et al. [13] have used bis-magnesium $[(C_2H_5)_2 Mg]$ as a p-type dopant for GaInP layers grown by MOCVD. They reported the similar behaviour for Mg as Zn in GaInP layers. They studied photo-luminescence properties of Mg doped GaInP layers as a function of hole concentration. The band-gap energy of Mg doped GaInP, grown under a condition in which undoped GaInP shows an anomalously low E_g, increased for $N_a - N_d > 1 \times 10^{18}\,\text{cm}^{-3}$. This anomalous behaviour was attributed to the Mg (or Zn) diffusion-enhanced randomization of the naturally formed monolayer $(1/2, 1/2, 1/2)$ superlattices on the column III sublattice.

2.2.3. Conclusion

The doping concentration as a function of dopant flow rate, growth temperature, and the V/III ratio can be expressed in an empirical form:

$$n = D_{\text{dopant}}^{\alpha} \cdot D_{\text{V}}^{\beta} \cdot D_{\text{III}}^{\gamma} \exp\left(-\frac{E_i}{kT}\right) \tag{14}$$

where α, β and γ are empirical coefficient for n- and p-type doping.

Table 2 lists the summary of the empirical coefficients for n and p-type doping in GaAs and GaInP. The doping concentration is always found to decrease slightly with the increasing V/III ratio. The temperature dependence of the doping concentration categories the dopants into two groups:

For group-IV elements (Sn, Ge, Si), the energy E_i is positive so that the incorporation efficiency of the dopant increases while increasing the temperature. E_i is weakly dependent on the doping element and is

TABLE 2
Summary of n and p-type doping behaviour in GaAs and GaInP.

Dopant	Materials	α	β	$E_i(\text{eV})$	γ
DEZn	GaAs	1	0.3	-2.6	
	GaInP	1	0.1	-2.5	
H_2S	GaAs	1	-1	-1.4	
	GaInP	1	-0.7	-2.1	
H_2Se	GaAs	1	-1	-2.8	1
SiH_4	GaAs	1	-0.5	$+1.2$	-1
	GaInP	1	-0.2	$+1.2$	
TMSn	GaAs	1	0.8	$+1$	
GeH_4	GaAs	1		$+1.5$	-1

close to 1 eV. In that case, the main limiting step for incorporation of the dopant is its cracking efficiency.

For group-II and -VI elements (Zn, S, Se), the energy E_i is negative so that the incorporation efficiency decreases with the increasing temperature. Its value is found to vary between -1 and -2 eV independent of dopant or the host material. In that case, the main limiting step for incorporation of the dopant is its desorption from the crystal surface.

3. BULK PROPERTIES OF GaInP

3.1. Structural Ordering in $Ga_xIn_{1-x}P$ Alloys Grown by MOCVD

GaInP alloys can be grown by MOCVD as disordered or ordered structure depending on the growth conditions. The disordered alloy has a bandgap of 1.92 eV at 300 K, while the ordered structure has a lower bandgap varying from 1.83 eV to 1.78 eV [14–15]. The existance of ordered and disordered structures has been found in various alloy systems such as GaInP [14, 16–18], AlInP [19], AlInAs [20], GaInAs [21], GaAsSb [22], GaAlAs [23], etc. Experimental results show that the kinetics of crystal growth plays an important role in the formation of an ordered structure. If growth kinetics generate ordered alloys, in this case, the CuPt-type ordered structure that has been found in a $Ga_{0.5}In_{0.5}P$ alloy may be also formed in a $Ga_{0.7}In_{0.3}P$ alloy.

$Ga_3In_1P_4$ ordered structure exists in either of two variations, "famatinite" or "luzonite" [24]. The famatinite-type ordered structure is a $(GaP)_3(InP)$ superlattice developing along the $(2\,1\,0)$ direction. If a famatinite-type ordered structure exists, a diffraction pattern with an electron beam incident along the $[1\,0\,0]$ crystal axis must show extra spots corresponding to the ordered structure [17]. However, the presence of a luzonite-type ordered structure cannot be proved by a diffraction measurement because this ordered structure does not have an equivalent superlattice.

Several studies have been reported on the structural ordering of $Ga_{0.5}In_{0.5}P$ [25], $Ga_{0.6}In_{0.4}P$ [17], $Ga_{0.7}In_{0.3}P$ [26], and InGaAlP [27] grown by MOCVD. The structural order has been studied by transmission electron diffraction (TED) technique.

A correlation has been established between the degree of structural ordering in GaInP and the bandgap energy measured by photoluminescence or electroreflectance which indicates that the disorder-order transition is reflected in the optical transition.

The most interesting structural ordering phenomena are observed in the $Ga_{0.5}In_{0.5}P$ alloy. The structural order of $Ga_{0.5}In_{0.5}P$ has been studied as a function of:

- growth temperature [25, 28]
- V/III vapor pressure ratio [29]

In all the cases, the $Ga_{0.5}In_{0.5}P$ samples were grown on (001) oriented GaAs substrates. Transmission electron diffraction (TED) studies have been performed along the $[110]$, $[1\bar{1}0]$ or $[001]$ directions [25–28]. These studies have shown that the disordered phase consists of a cubic face centered, zincblende-type crystalline lattice with a random distribution of Ga and In atoms on the group III sublattice. The TED pattern of the disordered phase shows diffraction spots at (hkl) positions with $h, k,$ and l being all even or odd integers. In this case, no superstructure was observed in the transmission electron diffraction pattern [30]. In the ordered phase, Ga and In atoms are distributed on group-III element sublattice of the zincblende structure in order.

$[110]$ beam direction is most often used for TED studies. Using this orientation, the TED pattern of the ordered phase showed $(h - 1/2, k + 1/2, l + 1/2)$ spots close to the (hkl) spots of the zinc-blende structure, corresponding to a double periodicity of the $(\bar{1}11)$ planes. Therefore, the ordered phase can be described as a regular alternative of GaP and InP monoatomic planes in the $[\bar{1}11]$ direction, or a CuPt type ordering [18, 27, 29, 30].

To confirm this assumption, transmission electron microscopic (TEM) studies have been performed by Suzuki et al. [30] on the ordered $Ga_{0.5}In_{0.5}P$, using a dark field imaging technique with an aperture including $(-1/2, 1/2, 1/2)$, $(-1/2, 1/2, 3/2)$, and (111) diffraction points. They observed at atomic resolution, a double periodicity of 6.5 Å of the $(\bar{1}11)$ planes, which can be related to TED observations. It proves the ordering of GaP and InP monoatomic planes along the $[\bar{1}11]$ direction.

Suzuki et al. [30], also observed by TEM imaging an intermixing of disordered and ordered phases as well as antiphase boundaries on partially ordered phases, with a variation of the relative extensions of the disordered phase, antiphase boundaries, and ordered phases corresponding to a short or long range structural order.

The degree of structural order can also be estimated by the TED technique by comparing the intensity and the sharpness of the superstructure spots $(h - 1/2, k + 1/2, l + 1/2)$ of the electron diffraction pattern.

The TED and TEM studies performed on $Ga_{0.5}In_{0.5}P$ alloys grown at different growth temperatures in the range of 550 to 750 °C showed that the degree of structural order was increasing with the growth temperature, from a very short range order at 550 °C to a long range order at 700 °C. At low growth temperatures, the density of antiphase boundaries is high. Intermixing of disordered and ordered phases can also be observed at low growth temperatures [27, 29, 30]. It is also observed that there is a negligible influence of the V/III vapaor pressure ratio on the structural ordering in $Ga_{0.5}In_{0.5}P$.

Photoluminescence measurements at 300 °K have shown that there is a negligible difference in energy gaps for $Ga_{0.5}In_{0.5}P$ alloys grown at the same growth temperature for different V/III ratios, (from V/III = 60 to V/III = 440) [28], although 4 K luminescence would yield more accurate bandgap values.

A correlation has been established between energy gap values and the degree of structural order in $Ga_{0.5}In_{0.5}P$. Photoluminescence results showed a minimum energy gap, which is 1.85 eV at 300 K, for the $Ga_{0.5}In_{0.5}P$ grown at 650°C. As a comparison, the energy gap of the disordered $Ga_{0.5}In_{0.5}P$ is 1.90 eV at 300°K. Electroreflectance studies also showed similar results [31].

Gomyo et al. [28] reported that the zinc diffusion through an ordered $Ga_{0.5}In_{0.5}P$ alloy increases its band gap by 50 meV, from 1.85 eV to 1.90 eV at 300 K: the effect of zinc diffusion is to produce an intermixing of the group III elements, creating structural disorder by randomizing Ga and In positions. This clearly proved the relation between the energy gap of $Ga_{0.5}In_{0.5}P$ and its structural order.

McDermott et al. [15] have grown $Ga_{0.5}In_{0.5}P$, at temperatures between 480 to 500°C, using atomic layer epitaxy. They found that all the samples grown on 2° misoriented substrates had CuPt structure, with typically two $1/2 \langle 111 \rangle$ variants. For samples grown on the (1 0 0) nominal substrates, no CuPt ordering was observed.

3.2. Defects in GaInP Layers Grown by MOCVD

Using deep-level transient spectroscopy (DLTS), the defects present in GaInP layers grown by MOCVD were characterized by Feng et al. [32]. The layers were grown on n^+-doped GaAs substrates in a horizontal cold wall reactor using trimethylindium, triethylgallium, and PH_3 as sources. The GaInP layers, $\sim 1 \mu m$ thick, lattice matched to GaAs are non-intentionally doped. Capacitance-voltage $(C\text{-}V)$ measurements show that the layers are n type with a carrier concentration of

2.4–$3.5 \times 10^{15} \mathrm{cm}^{-3}$ in the growth direction at room temperature. As shown in Fig. 9, a first peak is observed around 60 K corresponding to an ionization energy of 75 meV (Fig. 10) with a concentration of $3 \times 10^{13} \mathrm{cm}^{-3}$. This spectrum cannot be attributed to a possible emission over band discontinuities, in which case the emission rate should be strongly sensitive to the electric field and the width of the spectrum on the low-temperature side should increase with this field, i.e., with the applied reverse bias. Thus, there should be no correlation between this spectrum and the large photoluminescence shift versus excitation intensity commonly observed in similar layers [33]. A second peak, rather wide and exhibiting a double structure, appears between 350 and 450 K. The shape of this structure varies with the

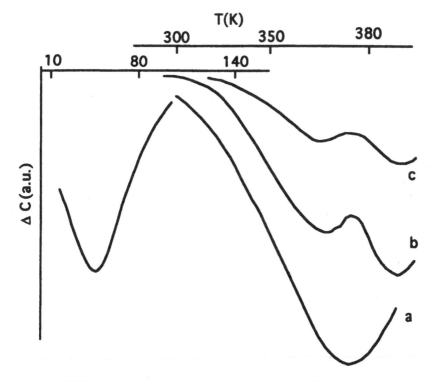

Fig. 9 DLTS spectrum obtained in the temperature range 4– 400 K. Low-temperature peak (60 K) measured at bias -1 V, bias pulse 1 V, emission rate 60 s^{-1}, pulse duration 1 ms. High-temperature peaks measured with an emission rate 30 s^{-1}, pulse duration 1 ms, bias pulse 1 V, and bias: (a) -3 V, (b) -2 V and (c) -1 V (Ref. 33).

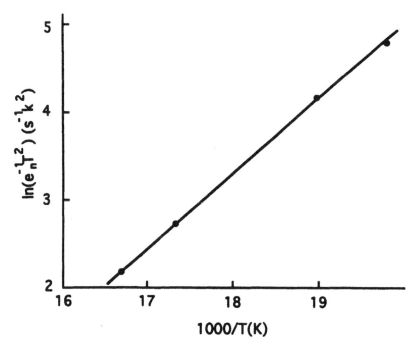

Fig. 10 Variation of the emission rate e_n versus temperature of the peak appearing around 60 K (Ref. 33).

filling pulse duration, suggesting that one of the traps is not complete-ly filled (see Fig. 9). The activation energy associated with the first maximum of this peak is 0.92 eV (see Fig. 11). Only the order of magnitude of the trap concentration, which is about 15% of the free-carrier concentration ($\sim 4.5 \times 10^{14}$ cm^{-3}), can be evaluated because it cannot be completely filled (see Fig. 12). The dependence of the high-temperature spectrum on the reverse bias, at constant pulse ampli-tude, indicates a nonuniform distribution of the associated defect: its concentration increases near the epi-substrate interface.

The GaInP layer studied above contains only one defect species emitting below room temperature (60 K) with a concentration of the order of 1% of the concentration of residual uncompensated donors. This defect cannot be attributed to a DX-like center since its concen-tration is saturated with short filling times (~ 50 μs). If it is DX-like center, it should have an energy level resonant in the conduction

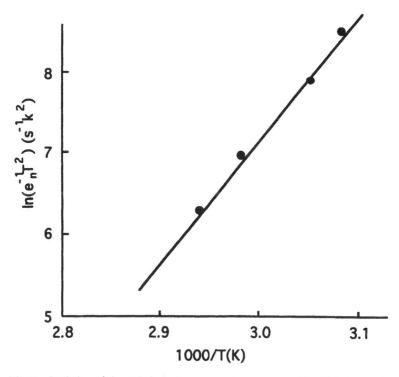

Fig. 11 Variation of the emission rate e_n versus temperature of the high-temperature peak under bias conditions (a) of Fig. 9 (Ref. 33).

band. According to Bourgoin [34], donor impurities are expected to introduce a DX-like energy level in the forbidden gap when the X or L band lies at an energy not larger than typically 0.2 eV from the bottom of the conduction band. In $Ga_{1-x}In_xP$ the condition is fulfilled only when $x = 0.7$.

Consequently, the persistent photoconductivity which was observed by Ben Amor et al. [35], should be ascribed to electron emission from deep traps emitting above room temperature. These traps, because of their relatively high concentration ($10^{14} cm^{-3}$), should dominate the electrical properties of these layers for low doping densities.

In conclusion, the nonintentionally doped $Ga_{1-x}In_xP(x = 0.51)$ layers grown by MOCVD exhibit an uncompensated electron concentration of a few $10^{15} cm^{-3}$. They contain one dominant deep level defect which is ionized just above room temperature with a

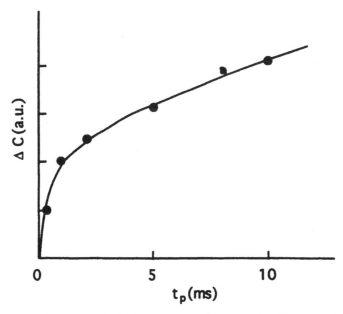

Fig. 12 Amplitude ΔC of the high-trmperature peak versus the filling pulse duration (Ref. 33).

concentration of the order of 10^{14}–10^{15} cm^{-3}. For the alloy composition considered, the DX center, if present, is not localized in the gap.

4. GaAs-GaInP HETEROSTRUCTURE

4.1. Band Offsets Measurements

4.1.1. Capacitance-Voltage (C-V) Measurement

Band offsets can be determined from C-V profiling [36]. Let's describe in detail the way to determine δE_V from C-V profiling through the p-GaInP/p-GaAs heterojunction by reverse-biasing the n^+-p junction between the substrate and the p-GaAs epilayer. The measurement frequency was 1 MHz and the magnitude of the applied differential voltage was 30 mV. The apparent majority carrier concentration profile, $p^*(x)$, was derived from the C-V profile according to the standard relation:

$$p^*(x) = \frac{2}{q\varepsilon}\left(\frac{d}{dV}\frac{1}{C^2}\right)^{-1}, \tag{15}$$

where

$$x = \varepsilon/C \tag{16}$$

is the width of the depletion region (the distance of the depletion region edge from the $p-n$ junction), $p^*(x)$ is the apparent majority-carrier (hole) concentration at position x, V is the reverse bias voltage, C is the capacitance per unit area, ε is the dielectric permittivity of the semiconductor.

The value of δE_V is obtained from the electrostatic dipole moment associated with the charge imbalance between a presumably known doping distribution $p(x)$ and the experimental $p^*(x)$ curve, using the relation given by Kroemer et al. [37], adapted to the p-p case:

$$\delta E_V = \frac{q^2}{\varepsilon} \int_0^\infty [p(x) - p^*(x)] \cdot (x - x_i)\,dx - kT \cdot \ln\left(\frac{p_2 N_{V1}}{p_1 N_{V2}}\right), \tag{17}$$

where $p(x)$ is the ionized acceptor distribution, assumed to be known and to level out far away from the heterojunction, $p_{1,2}$ are the asymptotic values of the doping concentrations in the GaAs (1) and the GaInP (2), x_i is the distance of the GaAs-GaInP interface from the n^+-GaAs substrate, $N_{V1,2}$ are the valence band density of states in GaAs and GaInP regions.

To evaluate δE_V, Kroemer et al. [37] assumed:

i) uniform doping on both sides of the interface at $x = x_i$
ii) an abrupt step in the doping at $x = x_i$,
iii) the existence of a localized interface defect charge density σ_i at $x = x_i$. For a given value of x_i, the value of σ_i is obtained from the requirement of overall charge neutrality:

$$\sigma_i = \int_0^\infty [p(x) - p^*(x)]\,dx. \tag{18}$$

In such a model an accurate knowledge of the interface position x_i, the thickness of the epilayers and the doping concentration in epilayers are essential. Kroemer et al. [37] found $\delta E_V = 0.238$ eV.

Kroemer et al. [37] also performed the additional check by performing a computer reconstruction of what the $p^*(x)$ profile should have been for an abrupt interface, with given values of the band offset and interface charge, using a computer program that Poisson's equation for incremental voltage steps applied to the heterojuction. Kroemer et al. [37] obtained a satisfactory agreement between experimental and theoretical results.

The largest source of uncertainty in above measurement is the quality of the samples. Another source of errors is the presence of deep levels near or at the interface, the ionization energy of which changes with changing bias. Errors in the capacitance measurements, such as errors in the diode area can occur, which may yield an incorrect apparent charge concentration.

4.1.2. DLTS Measurement

Deep level transient spectroscopy (DLTS) measurements were carried out on GaInP/Au Schottky diodes having a GaAs single quantum well (SQW) in the depletion region [38].

The emission energies of electrons and holes obtained from DLTS measurements are related to the band offsets δE_C and δE_V.

The emission rate of electrons from a quantum well can be derived from the thermionic current due to electrons emitted from the well to the barrier region. This emission rate e_n has been calculated by Martin et al. [39, 40] and is given by:

$$e_n = \left(\frac{kT}{2\pi n_W^*}\right)^{1/2} \frac{1}{L_W} \exp\left(-\frac{\delta E_C}{kT}\right), \tag{19}$$

where n_W^* is density of electrons in the well, L_W is well width, δE_C is conduction band offset, and T is the temperature of the sample.

It is also possible, by drawing an analogy between an electron in a quantum well and a deep level trap, to formulate a detailed balance between thermal capture and emission of electrons from the quantum well. From such detailed balance, the emission rate of electrons is given by:

$$e_n = \frac{16\pi^{3/2}}{3h^3} m_W^* X (kT)^{1/2} (\delta E_e)^{3/2} \exp\left(-\frac{\delta E_C}{kT}\right), \tag{20}$$

where δE_e is the electron emission energy from the conduction band well, m_W^* is the effective mass of electron in the well material, h is the Plank constant and X is a parameter related to the capture of carriers by the wells.

Equation (20) is also valid for hole emission from a valence band well from which δE_V can be measured. The thermal emission energy of carriers from a quantum well is related to the appropriate band offset.

Next, consider a single quantum well in the depletion region of a Schottky barrier. The existence of confined electrons in the well changes the depletion width W. Solution of the Poisson's equation in the well and barrier regions with the appropriate boundary conditions gives:

$$W^2 = W_0^2 \left(1 + \frac{2n_W L_W}{N_D W_0^2}\right), \tag{21}$$

where

$$W_0^2 = (2\varepsilon/qN_D)V, \tag{22}$$

is the depletion region width in the absence of the well, N_D is the net donor density in the barrier and $V = V_{app} + V_{b1}$, where V_{b1} is the build-in potential of the junction. The transient capacitance δC is then given by:

$$\frac{\delta C}{C(W)} = \frac{n_W L_W}{N_D W_0^2}. \tag{23}$$

The DLTS signal at the rate windows t_1 and t_2 is then given by:

$$C(t_2) - C(t_1) = \frac{C_0 n_{W_0} L_W}{N_D W_0^2} [\exp(-e_n t_1) - \exp(-e_n t_2)] \tag{24}$$

and

$$\delta E_C = \delta E_e + E_{e1} + L_W F \tag{25}$$

$$\delta E_V = \delta E_h + E_{h1} + L_W F, \tag{26}$$

where E_{e1} and E_{h1} are the electron and hole ground state subband energies, L_W is the well width and F is the electric field across the well region due to the applied reverse bias. The materials were grown by LP-MOCVD at 510 °C. The sample structure for δE_C measurements was as following:

First, a 0.2 µm thick n^+-doped GaAs buffer layer was grown on a n^+-GaAs substrate, followed by a 0.5 µm n-doped GaInP layer with a doping concentration of 2×10^{16} cm^{-3}. A 120 Å thick GaAs n-doped quantum well was grown, followed by a 0.5 µm thick n-doped GaInP layer with doping concentrations of 2×10^{16} cm^{-3}. Finally, an Au Schottky contact was deposited on the GaInP cap layer.

The sample structure for δE_V measurements was obtained by growing p-type doped materials on a p^+-doped GaAs substrate, with the same structure and doping concentrations used for δE_C measurements.

The values of δE_C and δE_V derived from this experiment are 0.198 eV and 0.285 eV, respectively. Also $\delta E_C + \delta E_V = \delta E_g = 0.48$ eV which agrees with the measured δE_g value from photoluminescence. The values of δE_C and δE_V estimated in this study agree reasonably well with the other measured values [36, 41].

4.1.3. I-V Curve of High Electron Mobility Transitors

According to the model proposed by Chen et al. [43], the current-voltage (I-V) characteristics of high electron mobility transitors can be described by the combination of two diodes in series: D_1 and D_2. D_1 represents the metal-semiconductor Schottky diode and D_2 is the heterojunction diode. Since the Schottky barrier (0.87 eV in this case) is much higher than the conduction band discontinuity, the gate current at low forward bias is determined by D_1. As the bias increases, especially when the potential drop across D_1 is larger than the Schottky barrier height, the change of gate current is determined by D_2. Two slopes are observed in the gate I-V characteristics. By extrapolating the I-V curve under D_2 operation it is possible to determine the reverse saturation current, I_{ss}, of D_2:

$$I_{ss} = A A^* T^2 \cdot \exp(-q\Phi(0)/kT), \tag{27}$$

where A is the diode area, A^* is the Richardson constant ($A^* = 4\pi q k^2 m_0/h^3$), and $\Phi(0)$ is the potential difference between the channel and the top of the conduction band discontinuity at zero gate bias. In order to reduce the influence of the series resistance, a large gate ($200 \times 50 \ \mu m^2$) diode was measured. The value of $\Phi(0)$ evaluated from I_{ss} is 196 meV.

Using a self-consistent solution of Schrodinger and Poisson's equations, the potential difference between the Fermi level in the channel and the bottom of the conduction band discontinuity is found to be 35 meV. Therefore the conduction band discontinuity of GaInP-GaAs is estimated to be $\delta E_C = 0.231$ eV, which is quite close to the previous results.

4.2. Properties of Two-Dimensional Electron Gas (2DEG)

4.2.1. Shubnikov-de-Haas Oscillation and Quantum Hall effect

The heterojunctions studied here were grown on a semi-insulating (1 0 0) GaAs substrate with the growth condition listed in Table 1. A nominally undoped, 3000-Å-thick GaAs layer with electron concentration of $2 \times 10^{16} \ cm^{-3}$ was covered by a sulphur or silicon doped

1000-Å-thick $Ga_{0.51}In_{0.49}P$ layer with an electron concentration of $2 \times 10^{17}\,cm^{-3}$.

From Auger profiling of phosphorus in a multiquantum well of GaAs/$Ga_{0.51}In_{0.49}P$ grown under identical conditions, the interface was estimated to be abrupt within two atomic layers.

Magnetotransport measurements were performed on long bar-shaped samples with current and potential contacts. Standard Hall bridges were photolithographically defined. Shubnikov-de Has (SdH) and quantum Hall effects (QHE) were measured in magnetic field up to 18 T [44].

The low field measurements performed on these heterojunctions gave electron mobilities of 49,000 cm^2/Vs at 4 K, 38,000 cm^2/Vs at 77 K, and 6000 cm^2/Vs at 300 K. SdH measurements were performed in normal and tilted field configurations.

Figure 13 shows SdH oscillations for various angles between the field direction and the normal to the interface. Extrema of resistivity

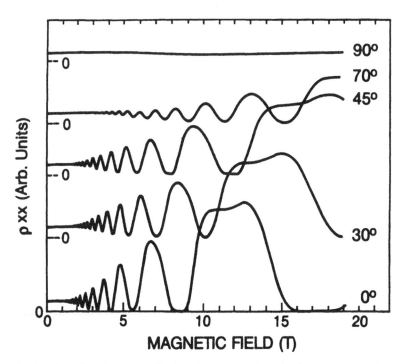

Fig. 13 Shubnikov-de Haas oscillations in a GaInP-GaAs heterostructure as a function of magnetic field for various tilt angles measured at 4.2 K (Ref. 44).

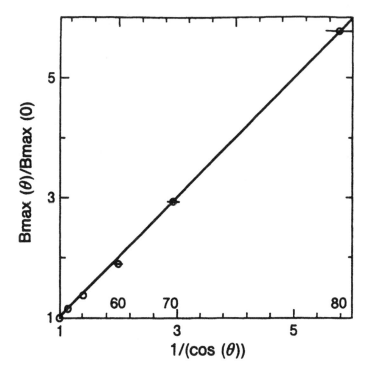

Fig. 14 Field value $B_{max}(\theta)$ of magnetoresistance extrema at angle θ divided by $B_{max}(\theta = 0)$ of corresponding extrema at $\theta = 0$ as a function of $[\cos(\theta)]^{-1}$ (Ref. 44).

are observed at constant values of the perpendicular component of the magnetic field $B_\perp = B \cos\theta$ (Fig. 14), giving evidence of two dimensionality of electrons at the interface. Figure 15 shows the diagonal resistivity ρ_{xx} and the Hall resistivity ρ_{xy} as functions of magnetic field at 4.2 K. In the vicinity of 5.2, 7.8, and 16 T ρ_{xx} approaches the zero-resistance state. At the same field positions ρ_{xy} develops plateaus at $h/i\,e^2$ for $i = 6, 4, 2$ characteristic of the QHE, hence giving further proof of the ideal 2D behaviour of this system.

Another important feature of this system is the persistent increase of the 2D electron concentration which was observed after illuminating the sample at low temperature, similar to that observed in GaAs/GaAlAs heterostructures [45–47].

Figure 16 shows the SdH oscillations before and after illuminating the sample. After illumination, a very clear shift towards higher fields is observed. The fundamental field $N \times B_N$ increases from 31.6 to 39.0 T

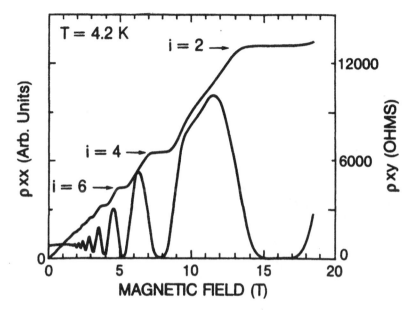

Fig. 15 Shubnikov-de Haas and quantum Hall effect as a function of magnetic field (Ref. 44).

giving a change of the total concentration from 7.64×10^{11} to 9.43×10^{11} cm^{-2}. The oscillations becomes more pronounced with increasing temperature because of the increased population of the first excited subband.

The population of the ground subband derived from the low field SdH periodicity changes from 7.51×10^{11} to 8.44×10^{11} cm^{-2}, leaving the difference $n_1 = n_{tot} - n_o$ to be 0.14×10^{11} and 0.98×10^{11} cm^{-2}, respectively, for the first excited subband's population.

Using the persistent photoconductivity (PPC) effect to tune the 2D electron concentration, one can measure the pupulation of each subband as a function of the total population (Fig. 17). The results are fitted well by linear dependences and very similar to the predictions of numerical calculations [49] for GaAs/GaAlAs heterojunctions. Extrapolation of this plot indicates that the second subbands starts to be populated at about 7.3×10^{11} cm^{-2}.

It is worth adding that the low field Hall concentration underestimated the total 2D electron concentration n_{tot}. The difference $\Delta n_H = n_{tot} - n_H$ clearly exceeded the experimental error and was obviously connected with a two conduction subband, each subband

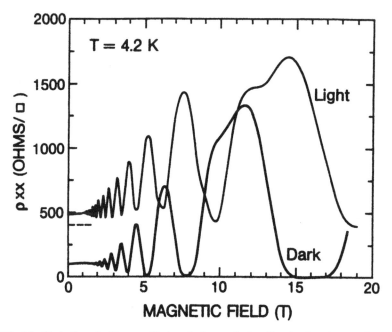

Fig. 16 Shubnikov-de Haas oscillations before and after illuminating the samples as a function of magnetic field (Ref. 44).

having a different scattering rate. The Hall mobility decreased slightly as the second subband was populated probably due to intersubband scattering.

4.2.2. High Electron Mobility GaInP-GaAs Heterostructures

Electron mobilities as high as 780,000 cm^2/Vs at a two-dimensional electron concentration $N_s = 4.1 \times 10^{11}$ cm^{-2} had been measured in GaAs-GaInP heterostructures grown by low-pressure metallorganic chemical vapor deposition (LP-MOCVD) [48].

The sample reported was grown on a semi-insulating GaAs substrate, oriented 2° off the (0 0 1) plane towards the (1 1 0) plane. On top of the substrate, a 1-µm-thick unintentionally doped GaAs layer followed by a 2000-Å-thick unintentionally doped layer of GaInP lattice matched to GaAs were grown. The GaInP layer was grown at a low pressure ($p = 76$ torr) at a substrate temperature in the range of 500 °C to 550 °C using triethylgallium, trimethylindium, and pure phosphine with hydrogen as the carrier gas. The optimum growth conditions are

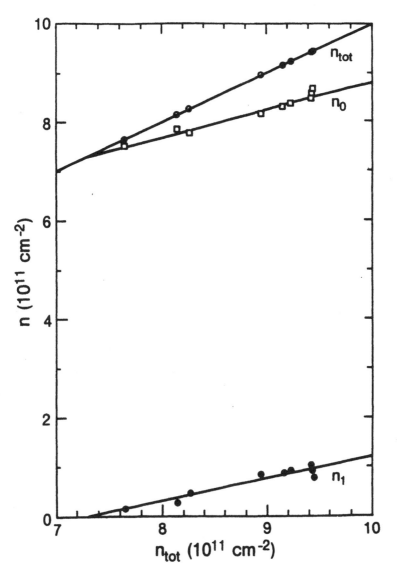

Fig. 17 Plot of electron populations in the first two electric subbands n_0 and n_1 as function of the total electron concentration, n_{tot} (Ref. 44).

summarized in Table 1. Corresponding layers of GaAs and GaInP with thicknesses of 2–3 µm, grown under identical conditions as the much thinner layers in the heterostructure, revealed $N_d - N_a = 5 \times 10^{14}\,\text{cm}^{-3}$ for GaAs and $N_d - N_a = 10^{15}\,\text{cm}^{-3}$ for GaInP (N_d-N_a: difference between donor and acceptor densities), as determined by Hall measurements. The sample has been structured into a standard Hall bar geometry, Au-Ge and Ni metallizations have been evaporated and then alloyed for 2 min at 420 °C in a hydrogen atmosphere to act as ohmic contacts.

Surprisingly, this nominally undoped heterostructure contains, depending slightly on the cooling procedure, about $1.7 \times 10^{11}\,\text{cm}^{-2}$ electrons as determined from the periodicity of the Shubnikov-de Haas oscillations. The fact that there are electrons and that they are on the GaAs side of the interface is clear from the cyclotron resonance experiments (Fig. 18), as the dependence on the cyclotron energy $\hbar\omega_c = AeB/m^*$ (see inset in Fig. 18) yields the effective mass expected for GaAs heterostructures $m^*/m_0 = 0.069\text{–}0.070$. (Here, $\hbar$ is Planck's constant, ω_c is the cyclotron frequency, e is the electron charge, B is the magnetic field, m^* is the electron effective mass, and m_0 is the free-electron mass, respectively). This m^* is clearly distinguishable from the estimated value $m^*/m_0 \sim 0.1$ in GaInP [49]. The question about where the electrons come from remains open. If they originate from donors in the barrier material, their concentration has to be at least of the order of $N_D \sim 2 \times 10^{16}\,\text{cm}^{-3}$, as a simple calculation (disregarding surface depletion) shows. It is possible that the 2000-Å-thick GaInP layer in the heterostructure has properties different from the 3-µm-thick reference layer ($N_D - N_A = 10^{15}\,\text{cm}^{-3}$) although these were grown under identical conditions. Such a big difference is rather surprising.

Successive illumination of the sample with light pulses from a red light-emitting diode (LED) at liquid-helium temperatures persistently increases the electron concentration. The photon energy of the LED is centered around $E = 1.87\,\text{eV}$, comparable to the band gap to $Ga_{0.51}In_{0.49}P$ (1.9 eV) [50] and above that of GaAs. The PPC is well known for GaAs-AlGaAs heterostructures [51] and was also recently reported for GaAs-GaInP heterostructures with intentionally n-doped barrier material. The mechanism for the PPC in the latter material combination, however, is not yet understood in detail, but it involves the ionization of a DX-center-like deep trap[35]. Using the PPC, the electron concentration could be increased up to $N_S = 3.9 \times 10^{11}\,\text{cm}^{-2}$ (N_S was determined from the periodicity of the Shubnikov-de Haas

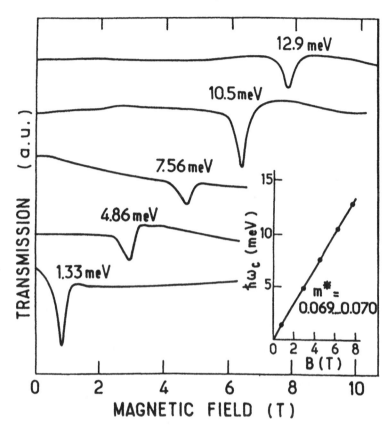

Fig. 18 Cyclotron resonance absorption in GaInP-GaAs for different photon energies (at $T = 1.6$ K). The far-infrared radiation originated from a CO_2-pumped molecular laser and a carcinotron. The inset shows the cyclotron energy, derived from the absorption peak position, as a function of the magnetic field (Ref. 48).

oscillations; a typical trace is shown in Fig. 19. Steady illumination increased N_S further on up to 4.1×10^{11} cm^{-2} (transient photoconductivity [52]). One cannot exclude that a stronger illumination might increase N_S even more since in the intensity range explored, there were no signs of saturation.

Razeghi et al. [53] do not see any evidence for parallel conduction through the barrier material after strong illumination. A current bypass would manifest itself in a nonzero magnetoresistivity ρ_{xx} in the plateau regions of the quantum Hall effect [54], especially also at filling factor $i = 2$ (see Fig. 19). The lack of parallel conduction might

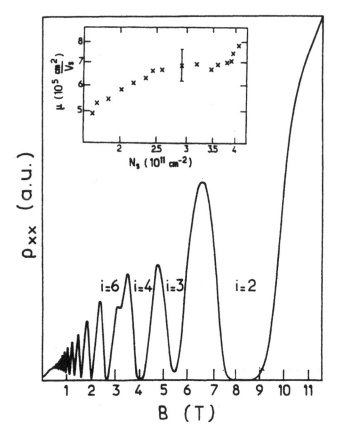

Fig. 19 Magnetoresistivity ρ_{xx} of a GaAs-GaInP heterostructure for an electron density of $N_S = 3.9 \times 10^{11}\,\mathrm{cm}^{-2}$ at a temperature $T = 1.4\,\mathrm{K}$. This N_S has been reached by illumination with red light using the persistent photoconductivity effect. Some filling factors $i = N_S h/(eB)$ are indicated. The inset shows the mobility as a function of the electron density varied by light pulses of a red light-emitting diode using the persistent photoconductivity effect. For clarity, the error bar is only shown on one of the points. Both axes are given in a logarithmic scale (Ref. 48).

find its explanation in the low doping concentration in the GaInP which is not intentionally doped.

After each successive illumination, the mobility μ was determined from the electron concentration N_S and the resistivity ρ at $B = 0$, according to the classical relation $\mu = (e\rho N_S)^{-1}$, as shown in the inset of Fig. 19. The sheet resistivity ρ deduced from the resistance R by using $r = (w/l)R$ (w is the width of the current carrying channel, l is the distance of the potential probes) involves an uncertainty of $\pm 10\%$ due to the finite extent of the potential probes. The mobility increases with N_S up to a concentration of about 2.5×10^{11} cm^{-2} and then remains constant upon further increasing N_S until 3.8×10^{11} cm^{-2} when μ again increases with N_S, leading in this sample to a final value of $\mu = 780,000$ cm^2/Vs at $N_S = 4.1 \times 10^{11}$ cm^{-2}. The half width at half amplitude $\Delta B_{1/2}$ of the cyclotron resonance line confirmed the high mobility if one used the semi-empirical expression $\Delta B_{1/2} = 0.63 \, \mu^{-1/2} B^{1/2}$ established in the case of GaAs-AlGaAs heterojunctions [55]. The value $\Delta B_{1/2} = 0.18 \, T$ measured for a resonance magnetic field $B = 6.5 \, T$ (photon energy 10.5 meV in Fig. 18 would lead to $\mu = 800,000$ cm^2/Vs), is in good agreement with the value deduced from transport experiments. This extremely high mobility proves the very high purity of the GaAs. For GaAs-AlGaAs heterostructures, such high mobilities are only reached by the growth of an undoped spacer layer, which separates the two-dimensional electrons from the ionized impurities in the barrier material. If we assume that the electrons are coming from the barrier material (residual donors in GaInP), there is nothing equivalent to a spacer layer, provided the growth of the barrier layer is uniform. The reduction of ionized impurity scattering in this heterostructure may be due to the fact that MOCVD-grown GaAs samples are slightly n-type, yielding an accumulation layer, contrary to the usual case for MBE-grown GaAs heterostructures having quasi-two-dimensional electron inversion layers. For accumulation layers, the penetration of the electronic wave function into the barrier is weaker, thus, reducing ionized impurity scattering in the barrier material. In addition, the density of ionized impurities is small, as the barrier is not intentionally doped. Therefore, the whole barrier might be assumed to act as a spacer layer.

The plateau-like behavior of the mobility beginning at $N_S^s = 2.5 \times 10^{11}$ cm^{-2} (cf. inset in Fig. 19), is due to the onset of intersubband scattering, and indicates that the first excited electric subband is starting to be populated. A comparison of this critical N_S with corresponding calculations [56] of the confining interface potential and the electronic wave functions leads to a depletion charge of the order of 10^9 cm^{-2}. These calculations are done for the GaAs-AlGaAs system. The results may be slightly different for GaAs-GaInP (e.g., due to a different conduction-band offset), but in any case, the order of magnitude of the depletion charge would be far too small for an inversion layer. This confirms again the well known fact that unintentionally doped MOCVD-grown GaAs is usually slightly n-type. A slight occupation of the second subband can be inferred from the positive magnetoresistance in ρ_{xx} as shown in Fig. 19. For some electron concentrations, additional structure in the Shubnikov-de Haas oscillations is observed, which is typical of second subband occupation.

Magnetotransport measurements showed that low pressure MOCVD-grown GaAs-GaInP heterostructures can reach extremely high electron mobilities, which up to now have been exclusively attained by MBE-grown GaAs-GaAlAs heterostructures. One can interpret this very high mobility as an indication of reduced ionized impurity scattering due to the high purity of the materials in the nominally undoped heterostructures.

4.2.3. Electron Spin Resonance

One possibility of the magneto-optic investigations of two-dimensional electron gases (2DEG) is electron spin resonance (ESR) [57]. The most frequently used method is cyclotron resonance [58] which yields the effective mass m^*. On the hand, ESR gives the g-factor. Both of these band-structure parameters (m^* and g-factor) are included in the energy spectrum of a 2DEG in a perpendicular magnetic field [47].

$$E_{Nms} = E_0 + (N + 1/2)(e\hbar/m^*)B + m_s g\mu_B B \qquad (28)$$

where E_0 is the energy of the lowest electric subband (higher ones are not considered), e is the elementary charge, $\hbar$ Planck's constant divided by 2π, μ_B Bohr's magneton and B the magnetic field. $N = 0, 1, 2, \ldots$ is the Landau level index and m_s is the magnetic spin quantum number, which takes the values $m_s = 1/2$ (spin up) and $m_s = -1/2$ (spin down). Strictly speaking, this energy spectrum is only valid in the case of parabolic bands, where m^* and g are really constants, independent of the magnetic field and the Landau level. In reality, however, the non-parabolicity of the conduction bands of the semi-conductors involved makes both of them vary with the magnetic field and the Landau quantum number [59].

The dependence of the g-factor on B and N has been studied systematically in the 2DEG of GaAs-AlCaAs heterostructures [60]. Both experiments and those reported here were done by looking at the change of the magnetoresistivity ρ_{xx}. Using the samples themselves as a detector for ESR. A basic requirement for this, however, turned out to be the following: the spin splitting must be resolved in ρ_{xx}, i.e., minima in the Shubnikov-de Haas oscillations corresponding to odd filling factors $i = N_s h/(eB)$ have to be visible (N_s is the concentration of 2DEG). This limits the measurability of ESR to high-mobility heterostructures.

The availability of very high mobility GaAs-GaInP heterostructures grown by low-pressure metallorganic vapor deposition (LP-MOCVD) also permits the study of the spin splitting of the Landau levels in the 2DEG of GaAs-GaInP heterostructures.

GaAs-GaInP heterostructure, grown by LP-MOCVD, consists of a 1 μm thick undoped GaAs layer on top of a semi-insulating GaAs substrate followed by a 200 nm thick, not intentionally doped layer of $Ga_x In_{1-x}P$ lattice matched to GaAs ($x = 0.51$). The concentration of 2DEG was about $N_s = 1.7 \times 10^{11}\,cm^{-2}$, depending slightly on the cooling procedure. By illumination with short light pulses of a red LED, N_s could be increased using the persistent photoconductivity effect. In order to see two resolved spin minima in the SdH oscillations within the available magnetic field range ($B = 12\,T$), N_s was increased to $N_s = 2.7 \times 10^{11}\,cm^{-2}$. The mobility at this concentration was $\mu = 650{,}000\,cm^2\,V^{-1}s^{-1}$. At this concentration, which remained constant for all the ESR experiments to be discussed, not only $i = 1$ minimum but also $i = 3$ minimum of the SdH oscillations are resolved (at $T = 1.3\,K$), thus, permitting the study of ESR in the two lowest Landau levels. At about $N_s = 2.5 \times 10^{11}\,cm^{-2}$ the second subband may begin to be populated. This might also be the case during the ESR study ($N_s = 2.7 \times 10^{11}\,cm^{-2}$). However, there is neither an additional oscillation in ρ_{xx} typical for second occupation nor a positive magnetoresistance at small magnetic fields (see Fig. 20). Therefore, the possibility of second subband occupation must be extremely small, and it will not be consider further.

The sample had a standard Hall bar geometry, and was located inside an oversized waveguide, immersed in liquid helium. In magnetic fields perpendicular to the plane of the 2DEG, ρ_{xx} was measured by a standard lock-in detection utilizing frequencies of the order of 10 Hz. Changes of the sample resistivity due to the chopped microwave radiation (of the order of 1 kHz) were detected using a second lock-in amplifier. In magnetic fields up to 12 T, using different klystrons and backward-wave oscillators to cover the frequency range up to 60 GHz, the output power of these microwave sources lay between a few and several hundred mW.

In Fig. 20, both ρ_{xx} and its change due to microwave radiation $\Delta\rho_{xx}$ are shown. The lower trace, ρ_{xx} reveals pronounced SdH oscillations already beginning at $B = 0.5\,T$. There are broad regions of vanishing

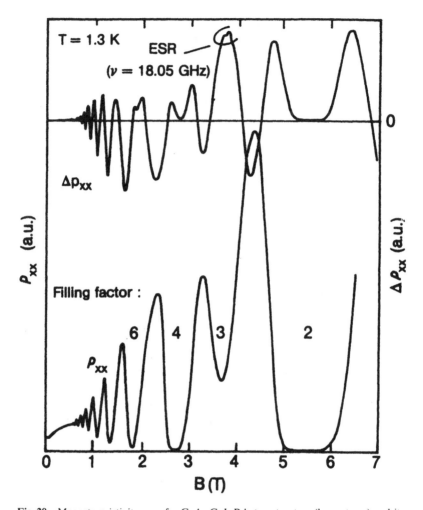

Fig. 20 Magnetoresistivity ρ_{xx} of a GaAs-GaInP heterostructure (lower trace) and its change due to microvave radiation $\Delta\rho_{xx}$ (upper trace) as a function of the magnetic field. ESR occurs at B 3.7 T. Note that the scale for $\Delta\rho_{xx}$ is greatly magnified compared with that of ρ_{xx} (Ref. 57).

ρ_{xx}, typical for the quantum Hall effect. At $B = 3.7$ T, the spin minimum corresponding to $i = 3$ is visible, whereas the next higher one ($i = 5$) is not resolved, because the magnetic field is too low. The upper trace shows $\Delta\rho_{xx}$. The broad spectrum reflects the periodicity of the SdH oscillation. It is due to a non-resonant heating of the sample

[61,62], and is related to the derivative of ρ_{xx} with respect to the temperature. On top of this broad spectrum, there is a sharp resonance structure, in this case ($\bar{\omega} = 18.05$ GHz) at $B = 3.7$ T. Whereas the broad non-resonant spectrum does not depend on the microwave frequency, the position of the ESR structure shifts with the applied microwave frequency on top of the non-resonant background. This can be seen in Fig. 21 which shows $\Delta\rho_{xx}$ for several microwave frequencies yielding ESR in the vicinity of filling factor $i = 1$. Here, the background changes because of a continuous variation of the bath temperature between $T = 2$ K and $T = 3$ K, and also due to different microwave powers at different frequencies; The width of the ESR structures is of the order of $\Delta B = 100$ mili-Tesla. This is a factor of 2 to 3 larger than has, up to now, been seen in the of case of GaAs-AlGaAs [63].

The ESR structures shown in Fig. 21 consist of an increase of ρ_{xx} due to ESR. Approaching the even filling factor $i = 2$, the effect of ESR on ρ_{xx}(and $\Delta\rho_{xx}$) turns into a decrease. The same observation has been made in the $N = 1$ Landau level: around $i = 3$, ESR results in an increase in ρ_{xx}; approaching the neighboring even filling factors, say around $i = 2.5$ and $i = 3.5$, ESR causes the decrease of ρ_{xx}. The same behavior was observed in GaAs-AlGaAs heterostructures [63], which might give some evidence to the assumption that this is an intrinsic property of a 2DEG.

A very weak hysteresis of the ESR structure was observed during a decrease of the magnetic field, which is an indication of an Overhauser shift of the ESR [64]. This hysteresis is due to a dynamic nuclear spin polarisation of the lattice nuclei via hyperfine interaction. This effect is only visible during very slow magnetic field sweeps (0.1 T/min), and even then it is very weak. Thus in practice, there was no complication for a determination of the electronic g-factors. Although one is not able to quantify properly the 'strength' of a dynamic nuclear spin polarization, it became obvious that it is much weaker in the present case of a GaAs-GaInP heterostructure compared with the GaAs-AlGaAs heterostructures [64], as all experiments were done with the same experimental set-up. Basically, there are two possibilities for the weakness of the Overhauser shift: either the nuclear spin relaxation times are very short so that a nuclear spin polarization, once created, immediately relaxes or more likely the electrons relax directly without polarizing the nuclei leaving the nuclear spin polarization unchanged from the beginning. The broader ESR structures in GaAs-GaInP indicate smaller electronic spin relaxation times. Therefore, the electrons

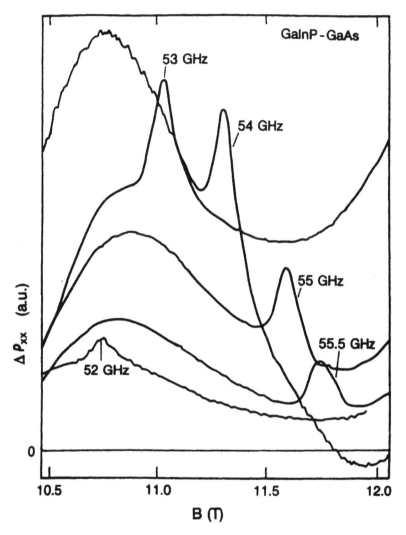

Fig. 21 Typical ESR structures in the change of magnetoresistivity due to microwave radiation $\Delta\rho_{xx}$ for different microwave frequencies. These ESR structures occur around filling factor $i = 1$ (Ref. 57).

feel more spin-flip scattering and the second interpretation seems to be preferable. Among the different electronic spin relaxation channels, the dynamic nuclear spin polarization is less important because all the others are more effective here.

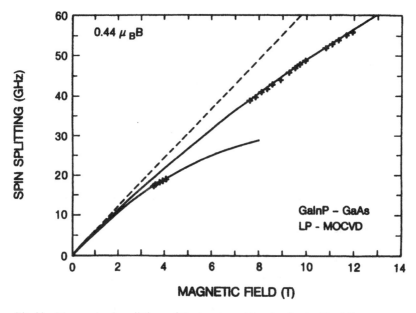

Fig. 22 Measured spin splittings of the two lowest Landau levels. The full curves are least-squares fits to the experimental data leading to the two coefficients in Eq. (29). For comparision, the spin splitting due to the bulk-GaAs conduction band edge g-factor, $g = 0.44$, is indicated as a broken line (Ref. 57).

From a series of experiments with different microwave frequencies as shown in Fig. 21 the magnetic field dependence of the spin splitting of 2DEG in GaInP/GaAs heterostructure can be deduced. This is shown in Fig. 22. The experimental results for a certain Landau level cover only a limited magnetic field range, as does the Fermi level. The spin splittings of the 2DEG are smaller than the spin splitting in bulk GaAs (broken line). Both of the two sets of data ($N = 0$ and $N = 1$) follow the quadratic magnetic field dependence:

$$\Delta E = 0.418 \, \mu_B B - 0.0133(1/T)(N + 1/2)\mu_B B^2 \,. \qquad (29)$$

The two coefficients, being determined by a common least-square fit of all the data. From Eq. (29), a magnetic-field and Landau-level-dependent g-factor can be deduced:

$$g(B, N) = 0.418 - 0.0133(1/T)(N + 1/2)B \,. \qquad (30)$$

This qualitative magnetic field dependence is predicted by theory [65], and confirmed by experiments [60, 63], in the case of

GaAs-AlGaAs. This dependence is also valid in GaAs-GaInP hetero-structures. The similarity of the spin splitting in GaAs-GaInP hetero-structures and those consisting of GaAs-AlGaAs is because the 2DEG in both cases is on the GaAs side of the interfaces. The extent of the electronic wave functions into the barrier material is only very small in either case. Therefore, an exchange of the barrier material does not alter the properties of the 2DEG very much.

MOCVD-grown GaAs is known to be slightly *n*-type. Thus, the 2DEG in MOCVD-grown GaAs heterostructures forms an accumula-tion layer, in contrast with MBE-grown heterostructures. The 2DEG, therefore, is confined at a slightly larger distance to the interface and enters to a lesser extent into the barrier material, an effect that is still supported by the high conduction band offset in GaAs-GaInP. The spin splitting, therefore, should be more 'GaAs-like'. There is slight evidence for this: according to Lommer et al. [65], the magnetic field dependence of the *g*-factor is simply an average of two corresponding quantities, one for each of the two materials, weighted with the prob-ability of finding the electrons on either of the two sides of the interface; if the electrons are more on the GaAs side, the coefficient in Eq. (31), constant *c* is bigger [65]. In fact, for the actual GaAs-GaInP hetero-structure, the magnetic field dependence of the *g*-factor is slightly stronger than found in GaAs-AlGaAs heterostructures [63], but unfor-tunately, not strong enough to verify the significance of this difference. Anyway, the experimental findings are consistent with the 2DEG being an accumulation layer at a higher conduction band offset, compared with that in GaAs-AlGaAs.

From these investigations, one can obtain the dependence of the *g*-factor on the magnetic field and the Landau level

$$g(B, N) = g_0 - c(N + 1/2)B, \tag{31}$$

which turned out to be quantitatively comparable with its dependence in GaAs-AlGaAs heterostructures.

5. GaAs-GaInP QUANTUM CONFINEMENT STRUCTURE

The quantum size effect and excitonic absorption in GaAs-GaInP multiquantum wells (MQW) and superlattices have attracted a great deal of interest because of the novel physical properties and the poten-tial usefulness in optoelectronic, such as quantum well infra-red photodetector (QWIP), and electronic devices [66]. Extensive studies on low-dimensional quantum confinement structures of GaInP-GaAs have been performed [66–69].

In this session, we will discuss the characterization tools used to obtain the important structural, optical parameters of the quantum confinement structures. In the end, we will present our recent results on the growth of quantum confinement structure on the Si substrate, which is imperative to the ultimate goal of QWIP: monolithically integrating QWIP focal plane arrays on a Si circuits.

5.1. Structural Characterizations

5.1.1. X-ray Diffraction and Dynamical Simulations

X-ray diffraction and reflectivity are the most important non-destructive tool to probe the structural parameters of superlattices. A four-crystal monochromator x-ray diffractometer (Philips MPD 1880/HP) with $K\alpha_1$ of Cu as x-ray source is used to measure diffraction patterns in this work. The diffractions are measured at a θ–2θ mode with a fine collimating of the incident and diffracted x-rays in order to track the evolution of the superlattice harmonics over several degrees. The more usual rocking curve measurements could not be employed for the measurement of superlattice satellites due to the large angular extent of superlattice satellites. Figure 23 shows (002) x-ray diffraction pattern of 10-period GaInP/GaAs superlattice grown by MOCVD. The nominal well thickness and barrier thickness of this superlattice are 90 Å and 600 Å, respectively. From a x-ray diffraction pattern, the following information about the superlattice sample can be extracted:

1) Period of superlattice

Superlattice, which is the periodic repetition of a well/barrier unit cell, introduces an extra period to the sample. As a result, the diffraction of x-ray from this extra period, called superlattice satellites, will present in the x-ray diffraction pattern. The separation of these satellites can be used to determine the superlattice period through the formula:

$$D_{sym} = \lambda/\Delta\theta \cos\theta_B, \tag{32}$$

where λ is the wavelength of the x-ray, $\Delta\theta$ is the separation of the neighboring peaks and θ_B is the Bragg peak of the epilayer. D_{sym} is

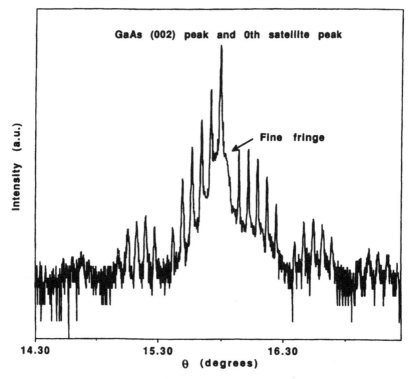

Fig. 23   The (002) x-ray diffraction pattern of 10-period GaInP/GaAs superlattice grown by low pressure MOCVD.

period of superlattice. This formula only works for the symmetric diffraction, or the diffraction with the diffraction plane parallel to the sample surface. For the sample with misorientation, only the diffraction pattern obained at the position, where the diffraction plane is perpendicular to the plane spanned by the diffraction plane direction vector and sample surface direction vector, will give the right structural data. Otherwise, an error as large as 10% may result for a sample grown on a 2 misoriented substate [70].

From Fig. 23, we can determined that the superlattice period is 612 Å, which is smaller than the nominal value 690 Å. Considering that nominal value are estimated through the known growth rate of the material, which may change slightly from growth to growth, the difference between nominal value and the value extracted from x-ray diffraction measurement is acceptable.

2) Quantum Well Thickness

Besides the strong superlattice satellites displayed in Fig. 23, the superlattice satellites are modulated with a period, that is about 6.5
times of the separation between the neighboring satellites. The modulation is owing to the interference of x-ray reflected from the
GaInP/GaAs and GaAs/GaInP interfaces. The period of the modulation implicit the information of the GaAs well thickness. Using Eq. 32,
we can determine the well thickness of the sample to be about 1/6.5 of
the superlattice period, or 94 Å.

3) Crystalline Quality

The separation between Bragg peak of substrate and that of epilayer
can be used to decide the mismatch between epilayer and substrate by
the formula:

$$\left(\frac{\Delta a}{a}\right)_\perp = -\cos\theta_B \Delta\theta_B \tag{33}$$

where $(\Delta a/a)_\perp$ is the mismatch in the growth direction and θ_B is the
Bragg angle.

The smaller mismatch will give better crystalline quality of the
epilayer. Mismatch is very important for the quality of the GaInP
epilayer while it is not a problem for AlGaAs, where AlGaAs is lattice
matched to GaAs at all of the composition. The measured mismatch
can be used to determine the composition of Ga in $Ga_xIn_{1-x}P$ by
Vegard's law, therefore, as a feedback to adjust the growth conditions.

No obvious mismatch is observed in the Figure 23. Higher order
diffraction is preferred to determine the small mismatch. Figure 24
shows the (004) x-ray diffraction of the same GaInP/GaAs superlattice.
In comparison to (002) x-ray diffraction, superlattice satellites in (004)
diffraction is less intense. For superlattice involving GaAs, (002) diffraction is a special position, at which, the scattering factor of GaAs is very
small; or the diffraction contrast is increased at this diffraction plane.
However, (002) is a diffraction which is at low θ angle, the small mismatch will not be distinguishable at this diffraction, higher order diffraction is required to obtain reliable mismatch value. In this figure,
mismatch between substrate and epilayer is clearly observed.

The quality of the crystal can also be qualitively evaluated by comparing the relative fullwidth at half maximum (FWHM) of the

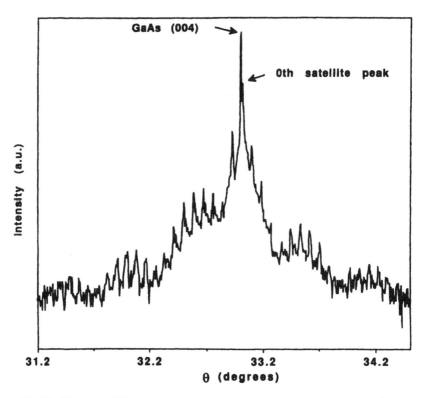

Fig. 24 (004) x-ray diffraction of the same GaInP/GaAs superlattice *a* in previous figure.

diffraction peak. The narrow FWHM indicates the better crystalline quality of the epilayer. The change in the FWHM of the superlattice satellites indicates the range of the fluctuate in the superlattice period. The larger change in the FWHM of the superlattice satelites reveals the larger fluctuate in the superlattice period.

The symmetry of the superlattice satellites hints the quality of the interface. For a lattice matched sample, the superlattice satellites should be completely symmetric around Bragg diffraction peaks. The asymmetry in the superlattice satellites implies the existence of the interfacial strain in superlattice sample. Figure 23 shows that there are interfacial strains for the measured sample.

To have a quantitative understanding of the interfacial strains and more precise determination of structure parameters of the epilayers, x-ray simulation is required. Two simulation models are now often

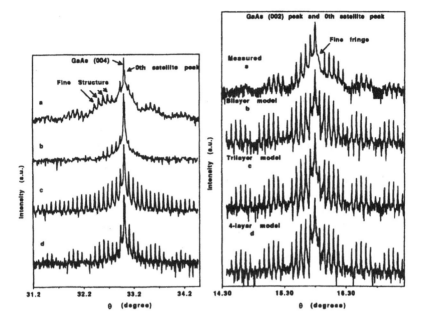

Fig. 25 The simulations of the (002) and (004) diffraction patterns shown in Figs. 23 and 24.

used: Kinematic model and dynamic model. Kinematic model treats the scattering of the x-ray inside the material by individual electrons in the crystal as the independent event and no multiple scattering is considered. The final x-ray intensity is the superposition of all these independent scattering. It succeeds for the thin layer where the total of scattered waves is sufficiently weak, but fails for the thick layer where the intercorrelation of the scattering from the crystals become stronger. The dynamic theory has to be used for the x-ray diffraction where scattering waves is strong. The basic equation for the x-ray simulation is Takagi-Taupin equations [71, 72]. Figure 25 shows the simulations of the (002) and (004) diffraction patterns shown in Figs. 23 and 24. They are performed as follows. The superlattice period is determined by fitting the observed superlattice satellites and the thickness of the well or barrier is decided by fitting the modulations of the superlattice satellites. The alloy composition of GaInP is determined by reproducing the observed mismatch. In Fig. 25(b) and (d) are the simulation results with the various choices of the superlattice unit cell shown in the lower right conner of the figure. Only 4-layer model, or a model includes a strained interfacial layer at both GaAs/GaInP and GaInP/GaAs interfaces, can fit (002) and (004) patterns well. The (002) pattern is apparently less sensitive. Therefore, the simulation of the (004) pattern is imperative to obtain more detail information about the sample structure.

The same superlattice sample was examined by transmission electron microscopy (TEM). A bright-field image of the complete superlattice is shown in Fig. 26. This image was taken by choosing [110] as the beam direction. Bright image is GaAs and the dark image is GaInP. The ten periods of GaInP/GaAs are clearly visible with average GaInP layer measured to be 623 Å and GaAs layer to be a 103 Å. The layer thicknesses measured from TEM picture is fairly close to the value measured from x-ray diffraction, which indicates that x-ray diffraction is a non-destructive, precise tool to determine the superlattice structure. The layers in the TEM picture are extremely regular and no local layer thickness variations were seen in the area examined. This demonstrates the good control of the layer thickness in the MOCVD growth process.

5.1.2. Auger Analysis on Chemical Bevels

Understanding the initial stage of epitaxial growth is essential for multilayer semiconductor materials, where interfaces between layers play a prominent part in their optical and electrical properties. The physical studies of the chemical species concentrations at the interfaces, can give valuable information on the initial steps of epitaxial growth.

The theoretical calculations of Auger currents given by Auger line scan measurements on chemically beveled heterostructure are performed by Olivier et al. [73], assuming exponentially varying concentrations at the interface and compared with experimental results.

Combined with ion etching, Auger Electron Spectroscopy (AES) can be used to get elemental concentration profiles within overlayers. However, several factors limit the resolution depth. There are different mechanisms which broaden the concentration profiles:

i) Auger electron escape depth and ion bombardment effects (ion knock-on mixing, preferential sputtering and ion-induced roughness).

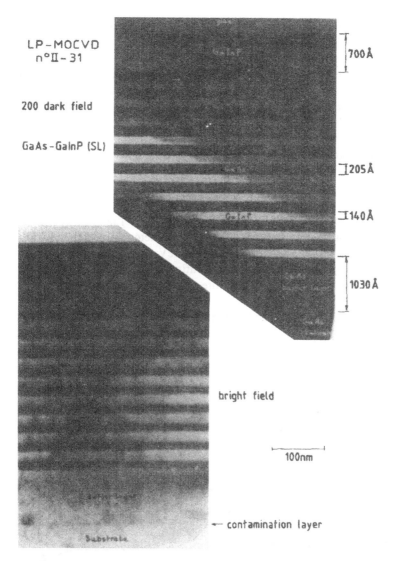

Fig. 26 TEM bright-field image of a complete superlattice. The x-ray diffraction spectra are shown in Figs. 23 and 24.

ii) In order to get round the disadvantages of ion milling, Bisaro et al. [73], developed a method of chemical beveling coupled with line scan Auger measurements to check interfaces of epitaxial III–V compounds.

The chemical bevels are produced by a technique in which the liquid-liquid interface between pure methanol and a bromine-methanol solution is raised progressively over the sample. The bevel angle α can be controlled by the speed of the etching solution flowing up the sample, and can be as small as 0.05 or less giving a magnification coefficient of $M = (\sin\alpha)^{-1} > 10^3$. In contrast to sputtering, the absolute depth resolution, R, of the bevel Auger profiling method is independent of the film thickness to be examined and is of the order of the electron beam size, s, plus the beam deflection resolution, d, divided by the lateral magnification of the bevel angle [74]:

$$R = (s + d)/M \qquad (34)$$

The schematic diagram of the bevel is shown in Fig. 27.

In the present experiment, the beam diameter was 2000 and 7500 Å (primary energy and current are 10 keV, 1 nA and 5 keV, 20 nA, respectively) and the deflection resolution 500 Å. With a magnification coefficient $M = 1000$, the absolute resolution R is 8 Å in the worst case.

The Auger current produced at x' by dx', $J(x')dx'$, is partially absorbed by the thickness $(x - x')$ before leaving the sample and

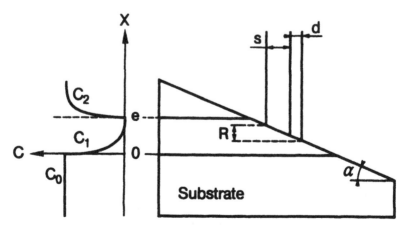

Fig. 27 Schematic diagram of the bevel. The mean parameters are the bevel angle, the electron beam size s, the beam deflection resolution d, and the depth resolution R. The phosphorous exponential composition profiles taken into account are $C_1(x) = C_0 \exp(-x/\Lambda_1)$ and $C_2(x) = C_0 [1 - \exp(-(x - e(/\Lambda_2))]$ (Ref. 73).

entering the analyzer with an angle α:

$$J(x')\,dx' = k\,n(x')\exp\left(-\frac{x-x'}{L_A\cos\alpha}\right)dx' \qquad (35)$$

where k is proportional to the product of the cross section for ionization of a level, of the Auger transition probability and of the secondary electron coefficient $(1+r)$ (Fig. 28).

$$n(x')\,dx' = N\exp\left(-\frac{x-x'}{L_p}\right) \qquad (36)$$

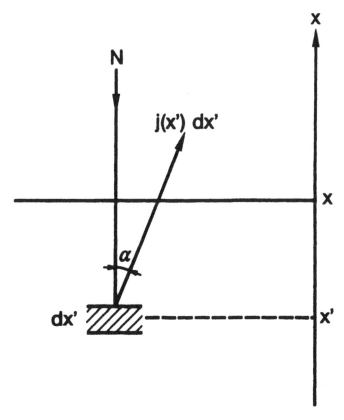

Fig. 28 Schematic diagram of the Auger current produced at x' by dx', which is partially
absorbed by the thickness $(x - x')$ before leaving the sample and entering the analyzer
with an angle a (Ref. 73).

is the attenuated primary electron beam, N is the incident beam intensity, L_A is the mean free path of the Auger electron. Defining $C(x') = C_B f(x')$ as the material atomic concentration, the Auger bulk current of a substrate of concentration C_0 is:

$$I_B = kNC_B L \tag{37}$$

with

$$(L)^{-1} = (L_p)^{-1} + (L_A \cos \alpha)^{-1} \tag{38}$$

The elementary Auger signal emerging from the surface x, normalized by the bulk value is expressed as:

$$J(x')dx' = \frac{f(x')}{L} \exp\left(-\frac{x-x'}{L}\right) dx' \tag{39}$$

In the case of a quantum well with exponentially varying concentrations at the interfaces, the normalized Auger currents are expressed by:

$$I(x) = 1 \quad \text{for } x < 0,$$

$$I(x) = \frac{\Lambda_1}{\Lambda_1 - L} \exp\left(-\frac{x}{\Lambda_1}\right) - \frac{L}{\Lambda_1 - L} \exp\left(-\frac{x}{L}\right) \quad \text{for } 0 < x < e, \tag{40}$$

and

$$I(x) = 1 + \left[\frac{L}{\Lambda_2 - L} - \frac{L}{\Lambda_1 - L} \exp(e - L) \right.$$

$$\left. - \frac{L(\Lambda_1 - \Lambda_2)}{(\Lambda_1 - L)(\Lambda_2 - L)} \exp\left(-\frac{e}{\Lambda_1}\right) \right] \exp\left(-\frac{x-e}{L}\right)$$

$$- \frac{\Lambda_2}{\Lambda_2 - L} \left[1 - \exp\left(-\frac{e}{\Lambda_1}\right) \right] \exp\left(-\frac{x-e}{\Lambda_2}\right) \quad \text{for } x > e, \tag{41}$$

where e is the escape depth. For abrupt interfaces: $\Lambda_1 = \Lambda_2$

$$I(x) = \exp(-x/L) \quad 0 < x < e, \tag{42}$$

and

$$I(x) = 1 - \exp[-(x-e)/L] \quad \text{for } x > e \gg L. \tag{43}$$

A log scale for $I(x)$ or $1 - I(x)$ gives straight lines, the slope of which allows the escape depth of Auger electron to be deduced.

The exact interface position does not correspond to relative intensity of about 50% as usually admitted: it corresponds either to the beginning of the decrease from unity ($0 < x < e$) or the beginning of the increase from zero ($x > e$). The quantum well profiles are asymmetric.

For gradual interfaces: Λ_1 and $\Lambda_2 \gg L$

$$I(x) = \exp\left(-\frac{x}{\Lambda_1}\right) \quad \text{for} \quad 0 < x < e, \qquad (44)$$

and

$$I(x) = 1 - \exp\left(-\frac{x - e}{\Lambda_2}\right) \quad \text{for} \quad x > e \gg \Lambda_1. \qquad (45)$$

Thus in a log-scale, the characteristic widths Λ_1 and Λ_2 govern the slope of the straight lines, when Λ_1 and $\Lambda_2 \gg L$.

Although one now has two interfaces of finite thickness, here too one can precisely determine their beginning that corresponds to starting or interruption of gas flow in the growth reactor. The distance between these two points corresponds to the aimed thickness as deduced from steady-state growth speed determined by measurements on thicker layers.

Figure 29 shows the experimental phosphorus Auger line scan of MOCVD grown GAInP/GaAs/GaInP quantum well. The good agreement between experimental and theoretical curves, a posteriori justifies the reasonable hypothesis of exponentially varying concentrations at the interfaces.

The experimental increase of the Auger signal that is less steep than the theoretical one at the two extremities of the well can be explained by a slight atomic mixing resulting from a short ion cleaning before the Auger line scan was taken. The two characteristic lengths are very different. The explanation is likely to come from the dynamics of the low-pressure MOCVD growth: in the neighborhood of the growth surface, there is a gas layer whose characteristics change gradually from those of the convective gas phase, to those of the growth surface-gas phase interface[75]. The evolution of that boundary layer by stopping or setting up again the phosphine flux in the reactor will give the key of the problem.

The GaAs QW of intended width $e = 100\,\text{Å}$ fits with a theoretical curve calculated using the following parameters $L = 4.75\,\text{Å}, e = 105\,\text{Å}$

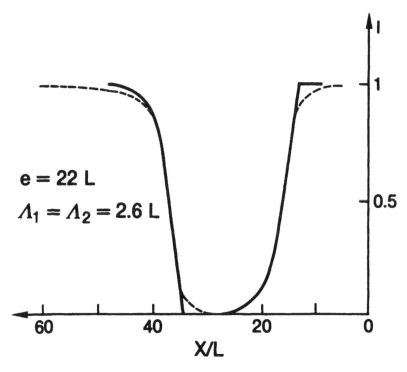

Fig. 29 Theoretical phosphorous Auger line scan of a GaInP/GaAs/GaInP QW. The dashed lines are the experimental curves (Ref. 73).

and $L_1 = L_2 = 12.5$ Å. The characteristic lengths L are equal, showing sharp interfaces and both interfaces are equivalent.

5.2. Optical Characterizations

The quantum well and superlattice samples grown for this study were nonintentionally doped. Residual impurity concentrations are assumed to be of the order of those determined for bulk layers. Corresponding layers of GaAs and GaInP with thicknesses of 3 μm, grown under identical conditions, revealed $N_D - N_A = 10^{14}$ cm^{-3} for GaAs and $N_D - N_A = 5 \times 10^{14}$ cm^{-3} for GaInP, and a uniform distribution of impurities in the direction perpendicular to the layers.

Figure 30 shows a photoluminescence (PL) spectrum of a three-well structure of GaAs-GaInP. The GaAs (well) and GaInP (barrier) indicated "nominal" thicknesses are those deduced from the measured

LP-MOCVD

MR : 1219

T = 4K

Laser Argon (2mW)

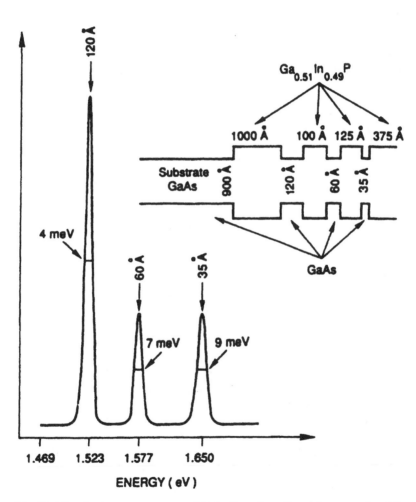

Fig. 30 Photoluminescence spectrum of GaAs-Ga$_{0.51}$In$_{0.49}$P sample measured at 4 K. Inset: Schematic representation of the sample structure grown by LP-MOCVD (Ref. 87).

growth rate of thick layers. The substrates were GaAs (1 0 0) mis-oriented 2° off axis towards $\langle 0\ 1\ \bar{1} \rangle$.

The PL measurements were done at 4 K using a 5435 Å argon laser operating at 2 mW, and dispersed on a HRS Jobin-Yvon mono-chromator equipped with a photomultiplier tube. The full width at half maximum of PL spectrum is 4, 7, and 9 meV for the nominal quantum well thicknesses of 120, 60, and 35 Å, respectively.

The quality of GaAs wells is proved by photoluminescence excita-tion (PLE) spectroscopy, using a tunable pyridine dye laser, pumped by an argon laser. The dye laser power is locked at 2 mW in the whole wavelength range. Figure 31 shows the PLE spectrum of GaAs-GaInP quantum well of 120 Å thickness at 4 K. The PLE spectrum exhibits a series of peaks, which can be attributed to the heavy hole exciton and light hole exciton as shown in Fig. 31. These results show that high-quality GaAs-GaInP MQW have been achieved. The intensity of PLE peaks indicate that the GaAs-GaInP system is of type 1, i.e., the GaAs layers are simultaneously quantum well for both the conduction and valence-band states.

By comparing PL and PLE spectra taken at 4 K in Fig. 31, we find that the subband transition energy from the first heavy hole subband to the first electron subband ($E_1 HH_1$) subband transition, as deter-mined by PLE, is 1524 meV and the peak energy of the $E_1 HH_1$ emission in PL spectrum is at 1522 meV, evidencing for quality of interfaces of GaInP-GaAs multiquantum well grown by low pressure MOCVD. The variation of PL intensity as a function of excitation power at low temperature is linear. X-ray diffraction measurements carried out on this sample showed a $\Delta a/a$ about 10^{-4}

Figure 32 shows the low temperature PL and PLE spectra of a GaAs-GaInP superlattice grown by LP-MOCVD. The superlattice consists of ten periods of approximately 100-Å-thick GaInP barriers and 90-Å-thick GaAs wells. The 4 K PL spectrum has a full width at half maximum of 4 meV evidencing a good layer-to-layer reproduci-bility. The PLE spectrum of the 90 Å quantum well exhibits a series of sharp peaks, which can be attributed to the electron to light-hole and electron to heavy bole transitions as indicated in Fig. 32. The energy of E1HH$_1$ subband transition, as determined by PLE, is 1545 meV and the peak energy of the (e,h) emission in PL spectrum is at 1542 meV, thus evidencing for high-quality interfaces and layer to layer homogeneity.

Room temperature photovoltage measurements were performed on the above ten period GaInP-GaAs superlattice. Figure 33 shows the

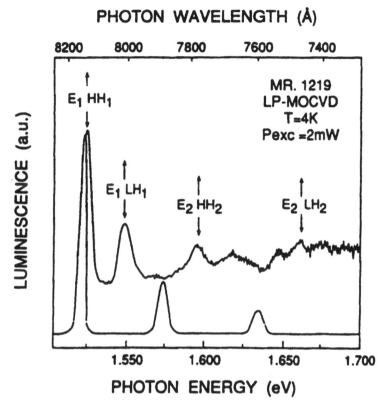

Fig. 31 Photoluminescence and photoluminescence excitation spectrum of the quantum well of 120 Å thickness at 4 K. Sample structure is shown in the inset of Fig. 30. (Ref. 87).

room temperature photovoltage spectroscopy (PVS) of a GaInP-GaAs superlattice. The well resolved peaks corresponding to transitions from the several sublevels. Also, the sharp peak, rather than step-like profile, of the photovoltage spectrum indicates a pronounced exciton absorption behavior at room temperature. In comparison with the PLE at 4K in Fig. 32, the intersubband transitions are more clearly revealed in the room temperature PVS. These results clearly demonstrate that photovoltage spectroscopy is a simple and powerful tool to study quantum confinement structures [76].

Using a standard quantum well calculation of energy levels, where we introduce, in a simple iterative form, the nonparabolicity of the energy bands. The transition energies for the nominal quantum well

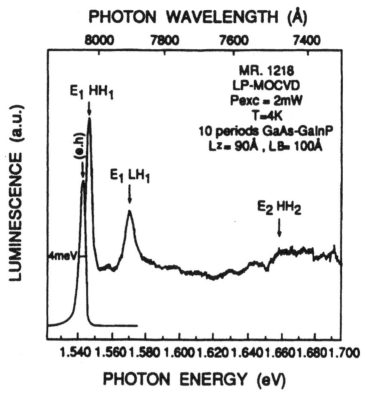

Fig. 32 Photoluminescence and photoluminescence excitation spectrum of a GaAs-GaInP superlattice consisting of ten periods of 90-Å-thick GaAs quantum wells and 100-Å-thick $Ga_{0.51}In_{0.49}P$ barriers (Ref. 87).

thicknesses have been calculated. The parameters used for this study are the following (all values taken at a temperature of 4 K) [50],

$$GaAs: E_g = 1518 \, meV, \quad m_0^* = 0.0665 \, m_0,$$
$$m_{hh} = 0.475 \, m_0, \quad m_{1h} = 0.087 \, m_0,$$
$$GaInP: E_g = 1984 \, meV, \quad m_0^* = 0.1175 \, m_0,$$
$$m_{hh} = 0.660 \, m_0, \quad m_{1h} = 0.145 \, m_0,$$
$$GaAs/GaInP: \Delta E_C/\Delta E_g = 0.4.$$

Table 3 indicates the experimental and theoretical values of the energy levels of GaAs-GaInP quantum wells of 90, 120, 60, and 35 Å thicknesses. A satisfying agreement between theoretical and experi-

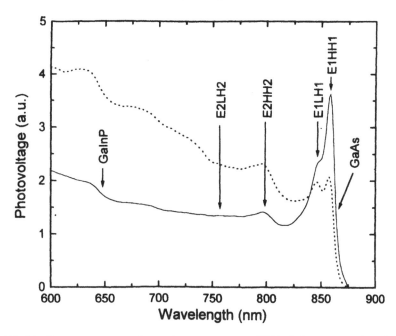

Fig. 33 Room temperature photovoltage spectroscopy (PVS) of the same sample as in Fig. 32 (Ref. 76).

TABLE 3
Experimental and theoretical values of energy levels for GaAs/GaInP MQWs.

Samples	$Lz(\text{Å})$	Energy Level	Calc. (meV)	Exp. (meV)	PL(e,h)(meV)
Superlattice	90	$E_1 HH_1$	1555	1546	1544
		$E_1 LH_1$	1576	1575	
		$E_2 HH_2$	1661	1664	
Multiquantum	120	$E_1 HH_1$	1541	1524	1522
Well		$E_1 LH_1$	1554	1550	
		$E_2 HH_2$	1608	1597	
		$E_2 LH_2$	1659	1661	
	60	$E_1 HH_1$	1582	1580	1574
		$E_1 LH_1$	1617	1618	
	35	$E_1 HH_1$	1637	1635	1634

mental values is observed for quantum wells of 90, 60, and 35 Å thicknesses. There is a noticeable discrepancy between the measured and calculated heavy hole-light hole splitting for the widest quantum well which could be attributed to compressive strain within this quantum well. However the fundamental transition should increase in this case, which is contrary to the observed trend. Such behavior is not clear at present but could be related to In segregation and subsequent localized strain. This could also be a source of uncertainty on the quantum well thickness. One could expect the widest quantum well to be the least sensitive to interface effects, but one should note that this widest well is the first of the whole structure: we carried out a mechanochemical bevel on this structure which showed that the first quantum well interface was not as good as the following ones.

Electrolyte electroreflectance is also used to study the optical transition in the quantum confinement structure. Electrolyte electroreflectance (EER) is the best and most direct technique for the determination of band profiles [77–79] in general and band offsets in particular. This is because EER yields especially sharp optical spectra and yields enhanced signals from interfacial surfaces. Thus, EER allows the determination of many more transition energies than does photoluminescencee. In particular, EER allows one to observe and measure the energies of transitions between states allowed only on one side of an interface and the tails of states allowed only on the other side; such transitions are called crossover transitions. For a type I quantum well or superlattice, these crossover transitions are between well states and the barrier band edges. Because of the ability to obtain a better signal-to noise ratio, to control better the modulating electric field and to apply bias voltages, EER also gives many more transition energies than does photoreflectance. In particular, the ability to apply a bias voltage allows one to observe and identify many transitions forbidden by symmetry in the absence of an electric field, as well as crossover transitions. This allowed one to measure directly intersubband, intraband transition energies without the use of band calculations. It also allowed one to measure almost directly the band offset in both samples, with the only theoretical input being the energy of the first heavy-hole level relative to the valence-band edge in the well.

Razeghi et al. [80] obtained approximately twenty electrolyte electro reflectance (EER) spectra, at different bias voltages. This result is the first reported direct optical measurement of either of these types of quantities for type I superlattices or quantum wells. Two GaAs/ $Ga_{0.51} In_{0.51} P$ based samples, a superlattice and a multiple quantum well have been used for this study. The EER spectra were fit using the generalized ER lineshape [81] to obtain the transition energies and linewidths, as well as the coefficients of the first-, second- and third-derivative terms in the generalized lineshape. Well-state-to-well-state transitions were assumed to have only a first-derivative term and crossover transitions were assumed to be primarily first-derivative in nature. This study of the spectra as a function of bias voltage helped to confirm the identification of the types of transitions observed. For example, for quantum wells and superlattices, the linewidth of cross-over transitions increases much more rapidly with internal electric field than does that of any other type of transition.

For the superlattice, 35 transitions were identified. This large number of observed transitions gives a multiplicity of self-consistency conditions and thus gives many checks both on the accuracy of our values for the transition energies and the identification of the transitions. All of the self-consistency conditions were well satisfied. For the multiple quantum well, which contained wells of three different widths, 30 transitions were identified. For both samples, the observed transition energies were fitted with a parameterized envelope-function band calculation in which the energy dependence of the effective masses was taken into account. The well widths obtained from these fittings were in excellent agreement with those obtained by calibration of the growth rate.

For both samples studies, a superlattice (SL) and a three-quantum-well structure, the linewidths of the well-state-to-well-state transitions were very narrow, from ~ 3 meV for the lowest-energy transitions up to ~ 8 meV for the highest-energy well-state-to-well-state transitions. These linewidths are narrower than we have seen before at room temperature, even on other quantum-well systems. These narrow linewidths are evidence of the unusually high quality of these samples. Another evidence of their exceptional quality was our ability to see many more transitions than is usual for such samples.

From a fitting of all the observed transition energies to a parameterized band calculation, the band offset of GaInP/GaAs is obtained. The fitting yields a 91 Å well width, in excellent agreement with the value of 90 Å found by calibration of the rate of epitaxial growth. The values found for the conduction-band, heavy-hole and light-hole masses in the barriers and the wells and for the split-off well mass are in excellent agreement with the literature.

5.3. GaInP/GaAs Quantum Confinement Structure on Si and InP Substrates

As have been mentioned in the introduction, one of the advantage of GaInP/GaAs over AlGaAs/GaAs is the possibility of monolithical integration of quantum well infrared photodetector (QWIP) focal plane array on Si circuit. As the first step, both multiquantum well structures and superlattices were grown by low pressure-MOCVD [82] on Si substrate.

Figure 34 shows the PL spectra a 4 K for a five-well structure of GaAs-GaInP grown on GaAs, Si and InP substrates. The excitation power is 10 mW. The nominal thicknesses of GaAs (well) and GaInP (barrier) indicated in the inset of Fig. 34 are those deduced from the measured growth rate of thick layers. A 2 μm GaAs buffer layer was grown on the substrate to block the defects originating from the mismatch between the GaAs and Si or InP substrate. The GaAs-GaInP interfaces were realized by turning off the AsH_3 flow and turning on both the TMI and PH_3 flows. The reversed procedure was used to obtain GaInP-GaAs interfaces. The purge of In is critical to the sharp GaInP-GaAs interfaces due to the memory effect of In. The growth rate was 100 Å/min and 200 Å/min for GaAs and GaInP respectively, and the gas flow stabilized to its new steady state value in less than a few seconds after switching.

PL measurements were done at various temperatures ranging from 4K to 297K. PL is performed with the Bio-Rad PL6100 Fourier transform spectroscope system. The 488 nm line of Ar laser is used as the excitation. The linear relation of the PL intensity with the excitation power is observed at low temperature which reveals that the peaks originates from the intrinsic exciton transitions. The five peaks around 1.52 eV, 1.55 eV, 1.62 eV, 1.69 eV and 1.78 eV in Fig. 34 are attributed to the transitions from the first electron sub level (e1) to the first heavy hole sub level (hh1) for the five-well structure. Two peaks at energies lower than GaAs band gap are observed. At nearly the same positions, two set of peaks are also observed in the GaAs epilayer on GaAs substrate. It is believed that the peak at 1.49 eV is due to donor-acceptor pair recombination and the peak at 1.513 eV is due mainly to the recombinations of impurity bounded excitons as that observed in the PL spectra of GaAs-AlGaAs MQW at low temperature [83].

Table 4 lists the peak positions, the full widths at half maximum (FWHM) and PL intensities of e1-hh1 transition peaks at excitation powers of 10 mW for multiquantum well grown on GaAs, Si and InP substrate as deduced from Fig. 34. The theoretical well thicknesses and nominal well thicknesses are also included in Table 4. The theoretical thicknesses of the wells are determined by fitting the experimental data at near the room temperature with the single quantum well

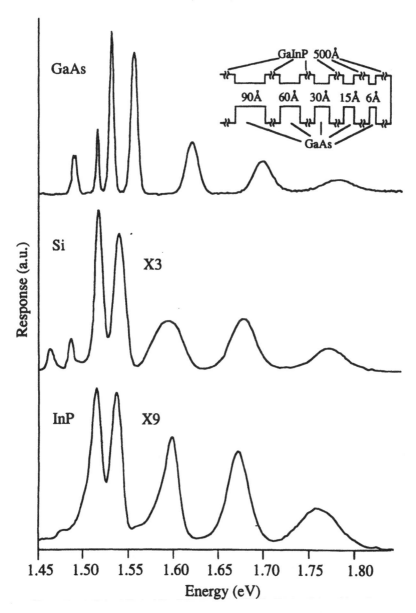

Fig. 34 The PL spectra of the GaAs-GaInP mutiquantum wells grown on GaAs, Si and InP substrates at 4 K. The structure is shown in the inset.

TABLE 4

The FWHM, peak positions and PL intensities of the MQW grown on GaAs, Si and InP substrates at 4K.

PL peaks (meV)			Intensities (a.u.)			FWHM (meV)			Well thickness (Å)	
GaAs	Si	InP	GaAs	Si	InP	GaAs	Si	InP	Calc.	Nominal
1529	1517	1515	3.673	1.113	0.238	5	9	16	110	90
1554	1539	1536	3.208	0.914	0.225	9	15	17	65	60
1615	1594	1598	1.176	0.317	0.151	16	36	24	31	30
1698	1679	1679	0.737	0.338	0.126	24	29	30	18	15
1782	1774	1760	0.320	0.148	0.054	38	37	52	10	6

model. Since the GaInP layers are 500 Å thick, the single quantum well model is a good approximation to each of the five wells. The parameters used in the calculations are as follows:

$$\text{GaAs: } m_c = 0.0665\, m_0, \quad m_{hh} = 0.475\, m_0$$
$$\text{GaInP: } m_c = 0.1175\, m_0, \quad m_{hh} = 0.660\, m_0$$
$$\text{GaAs/GaInP: } \Delta E_C / \Delta E_g = 0.41$$

The fitting results are shown in Fig. 35. The formula:

$$E_g(T) = 1.519 - 5.405 \times 10^{-4} T^2 / (T + 204)$$

is used for the temperature dependence of band gap for GaAs and a linear change of the band gap difference between GaAs from 0.466 eV at 4 K to 0.472 eV at 300 K is assumed. An approximate 15 meV discrepancy at low temperature between experimental value and calculation results is observed (Fig. 35). The discrepancy has two origins: One is the binding energy of quasi-2D excitons and another one is the strain-induced band gap change. The strain-induced band gap change is due to the different thermal expansion coefficient between GaAs and GaInP. Previous room temperature x-ray diffraction measurements indicated that all GaInP epilayer were perfectly lattice matched to GaAs. The difference of the thermal expansion coefficient between GaAs and GaInP multiplied by the temperature change gives the mismatch, which is 5×10^{-4} at low temperature. This mismatch results in about 3 meV decrease in the transition energy from the conduction band to the heavy hole valence band according to the

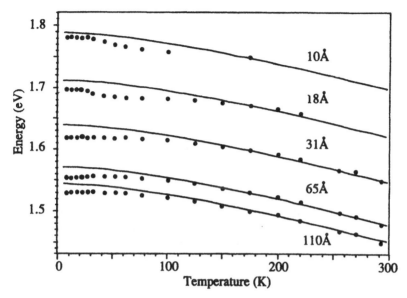

Fig. 35 The temperature dependence of the sub level transition energies (dots) for the five quantum wells and calculated transition energies (solid lines). The thickness shown in the figure are the calculated values.

formula. $(a = -9.8 \text{ eV}, \; b = -2.0 \text{ eV}, \; C_{11} = 11.8 \times 10^{10}, C_{12} = 5.36 \times 10^{10}$ for GaAs)

$$dE_{\text{heavy hole}} = -(2a(C_{11} - C_{12})/C_{11}$$
$$- b(C_{11} + 2C_{12})/C_{11}) \times \text{mismatch}.$$

The remaining 12 meV is the binding energy of the quasi-2D excitons. This value is the same as the results from the magneto optical experiments and similar to that from the other experiments (9–12 meV) on GaAs-GaAlAs quantum wells [83]. Although there are reports [84, 85] that the recombination of free excitons dominate even at room temperature and it is well accepted that free exciton absorption is clearly observed at room temperature for MQW, the temperature dependence of the PL energies in this work and the work done by Chiari et al. [86]. reveal that the excitonic recombinations dominate at low temperature while the free carrier recombinations dominate near room temperature.

The discrepancy on thicknesses of the wells between the nominal values and the calculated values is about 1–2 monolayer for the top

4 wells, but it is 20 Å for the first well which is immediately. on top of the buffer layer. A previous investigation [87] suggested that the interface of the first quantum well was not as good as the following ones and that the In segregation and subsequent localized strain would cause a relatively large energy shift. Our samples may suffer the same problem.

Figure 36 shows the temperature dependence of the PL spectra of the MQW grown on GaAs substrate. It indicates that the PL peak of the nominal thickness of 6 Å well disappears at 150 K and the PL peak of the nominal thickness of 15 Å well disappears at 297 K because of carrier heating and poor carrier confinement in such a narrow well.

Table 4 indicates that the PL peaks of GaAs-GaInP on Si and InP substrates are shifted about 15 meV toward the lower energy. There are two possible origins for this energy shift: One is the localized strain due to the defects and In segregations. The strain in the GaAs layer can cause the change of gap energy from the heavy hole band to conduction band. The observed energy shift can not be explained by the

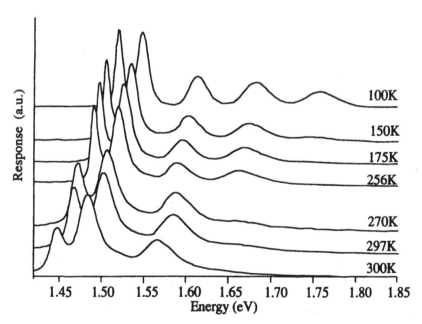

Fig. 36 The temperature dependence of the PL spectra of the MQW grown on GaAs substrate.

mismatch-induced uniform strain between substrate and epilayer. Theoretically speaking, there is hardly uniform strain in a 2 μm buffer layer, because no strain can be sustained in a 2 μm buffer layer for the mismatch as large as 4%. Most mismatch is eased by the dislocations instead of strain. Even though there is some remaining strain, considering the lattice constant of Si is 4% smaller than that of GaAs while the lattice constant of InP is 4% larger than that of GaAs, the strain induced energy shift should be the opposite value rather than the observed one. But the two impurity related peaks are shifted as well, thus the observed shift must be caused by the band gap shrinkage. We believe that the defects and In segregations may cause localized strain and as a result the energy shift. The second possible reason is that Indium diffuses to the epilayer through the dislocation threads from GaInP to GaAs. It has been observed that the diffusion of Si along the line of dislocation threads has caused the autodoping of GaAs epilayer grown on Si [88]. As were known, the band gap of GaInP is very sensitive to the In composition, it is easy to have 15 meV decrease in band gap of the well due to the diffusion of the In from GaInP to GaAs. The fact that the PL peaks of the top well are also shifted evidences that the defects are not effectively blocked by the buffer layers and they propagate to the utmost epilayer.

The FWHM and PL intensities of the samples listed in Table 4 indicate that the quality of the samples decreases in the order of MQW grown on GaAs, on Si and on InP, except the 30-well grown on the Si substrate which is broadened considerably. For all three samples, the PL peaks are relatively sharp. It evidences the high quality of the samples grown on the three different substrates. The sample grown on Si has better quality than the sample grown on InP, although they have similar mismatch with GaAs. In fact, the Si substrate is mechanically much more stronger than InP substrate. This fact indicates that Si is the better substrate for strained layer and mismatch epitaxy.

The PL peaks of MQW grown on Si and GaAs substrates are distinguishable until 300 K while the peaks from the sample grown on InP substrate disappear at about 200 K. The intense luminescence from the defects in GaAs is observed for the samples grown on Si and InP substrates when the temperature is higher than 125 K. The weak PL peaks from the MQW overlap on the luminescence from the defects in GaAs. This shows again that 2 μm buffer layer can not block the defects very effectively.

6. GaAs-GaInP QUANTUM WELL INFRARED PHOTODETECTORS

6.1. Introduction and Basic Principles

Quantum Well Intersubband Photodetectors were first proposed in 1983 by Smith, et al. [89] were first demonstrated in their modern form in 1987 by Levine, et al. [90] Since that time they have been steadily improved in both responsivity and detectivity [91]. They have also been used in focal plane arrays.

The operation of such devices is quite simple: carriers are trapped within the wells of the superlattice and cannot conduct unless excited out of the wells (See Fig. 37). The energy required to excite a carrier out of the well is on the order of the bandgap discontinuity, i.e. the barrier height of the quantum well. Therefore, in spite of the fact that QWIPs are grown with relatively wide bandgap material like GaAs, AlGaAs, and GaInP, narrow bandgap-like performance is achieved. This is particularly beneficial in light of the fact that these III−V semiconductors are well understood and easy to grow with extremely high uniformity relative to that HgCdTe and other narrow bandgap material (see Secs. 1 and 2).

In general, QWIPs are grown on a semi-insulating substrate; a notable exception to this is growth on Silicon substrates. The first layer of the structure to be grown is the bottom contact which is generally a 1 µm thick layer of heavily doped GaAs. On top of this is

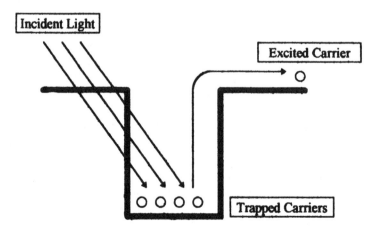

Fig. 37 Basic principles of QWIP function.

grown the actual QWIP superlattice – generally 50 periods of well and barrier material. Finally, a top contact is grown, typically 0.5 μm thick, heavily doped GaAs layer. This structure is the standard recipe, though it is certainly possible to use contacts of different size and a superlattice with a different number of periods. Once grown the sample is processed into mesa structures which allow for connection between each of the two contact layers and external circuitry.

Since GaAs forms a Type I superlattice with either GaInP or AlGaAs (i.e., the GaAs material is a potential well for both electrons in the conduction band and holes in the valence band), it is up to the designer to choose which well will be used for the QWIP. This choice made by doping the wells and contacts with the appropriate flavor of dopant (e.g. Zn for p-type or valence band QWIPs and Si for n-type or conduction band QWIPs). However, the choice is not without consequence because the two flavors of QWIPs are very different. n-type QWIPs tend to have a larger responsivity but they also have a higher dark current, and polarization dependent quantum mechanical selection rules make normal incidence detection impossible without external coupling systems. p-type QWIPs have much lower dark currents and band mixing in the quantum well overcomes the polarization dependent selection rules.

The majority of QWIP research thus far has concerned itself with n-type GaAs/AlGaAs structures. They have yielded good results (see Ref. 91) particularly with highly uniform arrays and improved Noise Equivalent Delta-Temperature (NEΔT) which is measure of the smallest change in temperature detectable by a system and the benchmark by which focal plane arrays are measured. However, for the devices to truly compete with existing technology, detectivity must be improved and background limited performance must be achieved at temperatures above that of liquid nitrogen. It is to achieve these improvements that investigations into the GaAs/GaInP system are worth- while. The superior electrical characteristics of GaInP (See Secs. 1 and 3) suggest improvements in the transport of photoexcited carriers. The superior surface recombination velocity at the GaAs/GaInP interface suggests further improvements.

n-type GaAs/GaInP has been tried once and the results were good enough to warrant further investigation. However, judging the reduction of dark current and capability for inherent normal incidence detection to be more important than improved responsivity, p-type QWIPs have been chosen for this investigation.

6.2. Characterization of GaAs/GaInP QWIPs

Three GaAs/GaInP QWIP structures were grown by MOCVD. All had 1.0 μm and 0.5 μm bottom and top contacts as depicted in Fig. 38. All had GaAs wells and GaAs top and bottom contacts doped with Zn to a concentration of 2×10^{18} cm^{-3}. Finally, all had undoped GaInP barriers 280 Å wide. The only differences between the three samples were the well widths which are indicated in Table 5.

The samples were fabricated into 400 μm × 400 μm mesas using photolithography and wet chemical etching. The etching procedure was a selective one which therefore guaranteed an etch down to the bottom contact and no further. A 5:1:1 mixture of $H_2O:H_3PO_4$: H_2O_2 was used as the GaAs etch and pure Hydrochloric Acid was used as the GaInP etch. Such a procedure is unavailable in GaAs/AlGaAs and it should ultimately help to improve the uniformity and yield of GaAs/GaInP focal plane arrays over their GaAs/AlGaAs counterparts, since the etch depth will be a constant. Once etched, these samples were given 100 μm × 100 μm Ti/Pt/Au contacts by electron beam evaporation. They were mounted on heat sinks and placed in a liquid nitrogen cryostat.

Figure 39 shows the normalized photoconductive response spectra for the three samples. The spectral response shifts towards longer wavelength as the well width is reduced, which is expected as the hole

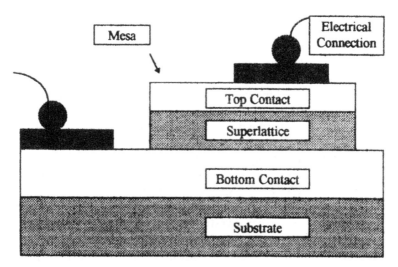

Fig. 38   The standard QWIP Mesa structure.

TABLE 5
Well widths of the three GaAs/GaInP QWIP Samples.

Sample	Well Width
A	25 Å
B	35 Å
C	55 Å

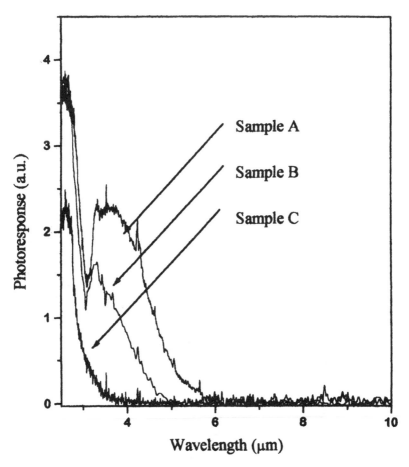

Fig. 39 Normalized photoconductive response spectra for the three GaAs/GaInP
samples. Note the increase in cutoff wavelength with decreasing well width.

ground state is pushed up towards the top of the barrier. For each of the three samples, the spectral response appears to peak outside the FTIR spectrometer wavelength range.

Dark current was measured versus bias at temperatures between 77 K and 300 K with an HP4155A Semiconductor Parameter Analyzer. Figure 40 shows a set of measurements for the 25 Å wide well sample. At low bias, tunneling is negligible and dark current

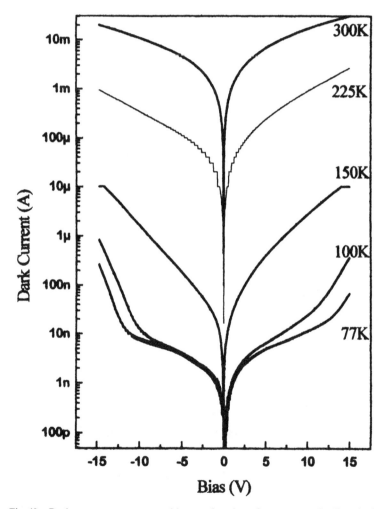

Fig. 40   Dark current curves versus bias as a function of temperature for Sample A.

originates from thermionic emission above the top of the barrier [89]. Arrhenius plots of the dark current measured at a fixed low bias of 0.10 V are shown in Fig. 41 for each of the three samples. The slopes of these plots reveal activation energies of 235 meV, 300 meV, and 330 meV for Samples A, B, and C respectively. These activation energies correspond to wavelengths of 5.28 μm, 4.13 μm and 3.75 μm, respectively. These values closely correspond to observed cut-off wavelengths in Fig. 39. This is expected since simple calculations place the Fermi level lies within a few meV of the heavy hole ground-state.

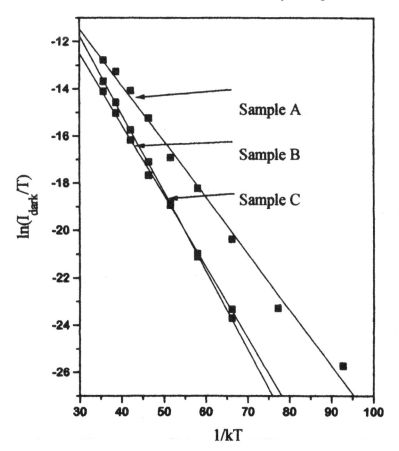

Fig. 41 Arrhenius plots of Dark current versus temperature for the three samples. The slopes of these plots indicate activation energies of 235 meV, 300 meV and 330 meV for Samples A, B, and C. Given the high doping level, the Fermi level is close to the ground hole state. Therefore, these slopes indicate cutoff wavelengths of 5.28 mm, 4.13 mm and 3.75 mm respectively.

Theoretical calculations of the cut-off wavelength were performed based on a modified Kronig-Penney model taking into account the effective mass discontinuity at the well/barrier interface, and reported parameters for GaAs and $Ga_{0.51}In_{0.49}P$ (see Sec. 4). The cut-off wavelength corresponds to the energy difference between the top of the barrier and the ground-state heavy hole energy level. Figure 42 shows measured and calculated cut-off wavelengths as a function of well width for a 280 Å barrier. Theoretical predictions closely match experimental observations, and further show that extremely narrow GaAs quantum wells must be used in order to reach the strategic 8–12 μm atmospheric window. Such narrow wells can be difficult to

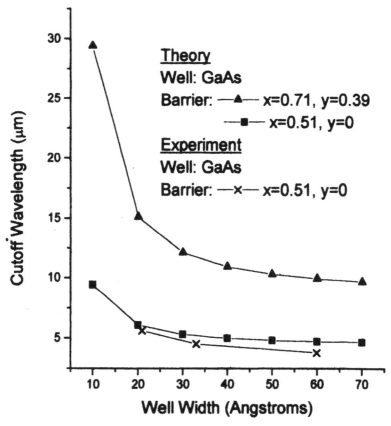

Fig. 42   Theoretical calculations of the cutoff wavelength versus well width for GaAs wells and A) GaInP barriers B) $Ga_{0.71}In_{0.29}As_{0.39}P_{0.61}$ barriers. Also shown is the experimental plot of cutoff wavelength versus well width for samples A, B, and C.

manufacture with high yield. This is the reason why we have also represented on Fig. 42 results of calculations for lattice-matched quaternary barriers. It can be seen that GaInAsP alloys should be used in order to reduce the valence band discontinuity thereby permitting operation at longer wavelength.

Finally, the samples were exposed to 800 K blackbody radiation and their absolute blackbody photoresponse was measured. Using the spectral shape and Planck's Law of Spectral Exitance, the peak responsivity was calculated. As an example, at 77 K, the current responsivity of Sample A (corrected for front reflection) at 2.5 μm wavelength was 0.5 mA/W at 5 V bias and increased linearly to 2 mA/W at 20 V bias. The other samples showed similar peak responsivities.

6.3. Characterization of Lattice Matched GaAs/$Ga_{0.71}In_{0.29}As_{0.39}P_{0.61}$ QWIPs

To test the validity of the theoretical predictions of Fig. 42, a GaAs/ $Ga_{0.71}In_{0.29}As_{0.39}P_{0.61}$ QWIP was grown, fabricated and tested. Like all of the GaAs/GaInP QWIPs, this structure consisted of a GaAs bottom contact 1.0 μm thick, 50 periods of well and barrier material, and a top contact of 0.5 μm thick GaAs. The barriers were all 280 Å thick $Ga_{0.71}In_{0.29}As_{0.39}P_{0.61}$ ($Eg = 1.75$ eV) and undoped. The wells were all 30 Å thick and doped 2×10^{18} cm^{-3}. Like the wells, the top and bottom contacts were doped with Zinc to a net acceptor concentration of 2×10^{18} cm^{-3}. The sample was processed into 400 μm × 400 μm mesa structures using standard photolithographic techniques and the same selective wet chemical etching technique. Finally, 100 μm × 100 μm Ti/Pt/Au contacts were formed by electron beam evaporation.

The effect of the quaternary alloy was to reduce the valence band discontinuity relative to lattice matched GaInP and thereby increase the cut-off wavelength. Photoresponse spectras were measured at 77 K for various biases using the Fourier Transform Infrared spectrometer. Measurements were taken across a 40 kΩ load resistor which was negligible compared to the QWIP differential impedance and the capacitive impedances of cables. Results are shown in Fig. 43. A pronounced photovoltaic effect was observed. A similar, though much less dramatic effect was observed in the GaAs/GaInP quantum wells. This arose from an asymmetric quantum well potential profile. Such asymmetry can be linked to either a structural difference in the two quantum well interfaces [89, 90], or to dopant migration during the growth [89]. The photoresponse magnitude increased with bias. The photoresponse spectra peaked around 4 μm with a broad maxima. Careful examination of the spectra revealed that the cut-off wavelength increased slightly with bias and that this effect was stronger for positive (mesa top positive) bias. We believe this increase originates in

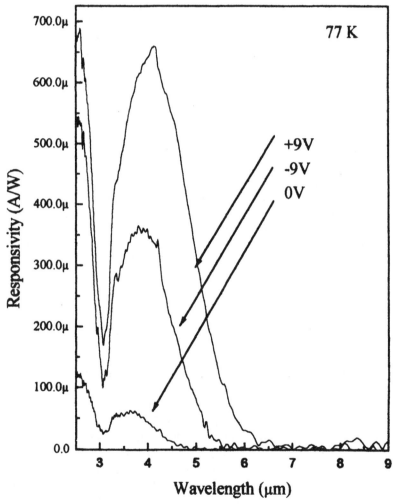

Fig. 43 Photoresponse spectra for 3 different biases for the GaAs/$Ga_{0.71}In_{0.29}$ $As_{0.39}P_{0.61}$ Sample. Note the slight change in spectral shape with bias and note the photovoltaic effect at 0 V bias.

part from the reduction of the effective barrier height (by one-half of the potential drop per well), and in part from easier tunneling out of the well through the triangular tip for the light hole state that is now close to the top of the barrier (see Band Structure in Fig. 44).

Again, the dark current was measured as function of bias for temperatures between 77 K and 300 K. However, this time the experiment was performed both with the sample cold shielded and with the sample exposed to a 300 K background. At a detector temperature of 300 K the current was identical with or without the cold shield. Figure 45 shows that at 77 K, the photodetector is clearly background-limited as the unshielded, 300 K background photocurrent exceeded the dark current by about two orders of magnitude. The device remains background-limited up to a detector temperature of 100 K for biases between -7.5 V and $+2.5$ V. To the best of our knowledge, there has been only one other report of BLIP operation in QWIPs at such high temperature [96]. The 77 K unshielded I(V) curve reached its minimum at around $+0.2$ V. This was a result of the photovoltaic response of the device. This was confirmed by increasing the level of infrared illumination using a blackbody source. At higher illuminatioin, the $I(V)$ curve moved up and reached a minimum at a voltage that approached $+1.5$ V asymptotically. Figure 46 shows Arrhenius plots of the dark current versus inverse temperature for different biases. They reveal activation energies which agree with the cutoff wavelengths observed in Fig. 43. In particular, the activation energy was lower and the cut-off wavelength is longer for $+9$ V than for -9 V, for reasons discussed at the end of the preceding paragraph.

Blackbody responsivity was measured as a function of bias using a chopped 800 K blackbody radiation source and conventional lock-in detection techniques, and is shown in Fig. 47. Here again the signal was measured across a 40 kΩ load resistor. Photoresponse goes to zero for approximately 1.5 V, consistent with the observation of a photovolatic effect. For negative biases between 0 V and -10 V, responsivity increased at a lower rate than for positive biases. For larger negative biases, the responsivity increased rapidly and reached a similar magnitude for $+18$ V or -18 V. The lower rate of increase indicates a larger photo-excited carrier capture probability, which results from the asymmetry in the quantum well profile. At higher biases, this asymmetry no longer plays a significant role because the photo-excited carriers gain enough energy from the electric field so that they are less sensitive to the actual well potential profile.

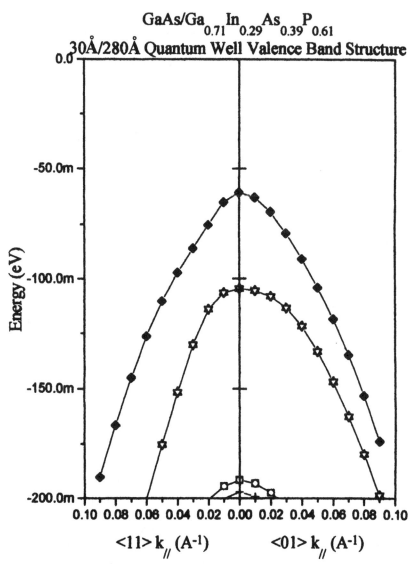

Fig. 44 Band structure calculated by an 8 × 8 envelope function approximation. The top of the barrier is approximately 195 meV. Note the existence of a light hole state just inside the well which would help contribute to high field dark current.

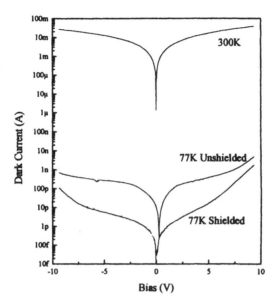

Fig. 45 The overwhelming difference between the unshielded and shielded dark currents at 77 K prove that the sample is BLIP. In fact, the sample remains BLIP up to 100 K.

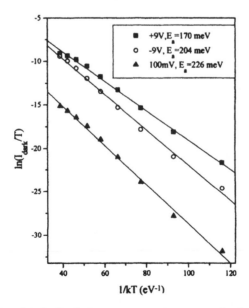

Fig. 46 Arrhenius plots for the dark current versus temperature for the biases used in the photoresponse spectra. The cutoff wavelengths agree with what is shown in Fig. 43.

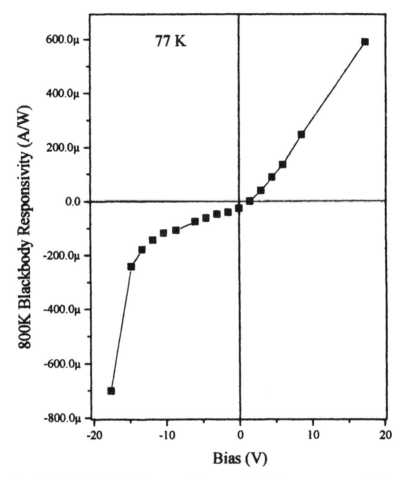

Fig. 47 A plot of the blackbody responsivity versus bias for the quaternary sample. Note the pronounced asymmetry in the responsivity due to the asymmetric quantum well profile.

7. CONCLUSION

We discussed in detail the growth conditions of GaInP and the doping behavior in GaInP. The properties of GaInP and GaInP/GaAs heterostructure have been reviewed as a guide to the understanding of the GaInP/GaAs system. The structural and optical characterization of the quantum confinement structure. i.e., quantum wells and superlattices, present an overview of the tools to understand the properties of these structures. The strong hole intersubband absorption has been observed at 9mm, and its dependence on light polarization has been investigated. As we described in the introduction part, we believe that GaInP/GaAs system has advantages over GaAs/GaAlAs system that currently dominates the quantum well infrared photodetector (QWIP) applications. More researchs on the development of GaInP/GaAs QWIP are worthwhile in order to explore the potential benefit of this promising material system.

Acknowledgments

The work at Northwestern University was supported by Air Force Contract No. F33615-93-C-5382 through KOPIN Corporation. The author would like to thank Dr. G. Brown from Wright Patterson, Dr. E. Bigan, X-He, J. Hoff, C. Jelen, S. Kim, S. Slivken, M. Erdtmann for their participation in this work.

References

1. B.F. Levine, *Semicond. Sci. and Technol.*, **8**, S400 (1993).
2. J.M. Olson, R.K. Ahrenkiel, D.J. Dunlavy, B. Keyes, and A.E. Kibbler, *Appl. Phys. Lett.*, **55**, 1208 (1989).
3. L.W. Molenkamp and H.F.J. van't Blik, *J. Appl. Phys.*, **64**, 4253 (1988).
4. H.C. Casey and M.B. Panish, *Heterostructure Lasers*, Part. B (Academic, New York, 1978), pp. 20.
5. C.C. Hsu, J.S. Yuan, R.M. Cohen, and G.B. Stringfellow, *J. Appl. Phys.*, **59**, 395 (1985).
6. M. Razeghi, F. Omnes, J. Nagle, M. Defour, O. Acher, and P. Bove, *Appl. Phys. Lett.*, **55**, 1677 (1989).
7. K. Kitahara, M. Hoshino, K. Kodama, and M. Ozeki, Jpn. *J. Appl. Phys.*, **25**, L534 (1986).
8. F. Omnes, Defour, and M. Razeghi, *M. Revue Technique*, Thomson-CSF, **23**, No. 3 (1991).
9. Y. Kawamura and H. Asahi, *Appl. Phys. Lett.*, **43**, 780 (1983).
10. I. Ladany, *J. Appl. Phys.*, **43**, 654 (1971).
11. A. Gomyo, H. Hotto, and I. Hino, *Jpn. J. Appl. Phys.*, **28**, L1330 (1989).
12. C.C. Hsu, J.S. Yuan, R.M. Cohen, and G.B. Stringfellow, *J. Appl. Phys.*, **59**, 395 (1985).
13. T. Suzuki, A. Gomyo, and S. Iijima, *J. Cryst. Growth*, **93**, 389 (1988).
14. A. Gomyo, T. Suzuki, and S. Iijima, *Phys. Rev. Lett.*, **60**, 2645 (1988).
15. B.T. McDermott, N.A. El-Marry, B.L. Jiang, F. Hyuga, and S.M. Bedair, *J. of Cryst. Growth*, **107**, 96 (1991).
16. O. Ueda, M. Takikawa, J. Komeno, and I. Umebu, *Jpn. J. Appl. Phys.*, Part. 2, **26**, L1824 (1987).
17. P. Bellon, J.P. Chevalier, G.P. Martin, E. Dupont-Nivel, C. Theibant, and J.A. Andre, *Appl. Phys. Lett.*, **52**, 567 (1988).

18. M. Kondow, H. Kakibayashi, S. Minagawa, Y. Inoue, T. Nishino, and Y. Hamakawa, *Appl. Phys. Lett.*, **53**, 2953 (1988).
19. S. Yasuami, C. Nozaki, and Y. Ohba, *Appl. Phys. Lett.*, **52**, 2031 (1988).
20. O. Ueda, M. Takikawa, M. Takechi, J. Komeno, and I. Umebu, *J. Cryst, Growth*, **93**, 418 (1988).
21. H. Nakayama and H. Fujita, in GaAs and Related Compounds–1985, edited by M. Fujimoto, IOP Conference Proceedings, **79**, (Hilger, Bristol, 1986), p. 289.
22. H.R. Jen, M.J. Jou, Y.T. Cerng, and G.B. Stringfellow, *J. Cryst, Growth*, **85**, 175 (1987).
23. T.S. Kuan, T.F. Kuench, W.I. Wang, and E.L. Wilkie, *Phys. Rev. Lett.*, **54**, 201 (1985).
24. L.D. Landau and E.M. Lifshitz, *Statistical Physics*, Chap. 14, (Pergamon, Oxford, 1969).
25. M. Kondow, H. Kakibayashi, S. Minagawa, Y. Inoue, T. Nishino, and Y. Hamakawa, *J. Cryst. Growth*, **93**, 412 (1988).
26. M. Kondow, H. Kakibayashi, T. Tanaka, and S. Minagawa, *Phys. Rev. Lett.*, **63**, 884 (1989).
27. T. Suzuki, A. Gomyo, and S. Iijima, *J. Cryst. Growth*, **93**, 389 (1988).
28. A. Gomyo, T. Suzuki, K. Kohayashi, S. Kawata, I. Hino, and T. Yuara, *Appl. Phys. Lett.*, **50**, 673 (1987).
29. T. Suzuki, A. Gomyo, I. hino, K. Kobayashi, S. Kawata, and S. Lijina, *Jpn. J. Appl. Phys.*, **27**, L1549 (1988).
30. T. Suzuki, A. Gomyo, S. Ijima, K. Kobayashi, S. Kawata, I. Hino, and T. Yuasa, *Jpn. J. Appl. Phys.*, **27**, 2098 (1988).
31. Y. Inoue, T. Nishino, Y. Hamakawa, M. Kondow, and S. Minagawa *Optoelectronics Devices and Technologies*, **3**, 61 (1988).
32. S.J. Feng, J.C. Bourgoin, F. Omnes, and M. Razeghi, *Appl. Phys. Lett.*, **59**, 941 (1991).
33. M.C. Delong, P.C. Taylor, and J.M. Olson, *Appl. Phys. Lett.*, **57**, 620 (1990).
34. J.C. Bourgoin, *Appl. Phys. Lett.*, **59**, 941 (1991).
35. S. Ben Amor, L. Dmowski, J.C. Portal, N.J. Pulsford, J. Singleton, R.J. Nicholas, and M. Razeghi, *J. Appl. Phys.*, **65**, 2756 (1989).
36. M.A. Rao, E.J. Caine, H. Kroemer, S.J. Long, and D. Babic, *J. Appl. Phys.*, **61**, 643 (1987).
37. H. Kroemer, W.Y. Chien, J.S. Harris, and D.D. Edwall, *Appl. Phys. Lett.*, **36**, 295 (1980).
38. D. Biswas, N. Debbar, P. Bhattacharya, M. Razeghi, M, and F. Omnes, *Appl. Phys. Lett.*, **56**, 833 (1990).
39. K.P. Martin, R.J. Higgins, J.J.L. Rascol, H.M. Yoo, and J.R. Arthur, *Surf. Sci.*, **196**, 323 (1988).
40. P.A. Martin, K. Meehan, P. Gavrilovic, K. Hess, N. Holonyak, and J.J. Coleman, *J. Appl. Phys.*, **54**, 4689 (1983).
41. O. Miyoko, N. Watanabe, and Y. Ohba, *Appl. Phys. Lett.*, **50**, 906 (1987).
42. Y.J. Chan, D. Pavilidis, M. Razeghi, and F. Omnes, *IEEE Trans. Electron. Devices*, **37**, 2141 (1990).
43. C.H. Chen, S.M. Baier, and D.K. Arch, M. Shur, *IEEE Trans, Electron. Devices*, **ED-35**, 570 (1988).
44. M. Razeghi, P. Maurel, F. Omnes, S. Ben Armor, L. Dmowski, and J. C. Portal, *Appl. Phys. Lett.*, **48**, 1267 (1986).
45. H.L. Stormer, A.C. Gossard, W. Wsiegmann, and K. Baldwin, *Appl. Phys. Lett.*, **39**, 912 (1981).
46. A. Kastalsky and J.C.M. Hwang, *Solid State Commun.*, **51**, 317 (1984).
47. T. Ando, A.B. Fowler, and F. Sterm, *Rev. Mod. Phys.*, **54**, 437 (1982).
48. M. Razeghi, M. Defour, F. Omnes, M. Dobers, J.P. Vieren, and Y. Guldner, *Appl. Phys. Lett.*, **55**, 457 (1989).

49. M.O. Watanabe and J. Ohba, *Appl. Phys, Lett.*, **50**, 906 (1987).
50. Madelung, *Series III*, 17a: *Semiconductors* (Springer, New York, 1982).
51. H.L. Stormer, R. Dingle, A.C. Gossard, W. Wiegman, and M.D. Sturge, *Solid State Commun.*, **29**, 705 (1979).
52. E. Schubert, A. Fisher, and K. Ploog, *Phys. Rev.*, B **31**, 7937 (1985).
53. M. Razeghi, F. Omnes, M. Defour, J. Nagle, and P. Bove, Unpublished.
54. K.V. Von Klitzing and G. Ebert, in Springer Series in Solid State Sciences, **53**, edited by G. Bauer, Kuchar, F, and H. Heinrich (Springer, Berlin, Heidelberg, New York, Tokyo, 1984), p. 242.
55. P. Voisin, Y. Gludner, J.P. Vieren, M. Voos, P. Deslecluse, and Nguyen T. Linh, *Appl. Phys. Lett.* **39**, 982 (1981).
56. G. Bastard, in Wave Mechanics Applied to Semiconductor Heterostructures (Les Editions de Physique, Les Ulis, France, 1988), p. 175.
57. M. Dobers, J.P. Vieren, M. Razeghi, M. Defour, and F. Omnes, *Semiconduc. Sci. Technol.*, **4**, 687 (1989).
58. U. Merkt, *Festkorperproblem* **27**, 109 (1987).
59. U. Rossler, F. Malcher, and G. Lommer, *Semiconductors in High Magnetic Fields*, Springer Series in Solid State Sciences, ed. Landwehr (Berlin: Springer, 1989).
60. M. Dobers, K. von Klitzing, and G. Weimann, *Phys. Rev.*, B, **38**, 5453 (1988).
61. D. Stein, G. Ebert, K. von Klitzing, and G. Weimann, *Surf. Sci.*, **142**, 406 (1984).
62. Y. Guldner, M. Voos, J.P. Vieren, J.P. Hirtz, and M. Heiblum, *Phys. Rev.*, B, **36**, 1266 (1987).
63. M. Dobers, F. Malcher, K. von Klitzing, U. Rossler, K. Ploog, and G. Weirmann, *Semiconductors in High Magnetic Fields*, Springer Series in Solid State Sciences, ed. G. Landwehr (Berlin.: Springer, 1989).
64. M. Dobers, K. von Klitzing, J. Schneider, G. Weimann, and K. Ploog, *Phys. Rev. Lett.*, **31**, 1650 (1988).
65. G. Lommer, F.F. Malcher, and U. Rossler, *Phys. Rev.*, B, **32**, 6995 (1985).
66. Y.J. Chan, D. Pavilidis, M. Razeghi, M. Jaffe, and J. Singh, *Inst. Phys. Conf. Ser.*, **96**, 459 (1988).
67. M. Razeghi, A. Machado, S.M. Koch, O. Acher, F. Omnes, and M. Defour, SPIE **1283**, 64 (1990).
68. M. Razeghi, and F. Omnes, *Appl. Phys. Lett.*, **59**, 1034 (1991).
69. M. Razeghi, F. Omnes, P. Maurel, J. Chan, and D. Pavalidis, *Semicond. Sci. Technol.*, **6**, 103 (1991).
70. X.G. He, and M. Razeghi, *J. Appl. Phys.*, **73**, 3284 (1993).
71. S. Takagi, *Acta Crystallogr.* **15**, 1311 (1962); *J. Phys. Soc. Jpn.*, **26**, 1239 (1969).
72. D. Taupin, *Bull. Soc. Fran. Miner. Cryst.*, **87**, 469 (1964).
73. J. Olivier, P. Etienne, M. Razeghi, and P. Alnot, *Nat. Res. Soc. Symp. Proc.*, **91**, 497 (1987); R. Bisaro, G. Lauzinich, A. Friederich, and M. Razeghi, *Appl. Phys. Lett.*, **40**, 978 (1982).
74. M.L. Tarng and D.G. Fisher, *J. Vac. Sci. Technol.*, **15**, 50 (1978).
75. M. Razeghi, *Semiconductors and Semimetals*, Edited by (Academic Press, New York 1985) **Vol. 22**, Part A. p. 305.
76. X.G. He and M. Razeghi, *Appl. Phys. Lett.*, **62**, 618 (1993).
77. L. Kassel, J.W. Garland, H. Abad, P.M. Raccah, J.E. Plotts, M.A. Haase, and H. Chang, *Appl. Phys. Lett.*, **56**, 42 (1990).
78. L. Kassel, J.W. Garland, P.M. Raccah, M.A. Haase, and H. Chang, *Semicond. Sci. Technol.*, **6**, A146 (1991).
79. L. Kassel, J.W. Garland, P.M. Raccah, M.C. Tamargo, and H. Farrell, *Semicond. Sci. Technol.*, **6**, A152 (1991).
80. M. Razeghi and Yang De, SPIE, 1992.
81. J.W. Garland and P.M. Raccah, SPIE, **Vol. 659**, 32 (1986).
82. M. Razeghi, *Prog. Crystal Growth and Characterization* 19 (1989).

144   M. RAZEGHI

83. R.C. Miller and D.A. Kleinman, *J. Lumin.*, **30**, 520 (1985).
84. S. Tarucha, H. Okamoto. Y. Iwasa, and N. Miura, *Solid State Commun.*, **52**, 815 (1984).
85. P. Dawson, G. Duggan, H.I. Ralph, and K. Woodbridge, *Phys. Rev.*, *B*, **28**, 7381 (1983).
86. A. Chiari, M. Colocci, F. Fermi, Y. Li, R. Querzoli, A. Vinattieri, and W. Zhuang, *Phys. Stat. Sol.*, (b) **147**, 421 (1988).
87. F. Omnes and M. Razeghi, *Appl. Phys. Lett.*, **59**, 1034 (1991).
88. S.F. Fang, K. Adomi, S. Iyer, H. Morkoc, H. Zabel, C. Choi, and N. Otsuka, *J. Appl., Phys.*, **68**, R31 (1991).
89. J.S. Smith, L.C. Chiu, S. Margalit, and A. Yariv, *Journ. Vac. Sci. Tech.*, **B1**, 376 (1983).
90. B.F. Levine, A. Zussman, S.D. Gunapala, M.T. Asom, J.M. Kuo, and W.S. Hobson, *J. Appl. Phys.*, **72**, 4429 (1992).
91. B.F. Levine, *J. Appl. Phys.*, **50**, R1 (1993).
92. M.A. Kinch and A. Yariv, *Appl. Phys. Lett.*, **55**, 2093 (1989).
93. Xiaoguang He and Manijeh Razeghi, *J. Appl. Phys.*, **73**, 3284 (1993).
94. K.L. Tsai, C.P. Lee, K.H. Chang, D.C. Liu, H.R. Chen, and J.S. Tsang, *Appl. Phys. Lett.*, **64**, 2436 (1994).
95. N.C. Liu, Z.R. Wasilewski, M. Buchanan, and Hanyou Chu, *Appl. Phys. Lett.*, **63**, 761 (1993).
96. Y.H. Wang Sheng S. Li, J. Chu and Pin Ho, *Appl. Phys. Lett.*, **64**, 727 (1994).

CHAPTER 3

Metal Grating Coupled Bound-to Miniband Transition III–V Quantum Well Infrared Photodetectors

SHENG S. LI

Department of Electrical Engineering, University of Florida, Gainesville, Florida, 32611 USA

We present three types of metal grating-coupled III – V quantum well infrared photodetectors (QWIPs) using bound-to-miniband (BTM) intersubband transition schemes for the 8–14 μm long-wavelength infrared detection. The BTM QWIPs described here used an enlarged quantum well with a well width ranging from 8.8 to 11 nm to enhance optical absorption and a short-period superlattice barrier layer to form a miniband inside the quantum well. These BTM QWIPs exhibit both the photovoltaic (PV) and photoconductive (PC) dual-mode detection characteristics. The BTM QWIP structures were grown by the MBE technique using GaAs/AlGaAs and InGaAs/InAlAs material systems. The intersubband transition mechanism for the BTM QWIPs is due to the electron transition from the highly populated ground bound state to the miniband states inside the quantum well/superlattice barrier layers followed by the coherent and resonant tunneling conduction via the global miniband states. A significant reduction of the dark current and improvement of the optical and electrical properties have been achieved in these BTM QWIPs. Detectivity in the high 10^9 to low 10^{10} cm-Hz$^{1/2}$/W has been obtained for these BTM QWIPs around 9–10 μm wavelengths and at 77 K operation.

To obtain an efficient coupling of the normal incident IR radiation into the quantum well, we developed a two-dimensional (2-D) planar square mesh metal grating coupler for the BTM QWIPs. The 2-D planar square mesh metal grating structure was deposited on top of the QWIP to couple the normal incident IR radiation into an absorbable angle in the quantum wells and to provide good ohmic contacts for the QWIPs. The main feature of this grating structure is that the coupling of normal incident IR radiation is independent of its polarization direction. Theoretical and experimental studies of the dependence of coupling quantum efficiency on the grating period and aperture size for the GaAs/AlGaAs and InGaAs/InAlAs BTM QWIPs are depicted in this chapter. The method of moment was employed to analyze the 2-D square mesh metal grating structure for both the conventional and BTM GaAs QWIPs. A high coupling quantum efficiency (i.e., 30–50%) has been achieved in the grating coupled GaAs BTM QWIPs in the 8–14 μm wavelengths.

1. INTRODUCTION

Quantum well infrared photodetectors (QWIPs) based on intersubband transitions have been extensively investigated for a wide variety of long wavelength infrared detection and focal plane array image sensor applications in recent years [1–15]. Most of the QWIPs are fabricated from III–V semiconductor materials grown by using molecular beam epitaxy (MBE) and metalorganic chemical vapor deposition (MOCVD) techniques. The large electric dipole matrix element [1] existing between the subbands of quantum well makes QWIP structure very attractive for long wavelength infrared (LWIR) detection, especially, in the atmospheric window spectral range of 8–14 μm wavelength.

Studies of heterojunction superlattices and their transport properties were first reported by Esaki and Tsu in 1970 [16]. Due to the coupling effect between the quantum wells, Kazarinov and Suris [17] predicted in 1971 that the existence of current-voltage peaks is a negative differential conductance characteristic of the superlattice corresponding to the resonant tunneling between the different states of adjacent wells. In 1973, Tsu and Esaki[18] reported their theoretical calculations of the resonant transmission coefficients for the double, triple, and quintuple barrier structures based on the tunneling process. The oscillatory conductance along the superlattice axis in the AlAs/GaAs system was observed by Esaki in 1974 [19]. The voltage period of the oscillations was comparable to the energy separation between the first two conduction minibands. In 1985, Capasso et al. [20] demonstrated a novel effective mass filtering phenomenon by using superlattice miniband transport. A new quantum conductivity and giant quantum amplification of photocurrent in a semiconductor superlattice was obtained from this study.

The main advantage of the intersubband transition QWIPs lies in the fact that they can be fabricated from the more mature III–V semiconductor material systems that have fewer growth and processing problems than that of HgCdTe (MCT) material to achieve monolithic integration of QWIPs with readout circuits on the same chip. In addition, flexibility associated with the control of alloy compositions and dopant densities during the epitaxial growth can be used to tailor the spectral response of the QWIPs to a single or multi-color infrared detection. In fact, peak detection wavelengths ranging from 2 to 18 μm have been demonstrated in a wide variety of III–V QWIPs [15]. Many infrared applications using QWIPs are possible including low loss space communication systems, remote astronomy telescopes, space-borne surveillance radars and environmental thermal imaging systems. Furthermore, QWIPs can also be used in a wide variety of high-speed detection such as picosecond CO_2 pulse experiments [21], high frequency heterodyne experiments [22], and broadband terrestrial spectroscopy and astrophysics. The uniformity of III–V epitaxial growth over large areas could provide opportunity for fabricating large area quantum well infrared focal plane arrays (FPAs) for infrared image sensor systems. In many respects, recent experimental work has confirmed the potential advantages of QWIPs for FPAs applications [15]. To predict the intersubband optical absorption and photoresponse in QWIPs, theoretical models for computing the energy levels of the bound states and continuum states as well as transport properties in the quantum well and superlattices have been developed [23–25]. Recent works have also dealt with improved means of coupling the normal incidence infrared radiation into the quantum wells with strong intersubband absorption by using planar metal or dielectric grating couplers for various n-type QWIPs [4, 15, 27, 28].

A majority of the studies on QWIPs so far have been focused on GaAs/AlGaAs material system [15]. However, other material systems such as n-type InGaAs/InAlAs, InGaAsP/InP, GaAs/GaInP, type-II AlAs/Al$_{0.5}$Ga$_{0.5}$As, and SiGe/Si have also been investigated for QWIPs applications [15]. In general, the hetero-interface quantum well/superlattice structures may be classified into four categories: type-I, type-II staggered, type-II misaligned, and type-III, as shown in Fig. 1. Type-I is the most widely used structure for QWIPs, which may be fabricated from n-type GaAs/AlGaAs, InGaAs/InAlAs, GaSb/AlSb, GaAs/GaInP material systems. In a type-II quantum well structure, electrons and holes are confined in different semiconductor layers at their heterojunctions and superlattices. A type-II QWIP using indirect bandgap AlAs/Al$_{0.5}$Ga$_{0.5}$As material system has been reported recently [14]. The type-III QWIP is another interesting member of the family which involves the use of a zero bandgap material system such as HgCdTe. Since the main objective of this chapter is to present a new class of metal grating coupled III–V semiconductor QWIPs using bound-to-miniband (BTM) transition mechanisms for long wavelength infrared (LWIR) detection, we shall focus our discussions only on the type-I QWIPs fabricated from n-type GaAs/AlGaAs and InGaAs/InAlAs material systems. For n-type QWIPs, a metal or dielectric grating coupler is needed in order to

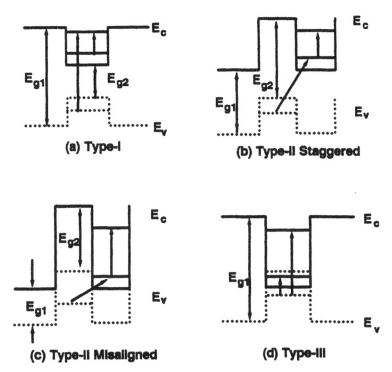

Fig. 1 Energy band diagrams for the type-I, type-II staggered, type-II misaligned, and type-III hetero-interface quantum well/superlattice structures. Type-I is the most widely used structure for QWIPs, which may be fabricated from n-type GaAs/AlGaAs, InGaAs/InAlAs, GaSb//AlSb, GaAs/GaInP material systems.

couple the normal incidence IR radiation into the quantum wells with high coupling quantum efficiency [4, 27, 28].

In this chapter we present three types of metal grating coupled bound-to-miniband (BTM) transition III–V semiconductor QWIPs for 8–14 μm infrared detection. The detectors are fabricated from n-type GaAs/AlGaAs and InGaAs/InAlAs material systems grown by the MBE technique. These BTM QWIPs use an enlarged quantum well with a well width ranging from 88 to 110 Å and a short-period (i.e., 6–9 periods) superlattice barrier structure. Due to the use of the wider well and superlattice barrier layers, an increase in the potential barrier height and the creation of two bound states and a superlattice miniband inside the quantum well are achieved in the BTM QWIPs. The infrared detection process in a BTM QWIP is based on the intersubband transition from the highly-populated bound ground state to the global miniband states inside the quantum well. The superlattice miniband states are resonantly aligned with the first bound excited state in the quantum well. The shape and peak position of the IR response of a BTM QWIP depend on the relative position of the bound excited state within the miniband, as will be discussed later. To couple the normal incidence IR radiation into a BTM QWIP, a planar two-dimensional (2-D) square mesh metal grating coupler structure [28] is used to achieve high coupling quantum efficiency in such a QWIP.

To study the performance and characteristics of a BTM QWIP, three different types of BTM QWIPs were grown by the MBE technique using GaAs/AlGaAs and lightly strained InGaAs/GaAs/AlGaAs on GaAs substrate, and lattice matched InGaAs/InAlAs on InP substrates. This is depicted in Sec. 2.

Section 3 describes the calculations of energy states and intersubband transition physics of BTM QWIPs. We use the transfer matrix method (TMM) [24] to calculate the energy levels of the bound states and transmission coefficients in the BTM QWIPs. The TMM offers several advantages over the envelop wavefunction method. It provides a relatively simple and direct means for solving the coupling effects, the energy levels of the bound and miniband states in the quantum well, the carrier lifetimes, and the transmission coefficient of electrons tunneling through the miniband states.The calculated energy states and transmission coefficients for the three BTM QWIPs are described in this section. The results of the intersubband absorption study on the BTM QWIPs are also presented in this section.

In Sec. 4 we present the fundamental principles of the conduction processes and analysis of the dark currents in the BTM QWIPs. A thermionic-assisted tunneling model was used to calculate the dark currents in the BTM QWIPs, and the results are compared with the measured dark currents for the GaAs/AlGaAs and InGaAs/InAlAs BTM QWIPs.

Section 5 is concerned with the spectral responsivity measurements on the BTM QWIPs. The effects of bias voltage and the relative alignment of bound excited state to the miniband states on the spectral response characteristics are also discussed in this section.

The noise theory and noise characterization of the BTM QWIPs are depicted in Sec. 6. Discussion of different noise sources such as Johnson noise, g-r noise, and flicker noise, and the measurements of noise current density versus frequency for the BTM QWIPs are given in this section. In addition, calculations of the detectivities for the BTM QWIPs are also presented in' this section.

Section 7 deals with the design of 2-dimensional (2-D) square mesh metal grating couplers for efficient coupling of the normal incidence IR radiation into the quantum wells with high quantum efficiency. Theoretical simulations of the coupling quantum efficiency as a function of grating parameters and wavelength have been carried out for the BTM QWIPs using method of moment (modal expansion) technique. A comparison of the theoretical calculations with the experimental data for the 2-D square mesh grating coupled GaAs BTM QWIP and a standard GaAs QWIP are discussed in this section. Conclusions drawn from the study of BTM QWIPs are given in Sec. 8.

2. FABRICATION OF BTM QWIPs

Recent advances in the molecular beam epitaxy (MBE) and metalorganic chemical vapor deposition (MOCVD) techniques have made possible the growth of high quality III–V semiconductor epitaxial layers routinely for a wide variety of device fabrications. Since both the MBE and MOCVD techniques allow the growth of alternate monatomic epilayers of different bandgap materials, a wide variety of quantum effect devices has been fabricated for high-speed and optoelectronic device applications. High quality quantum well structures have been grown on alternate layers of GaAs as quantum well and $Al_xGa_{1-x}As(0 \leqslant x \leqslant 1)$ as barrier layer. The confinement of electrons occurs because the bandgap of $Al_xGa_{1-x}As(E_g = 1.424 + 1.247x(eV)$ for $x \leqslant 0.45)$ is much larger than that of GaAs, and the well width is smaller than the de Broglie wavelength of electron. The GaAs/AlGaAs heterojunction system creates a 65% of the band offset in the conduction band, and a 35% band offset in the valence band. Thus, the potential barrier height between the GaAs and AlGaAs for the electrons is given by $\Delta E_c = 0.81x(eV)$, where x is the aluminum mole fraction. Other high quality III–V heterojunction structures such as the lattice matched $In_{0.53}Ga_{0.47}As/In_{0.48}Al_{0.52}As$ on InP and the InGaP/GaAs on GaAs substrates have also been grown using both the MBE and MOCVD techniques.

In this chapter, we describe three types of BTM QWIPs which are fabricated using the lattice-matched n-type GaAs/AlGaAs, $In_{0.53}Ga_{0.47}As/In_{0.52}Al_{0.48}As$, and a lightly-strained $In_{.07}Ga_{.93}As/GaAs/AlGaAs$ material systems grown on the semi-insulating (SI) GaAs or InP substrates. Detailed layer structures and device parameters for these BTM QWIPs are summarized in Table 1. The unique features of the BTM QWIPs include: (i) using the enlarged quantum well with a well width ranging from 88 to 110 Å, and (ii) replacing the bulk barrier layer of a standard QWIP with a short-period of undoped GaAs/AlGaAs or InGaAs/InAlAs superlattice barrier layer for the BTM QWIPs. By replacing the bulk barrier layer with a short-period superlattice barrier layer, a global miniband is formed inside the quantum well which is resonantly lined up with the first bound excited state of the quantum well. The miniband transport QWIPs offer advantages such as reduction of interface recombination and dark current, elimination of deep-level related phenomena, and a significant enhancement of intersubband absorption in the quantum well. Furthermore, the increase of potential barrier height in the quantum well offers additional design flexibility for selecting the peak response wavelength of the detection band. The transport mechanism for the BTM QWIPs is mainly due to the transition of electrons from the highly populated ground state to the global miniband via the thermionic-assisted tunneling conduction by the strong coupling of electron wave functions between the adjacent quantum wells and the thin superlattice barrier layers. If the miniband is sufficiently broad (e.g., a few tens of meVs), then eletrons making transition from the heavily populated bound ground state to the miniband states via thermal (dark current) or optical (photocurrent) excitation could attend high mobility under the applied electric field and tunnel through the global

TABLE 1
Device parameters for the three BTM QWIPs used in this study.

QWIP	A	B	C
SL barrier	$Al_{0.4}Ga_{0.6}As$	$Al_{0.4}Ga_{0.6}As$	$In_{0.52}Al_{0.48}As$
Width (Å)	58	30	35
SL well	GaAs	GaAs	$In_{0.53}Ga_{0.47}As$
Width (Å)	29	59	50
L_b(Å)	493	478	460
QW	GaAs	$In_{0.07}Ga_{0.93}As$	$In_{0.53}Ga_{0.47}As$
Width(L_z)(Å)	88	106	110
$N_D(10^{18}cm^{-3})$	2.0	1.4	0.5
Periods	40	40	20
Substrate	GaAs	GaAs	InP
Mesa (μm^2)	4×10^4	4.9×10^4	1.92×10^5
Transition	BTM	SBTM	BTM

miniband, which are then collected by the external ohmic contacts. To facilitate dark current and photoresponse measurements, the detectors are mesa-etched down to the bottom contact layers. The typical mesa-etched device area is around $200 \times 200 \, \mu m^2$. Finally, a 2-D square mesh metal grating coupler structure is deposited on the QWIPs to couple the normal incidence IR radiation into these devices.

3. PHYSICS OF THE BTM QWIPs

In this section, we discuss the physics of bound-to-miniband (BTM) transition QWIPs. Theoretical calculations of the eigenvalues for the bound states and miniband states in the quantum well and the continuum states above the barrier layer as well as the intersubband optical absorption in the BTM QWIPs are depicted in this section. The common approaches used to explain the resonant tunneling phenomena in quantum well and superlattice structures are (i) the Wentzel-Kramers-Brillouin (WKB) approximation [29, 30], which is valid if the barrier energy varies slowly compared to the scale of electron wavelength, (ii) the Monte Carlo solution [31, 32] of the semiclassical Boltzmann transport equation which assumes a quasi-particle picture of the electron, and (iii) the transfer matrix method which calculates the transmissivity of the quantum well/superlattice structures as a function of energy directly. The WKB approximation is not valid for quantum well structures with narrow barriers due to the rapid potential changes in these structures. To employ the WKB approximation, the electron wavelength must be small compared to the distance in which the momentum or potential has a significant change. This condition can not be satisfied in the quantum well structures in which the barriers are narrow, especially at low incident carrier energies. The Monte Carlo approach is useful because it considers the phonon scattering but does not easily lend itself to the calculation of the structure's transparency. Even with the popular variational method, it does not give the correct value of the wave function outside the quantum well, nor predicts the lifetimes of the quasi-bound states. Solutions of the energy levels and wave functions using variational method depend on the choice of trial function, which becomes very difficult for structures involving more than one quantum well and also for the graded potential structures.

In view of the limitations cited above, we have applied the transfer matrix method (TMM) to calculate the energy levels of the bound

states and the transmission coefficient in the quantum well and superlattice structures. The TMM is a simple and straightforward technique
for investigating the quantum well and superlattice structures, and
hence can be applied directly to the design of a wide variety of
QWIPs. It overcomes all the problems cited above and can be used to
analyze arbitrary potential profiles, including asymmetric, graded, and
parabolic quantum well structures. Unlike the conventional methods,
the TMM can be used to calculate the energy levels of the bound
states, the lifetimes of the quasi-bound states, and to predict the influence of electric field on the lifetimes as well. It is capable of predicting
not only the wave functions inside the quantum wells, but also outside
the wells, either with or without the external electric fields. We next
present the TMM for the design of BTM QWIPs.

3.1. Transfer Matrix Method

In this section the transfer matrix method (TMM) for calculating the
eigenvalues of bound states and transmission coefficients of a BTM
QWIP are discussed. The time-independent one-electron Schrödringer
equation in a quantum well can be expressed as

$$\frac{d^2\psi}{dz^2} + \frac{2m^*}{\hbar^2}[E - V(z)]\psi = 0 , \tag{1}$$

where m^* is the electron effective mass, E and ψ denote the eigenvalues and eigenfunctions of the bound states inside the quantum well,
respectively. The potential function, $V(z)$, under bias condition is given
by

$$V(z) = V_0 + q\varepsilon z , \tag{2}$$

where ε is the electric field, and q is the electronic charge.

The electron waves in each of the quantum well and superlattice
potential regions consist of two components propagating in the forward and backward directions respectively. This can be expressed by

$$\psi_i = \psi_i^+ e^{-i\Delta_i} e^{+ik_i} + \psi_i^- e^{+i\Delta_i} e^{-ik_i} \tag{3}$$

and

$$\Delta_1 = \Delta_2 = 0, \quad \Delta_i = k_i(d_2 + d_3 + \cdots + d_i) \quad i = 3, 4, \ldots, N \tag{4}$$

$$k_i = \left[\frac{2m^*}{\hbar^2}(E - V_i)\right]^{1/2} \tag{5}$$

where ψ_i^+ and ψ_i^- represent the electron wave functions propagating along the $+x$ and $-x$ directions, respectively. Since both ψ and $d\psi/dx$ are continuous at the boundaries, we obtain

$$\psi_i^+ = (e^{i\delta_i}\psi_{i+1}^+ + r_i e^{i\delta_i}\psi_{i+1}^-)/t_i \tag{6}$$

$$\psi_i^- = (r_i e^{-i\delta_i}\psi_{i+1}^+ + e^{-i\delta_i}\psi_{i+1}^-)/t_i \tag{7}$$

Here the recurrence relation may be written in matrix form as

$$\begin{pmatrix} \psi_i^+ \\ \psi_i^- \end{pmatrix} = \frac{1}{t_i}\begin{pmatrix} e^{i\delta_i} & r_i e^{i\delta_i} \\ r_i e^{-i\delta_i} & e^{-i\delta_i} \end{pmatrix}\begin{pmatrix} \psi_{i+1}^+ \\ \psi_{i+1}^- \end{pmatrix} \tag{8}$$

where

$$r_i = \frac{(m_{j+1}^*/k_{i+1}) - (m_j^*/k_i)}{(m_{j+1}^*/k_{i+1}) + (m_j^*/k_i)} \tag{9}$$

$$t_i = \frac{2(m_{j+1}^*/k_{i+1})}{(m_{j+1}^*/k_{i+1}) + (m_j^*/k_i)} \tag{10}$$

$$\delta_i = k_i d_i \tag{11}$$

Thus, we have

$$\begin{pmatrix} \psi_1^+ \\ \psi_1^- \end{pmatrix} = S_1\begin{pmatrix} \psi_2^+ \\ \psi_2^- \end{pmatrix} = S_1 S_2\begin{pmatrix} \psi_3^+ \\ \psi_3^- \end{pmatrix} = \cdots = S_1 S_2 \cdots S_N\begin{pmatrix} \psi_{N+1}^+ \\ \psi_{N+1}^- \end{pmatrix} \tag{12}$$

Since there is no forward propagating component in the last medium, i.e., $\psi_{N+1}^- = 0$, one can find $\psi_j^+ (j = 2, 3, ..., N+1)$ in terms of E_1^+, where j represents the layer region to be investigated. By calculating the quantity $\eta(\beta) = \psi_j^+/\psi_1^+$ as a function of energy E, one obtains certain resonance peaks. It can be shown that the resonance curve has a Lorentzian distribution which corresponds to each bound state and quasi-bound state of the form [33, 34]

$$\left|\frac{\psi_j^+}{\psi_1^+}\right|^2 = \frac{\sigma}{(E - E_b')^2 + \Gamma^2}, \tag{13}$$

where $E = E_b'$ is the energy level of the bound state; 2Γ is the full width at half maximum (FWHM) of the bound or miniband states. It is noted that Γ will approach zero for the bound states and have a finite value for the miniband states. $\tau = \hbar/2\Gamma$ is the lifetime of the quasi-bound states. By calculating the ratio $|\psi_j^+/\psi_1^+|$ one obtains all the energy levels and related properties inside each of the wells. Note that σ/Γ^2 represents the peak value of the relative electron density distribution. Similar results can also be obtained outside the well.

The 2×2 matrix method discussed above is a simple and straight-forward technique for solving the 2-D quantum well and superlattice structures. It can simultaneously give the transmission coefficient, relative density of states, energy levels of bound states, and lifetimes of the quasi-bound states in the quantum well and superlattice struc-tures. This method has been applied to the design of three III–V semiconductor QWIPs using bound-to-miniband (BTM) transition scheme, and the results are discussed next.

Figure 2 shows the energy band diagrams for the three BTM QWIPs fabricated from (a) the lattice-matched GaAs/AlGaAs (QWIP-A) on the semi-insulating (SI) GaAs substrate, (b) lightly-strained $In_{0.07}$ $Ga_{0.93}$As/GaAs/AlGaAs (QWIP-B) on the SI GaAs, and (c) the lat-tice-matched InGaAs/InAlAs/InGaAs (QWIP-C) on SI InP substra-tes. The device structures and parameters used in these BTM QWIPs are summarized in Table 1.

In a BTM QWIP, the device parameters are chosen so that there are two bound states and one miniband formed inside the quantum well. The first bound excited state in the quantum well is lined up with the global miniband created by the superlattice barrier layer to achieve a large oscillation strength f and to enhance the intersubband optical absorption. As shown in Fig. 2(a), the energy separation (~ 140 meV) between the ground bound state and the miniband states inside the quantum well/superlattice layer determines the peak spec-tral response of this QWIP, which has a peak response wavelength at $\lambda_p = 8.9$ μm. Figure 2(b) shows a step-bound-to-mininband (SBTM) InGaAs QWIP structure, which uses the same superlattice barrier layer as QWIP-A but the GaAs well layer is replaced by a lightly-strained $In_{0.07}Ga_{0.93}$As (106 Å) as the quantum well so that a poten-tial step below the bottom of the barrier well is created. As a result, the ground state of the enlarged InGaAs quantum well is pushed below the barrier well, which eliminates the direct tunneling current flowing from the highly populated bound ground state through the quantum well/superlattice layers. Figure 2(c) shows the third BTM QWIP fabricated from the lattice-matched $In_{0.53}Ga_{0.47}$As/$In_{0.52}$ $Al_{0.48}$As on the InP substrate.

Theoretical calculations of the energy levels E_{Wn}, E_{SLn} ($n = 1, 2, ...$) for the bound and miniband states and the transmission coefficient $|T * T|$ for the three BTM QWIPs were carried out by using the TMM described above. The energy eigenvalues of the bound and miniband states inside the quantum well and superlattice regions of this GaAs BTM QWIP (see Fig. 2(a)) were calculated by the TMM.

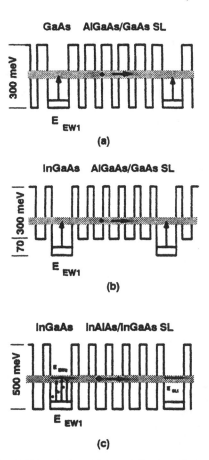

Fig. 2 The energy band diagrams for the three BTM QWIPs fabricated from (a) the lattice-matched GaAs/AlGaAs (QWIP-A) on GaAs, (b) lightly-strained $In_{0.07}Ga_{0.93}As$/GaAs (QWIP-B) on GaAs, and (c) the lattice-matched InGaAs/InAlAs (QWIP-C) on InP substrates. The device structures and parameters used in these BTM QWIPs are summarized in Table 1.

The results reveal that only two localized states create inside the GaAs quantum well, namely the ground state with energy $E_{W1} = 47$ meV above the well edge (taking into account the Hartree effect and exchange term of E_{exch}), and the first excited state $E_{W2} = 174$ meV, which is merged with the superlattice miniband E_{SL1} (179–189 meV above the well edge). The Fermi level E_F was found to be 42 meV above the bound ground state E_{W1} in the well. Outside the quantum wells, the first continuum state is estimated to be 4 meV above the barrier ($E_{CN1} = 326$ meV), and the second continuum state E_{CN2} is 48 meV higher than the first continuum state. The highly localized and heavily populated ground state E_{W1} is located deep inside the quantum well. Therefore, the dark current due to thermionic emission of electrons out of the quantum wells into the continuum states in QWIP-A can be greatly reduced. On the other hand, the thermionic-assisted tunneling current through the E_{W2} and the miniband states is also reduced due to the narrow bandwidth of the miniband E_{SL1}. Figure 3 shows the normalized transmission probability ($|T * T|$) versus energy for this BTM QWIP, calculated at bias voltages, $V_b = 0, 0.5$

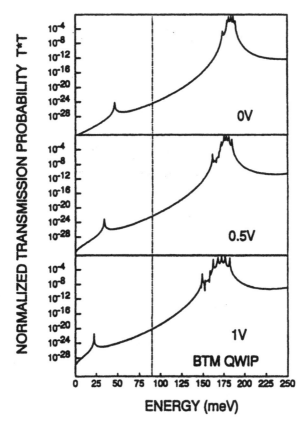

Fig. 3 The normalized transmission probability ($|T * T|$) versus energy for the GaAs BTM QWIP (QWIP-A), calculated by TMM at bias voltages, $V_b = 0$, 0.5 and 1V; dashed line is the Fermi level.

and 1 V; dashed line is the Fermi level. The results clearly show the shift of bound ground state to lower energy and the broadening of the superlattice miniband and the first bound excited state with increasing bias voltage. This effect becomes more pronounced for the InGaAs SBTM QWIP, as shown in Fig. 4.

A similar calculation of the energy values of the bound and miniband states, Fermi-level, and the transmission probability as a function of bias voltage was also carried out for the InGaAs SBTM QWIP, and the results are shown in Fig. 4 under different bias conditions. For the InGaAs SBTM QWIP (QWIP-B), only two localized

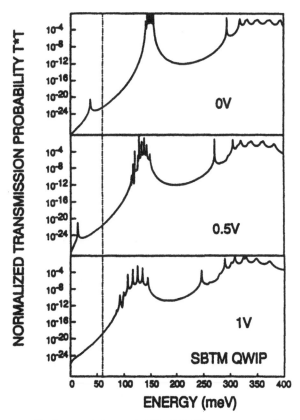

Fig. 4 The normalized transmission probability ($|T * T|$) versus energy for the InGaAs SBTM QWIP (QWIP-A), calculated by TMM at bias voltages, $V_b = 0,0.5$ and 1V; dashed line is the Fermi level.

states were found in the InGaAs quantum well: the bound ground
state $E_{W1} = 37$ meV and the first excited state $E_{W2} = 143$ meV, which
is merged with the superlattice miniband E_{SL1} ($= 144$–155 meV). The
Fermi level E_F was found to be 24 meV above the ground state E_{W1}.
The peak detection wavelength for this QWIP is 10.5 μm.

For the lattice matched InGaAs/InAlAs BTM QWIP (QWIP-C)
grown on the InP substrate, a broad and highly degenerated
miniband was formed by using a 6-period of undoped InGaAs/InAlAs
superlattice barrier layers. Figure 5 shows the calculated energy levels
of the bound and miniband states, Fermi-level, and the transmission
coefficient for this BTM QWIP at bias voltages $V_b = 0, 0.5$V. The

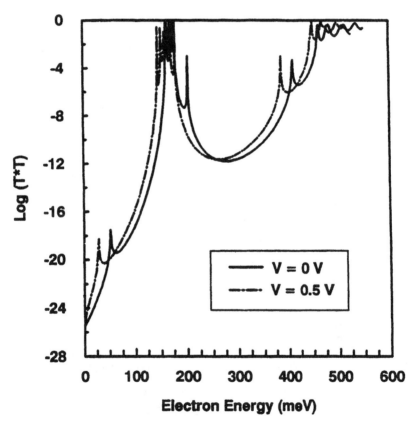

Fig. 5   The energy levels of the bound and miniband states, Fermi-level, and the transmission coefficient for the InGaAs BTM QWIP (C) calculated at bias voltages, $V_b = 0$,
0.5 V.

center energy position of the miniband states was found to be 163 meV above the conduction band edge of the InGaAs quantum well with a bandwidth of $2\Gamma \sim 20$ meV. In order to accurately determine the intersubband transition energy values, the effects of band nonparabolicity [35], electron-electron interaction [36], and depolarization [37] on the subband energy levels should be taken into account in the calculations of subband energy differences. In our calculations of the energy levels for the BTM QWIPs, both the electron-electron interaction (exchange energy: E_{exch}) and the depolarization (E_{dep}) effects are considered. The net result of these effects is a lowering of about 5 meV for the heavily populated ground bound state E_{W1} in the quantum well. The peak absorption wavelength can be found from the relation

$$\lambda_p = \frac{1.24}{E_{\text{SL1}} - E_{W1} + E_{\text{exch}} - E_{\text{dep}}}\ \mu\text{m} \tag{14}$$

Now, substituting values of $E_{\text{SL1}} = 163$ meV, $E_{W1} = 51$ meV, and $(E_{\text{exch}} - E_{\text{dep}}) \sim 5$ meV into Eq. (14), the peak wavelength of absorption was found to be $\lambda_p = 10.6\ \mu$m. The effects of electron-electron interactions on the energy shift in the quantum well are discussed next.

In calculating the electronic states in a quantum well/superlattice structure, the electron-electron interactions should be considered when the quantum well is doped to 10^{18} cm^{-3} or higher. The interaction includes two components: direct Coulomb force and quantum exchange interaction, which have opposite effects on the energy level shift [38]. The Coulomb interaction shifts the subband up while the exchange interaction shifts the energy level down. In type-I QWIPs, the doping in the quantum well can give rise to charge neutrality within the well, and the exchange energy is more significant than that of Coulomb interaction.

The exchange interaction energy term associated with electrons in the bound ground state of a QWIP is approximately given by [36]

$$E_{\text{exch}}(k=0) \approx -\frac{e^2 k_F}{4\pi\varepsilon}\left[1 - 0.31\frac{k_F}{k_1}\right], \tag{15}$$

$$E_{\text{exch}}(k_F) \approx -\frac{e^2 k_F}{4\pi\varepsilon}\left[\frac{2}{\pi} - 0.31\frac{k_F}{k_1}\right], \tag{16}$$

where $k_1 = \pi/L_a$, L_a is the quantum well width, $k_F = (2\pi\sigma)^{1/2}$, and $\sigma = L_a N_D$ is the 2-dimensional electron density in the quantum well. For the unpopulated excited states, the exchange-induced energy shift is very small, and hence the dominant contribution to the energy shift is due to the electron-electron interaction in the highly populated ground bound state.

The energy shift in the ground state attributed to direct Coulomb interaction is [38]

$$E_{\text{direct}} = \frac{\sigma e^2}{8\varepsilon k_1^2 L_a}. \tag{17}$$

This term has a small contribution to the energy shift compared to the exchange-induced energy shift discussed above for $N_D \geqslant 10^{18}$ cm^{-3}

When IR radiation is impinging on a QWIP, resonant screening of the infrared field by electrons in the quantum well generates depolarization field effect, which can cause the subband energy shift (also called the plasmon shift) in the quantum well. The depolarization effect arises because each electron experiences a field which is different from the mean Hartree field due to other electrons polarized by the external field. The energy shift due to the depolarization field effect is given by [37]

$$E_{\text{dep}} = \sqrt{\frac{2\sigma e^2 (E_0 - E_1) S_{01}}{\varepsilon L_a}}, \tag{18}$$

where S_{01} is the Coulomb matrix element given by

$$S_{01} = \int_0^\infty dz \left[\int_0^z \phi_0(z') \phi_1(z') dz' \right]^2, \tag{19}$$

It is noted that depolarization effect increases as dopant density is increased. Figure 6 shows the calculated energy shifts due to direct Coulomb interaction, electron-electron interaction, and the depolarization effect for a GaAs QWIP with $N_D = 10^{18}$ cm^{-3} and $L_a = 100$ Å.

3.2. Intersubband Absorption

3.2.1. Theory

The intersubband transition in a QWIP takes place between the subband levels of either conduction band or valence band. It has some unique features which include: (i) large absorption coefficient [1], (ii) narrow absorption bandwidth [39], (iii) large optical nonlinearity [40], (iv) fast intersubband relaxation [41], (v) reduced Auger effect [42], (vi) wavelength tunability [43], and (vii) large photocurrent gain.

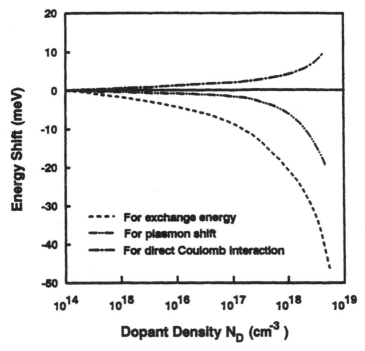

Fig. 6 The calculated energy shifts due to direct Coulomb interaction, electron-electron interaction, and the depolarization effect for a GaAs QWIP with $N_D = 10^{18}$ cm^{-3} and $L_x = 100$ Å.

The intersubband transition process can be analyzed by using the dipole transition model [44]. The transition rate W from the initial state ψ_i to the final state ψ_f can be described by

$$W_{i \to f} = \frac{2\pi}{\hbar} \sum_f |\langle \psi_f | V_p | \psi_i \rangle|^2 \delta(E_f - E_i - \hbar\omega) \tag{20}$$

where V_p is the interaction potential between the incident IR radiation and the electrons, which is given by [42]

$$V_p = \frac{eA_o}{m_o c} \hat{\varepsilon} \cdot \mathbf{P}, \tag{21}$$

where A_o is the vector potential, m_o the free-electron mass, $\mathbf{P}$ the momentum operator of electron, and $\hat{\varepsilon}$ the unit polarization vector of the incident photons.

Since the electron wavefunction $\psi_n(\mathbf{r})$ in the quantum well is the product of Bloch function $\phi_B(=U_n(\mathbf{r})e^{(ik_x x + ik_y y)})$ and the envelope function $\phi_n(z)$, the transition matrix element can be approximated by

$$H_{fi} = \langle (\phi_B \phi_n)_f | V_p | (\phi_B \phi_n)_i \rangle$$
$$\sim \langle (\phi_B)_f | V_p | (\phi_B)_i \rangle_{\text{cell}} \langle \phi_{nf} | \phi_{ni} \rangle + \langle (\phi_B)_f | (\phi_B)_i \rangle_{\text{cell}} \langle \phi_{nf} | V_p | \phi_{ni} \rangle. \quad (22)$$

In the interband transition scheme, the dipole transition occurs between the Bloch states while keeping the envelope states (or momentum vector) constant, and hence the first term on the right-hand side of Eq. (22) tends to vanish. However, in the intersubband transition scheme such as for the case of QWIPs, the dipole transition is between the envelope states while the Bloch states remain nearly constant, thus the second term on the right hand side of Eq. (22) becomes zero. From the calculations of transition matrix element $M_{fi} = \langle \psi_f | V_p | \psi_i \rangle$, the transition selection rules and incident polarization requirements for the intersubband transition can be determined.

Finally, by solving W_{fi} from Eq. (20), the absorption coefficient α can be calculated by using the expression [45]

$$\alpha_{i \to f} = \frac{2\pi \hbar c W_{i \to f}}{n_r A_o^2 \omega}, \quad (23)$$

where n_r is the refractive index of the medium. The absorption coefficient curve can be fitted by Lorentzian function. The integrated absorption strength I_A for the polarized incidence radiation at the Brewster angle is given by

$$I_A = \sigma NS \frac{e^2 h}{4\varepsilon_o m^* c} \frac{f_{\text{os}}}{n_r^2 \sqrt{n_r^2 + 1}}, \quad (24)$$

where N is the number of quantum wells, S is the quantum well structure factor, and f_{os} is the dipole oscillator strength given by

$$f_{\text{os}} = \frac{4\pi m^* c}{\hbar \lambda} \left(\int_{-L_a/2}^{L_a/2} \psi_f z \psi_i dz \right)^2 \quad (25)$$

When the incident radiation is perpendicular to the quantum well surface, the transition matrix element M_{fi} is zero if the shape of constant energy surface of the material is spherical. A nonzero transition rate can be obtained by using either a 45° polished facet illumination or a grating coupler [46]. For a transmission grating coupler, the grating equation is given by

$$n_r \sin \theta_m - \sin \theta_i = m\lambda/\Lambda, \quad (26)$$

where λ is the resonant incident wavelength, Λ is the grating period, $\theta_{i,m}$ denote the incident and m th order diffracted angle with respect to the superlattice axis, respectively. In a grating coupled QWIP, the integrated absorption strength I_A in Eq. (24) should be multiplied by a factor of $\sin^2 \theta_m / \cos \theta_m$. We shall next discuss the results of absorbance measurements on the BTM QWIPs using Fourier transform infrared spectroscopy (FTIR).

3.2.2. Results and Discussion

The intersubband absorption measurements were performed on the three BTM QWIPs at 300 K using a Perkin-Elmer 1640 Fourier transform infrared (FTIR) spectroscopy equipment. Figure 7 shows the absorption coefficient versus wavelength curves for the three BTM

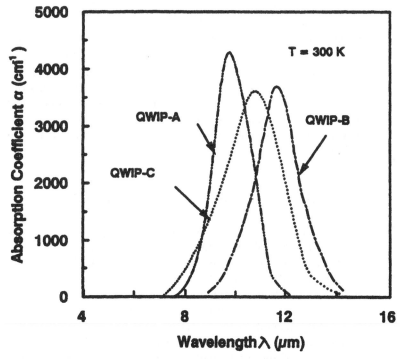

Fig. 7 The absorption coefficient versus wavelength curves for the three BTM QWIPs (A, B, C) as calculated from the absorbance data measured by FTIR spectroscopy at $T = 300$ K.

QWIPs as calculated from the absorbance data by the FTIR spectroscopy measured at Brewster angle and $T = 300$ K. The absorption peak is centered at 1060 cm^{-1} for the GaAs BTM QWIP (QWIP-A). The measured peak absorbance $A = -\log_{10}$ (transmission) for this QWIP was found to be about 7×10^{-2}. The bandwidth broadening, slightly larger than the expected value of 12 meV, might be attributed to problems associated with the interface roughness and spatial inhomogeneities in the well and superlattice width [47], high temperature influence, and higher subband lifetime broadening effect. The room temperature absorption coefficient versus wavelength for the InGaAs SBTM QWIP (B) was calculated from the absorbance data measured by the FTIR at Brewster angle ($\theta_B = 73°$) and $T = 300$ K. The measured peak absorbance $A = -\log_{10}$ [transmission] for this QWIP (B) was found to be about 40 mAbs. The absorption peak is centered at 11.4 μm. The full width at half-maximum (FWHM) of the absorption peak is about 220 cm^{-1}. The room temperature absorption coefficient versus wavelength for the InGaAs BTM QWIP (C) formed on InP was also calculated from the measured absorbance at Brewster angle ($\theta_B \sim 73°$) by FTIR spectroscopy. The main absorption peak for this BTM QWIP is centered at $\lambda_p = 10.7$ μm with a spectral linewidth of $\Delta v = 500$ cm^{-1}, which is in excellent agreement with the theoretical prediction. Note that the large absorption coefficients observed in these BTM QWIPs can be attributed to the use of enlarged quantum well and miniband structures as compared to the conventional QWIPs, which greatly enhance the intersubband absorption in these QWIPs.

4. CONDUCTION PROCESSES AND DARK CURRENT

In this section we discuss the carrier conduction process and dark currents in the BTM QWIPs. Similar to a standard bound-to-continuum (BTC) transition QWIP, the performance of a BTM QWIP is determined by the physical and device parameters such as absorption coefficient associated with intersubband transitions, dark current, spectral responsivity, and noise equivalent power (NEP). These parameters not only depend on the materials used but also closely relate to the quantum well and barrier layer structures of the QWIPs. In order to understand the fundamental limitations on the performance of BTM QWIPs we shall first discuss the conduction process and the device dark current in the BTM QWIPs.

4.1. Carrier Conduction in a BTM QWIP

In general, the carrier conduction processes in a quantum well/super-lattice structure are quite complicated. They can be divided into three different conduction processes, namely, the continuum state conduction, the miniband conduction, and the hopping conduction. We shall discuss each of these conduction processes as follows.

4.1.1. Continuum State Conduction

When the excited states of a QWIP lie above the quantum well barrier, the states become unbounded continuum states which have 3-dimensional (3-D) conduction properties. Charge carriers (i.e., either dark or photo-excited carriers from the ground bound states in the quantum well) transporting through the continuum states generally have high mobility under applied bias conditions. If the electric field is high enough, then hot carrier conduction through the 3-D continuum states is expected. This type of conduction has the advantages of high efficiency, high optical gain, and long mean free path. In fact, if the excited states are placed just above the barrier, a resonant infrared absorption and an optimized oscillator strength can be achieved. However, the interface defects and the image lowering effects will give rise to large dark current, and hence lowering the operating temperature of the QWIP. This type of conduction is usually the dominant transport process in a conventional bulk barrier QWIP.

4.1.2. Miniband Conduction

The miniband conduction is a coherent resonant tunneling process in which the photoexcited carriers are phase-coherent to the incident IR radiation. This coherent conduction can lead to much higher carrier transmission probability through the quantum well and superlattice. Resonant transmission modes build up in the miniband to the extent that the scattering reflected wave is cancelled out and conduction transmitted wave is enhanced. The miniband conduction depends strongly on the miniband bandwidth, heterointerface quality, and layer thickness fluctuation. For example, it has been demonstrated that the morphological quality of the heterointerface can be greatly improved by using interruption growth technique for a few tens of seconds [48]. The interruption growth allows one to reduce the density of monolayer terraces in the plane of heterointerface. As a result, the interface improvement can enhance the coherence of interfacing electron wave overlapping and resonant coupling. In the miniband conduction, the effective mass of the photoexcited electrons can be modulated by superlattice structure parameters, given by $m_z^* = (2\hbar^2)/(\Gamma L^2)$ (where $L = L_a + L_b$ is the superlattice basis). An effective mass m_z^* for the miniband smaller than that of both the well and barrier may be obtained. As a result, the electron transport in the miniband will have a high mobility, which leads to a large oscillator absorption strength, high quantum efficiency, and high response speed. On the other hand, increasing the miniband bandwidth will reduce the tunneling time constant (i.e., $\tau_0 = \hbar/\Gamma = 6.6 \times 10^{-16}/\Gamma$ (in eV)). The value of τ_0 in a QWIP is estimated to be about 20 fs (for $\Gamma = 30 \sim 70$ meV), while a typical scattering time constant τ_S is about 0.1 ps. Thus, for $\tau_0 \ll \tau_S$, the coherent resonant tunneling can be built-up in the miniband conduction process. The photocurrent strongly depends on the tunneling time constant τ_0, while the intersubband relaxation time constant τ_R is about 0.4 ps. From the theoretical calculation, τ_0 was found to be about 20 to 100 fs, and hence $\tau_0 \ll \tau_R$. Thus, the photoexcited electrons can tunnel resonantly out of the quantum well/superlattice barrier via global miniband states in a BTM QWIP.

In the miniband conduction, charge carrier transport through the miniband states inside the quantum well has an average wave vector $k_z = q\varepsilon\tau_R/\hbar$, where ε is the applied electric field. The drift velocity v_d along the superlattice axis can be expressed as

$$v_d = \frac{\Gamma L}{2\hbar}\sin\left(\frac{q\varepsilon\tau_R L}{\hbar}\right), \tag{27}$$

At low electric field, the carrier mobility along the superlattice axis is given by

$$\mu_z = \frac{e\Gamma L^2 \tau_R}{2\hbar^2}, \tag{28}$$

It is noted from Eq. (28) that the carrier mobility is proportional to the miniband width Γ and the relaxation time τ_R if the superlattice basis L is kept constant. Since the miniband bandwidth is an exponential function of the superlattice barrier thickness, the carrier mobility is also sensitive to the thickness of superlattice barrier layer. A similar conclusion can also be drawn from the Boltzmann equation using the relaxation time approximation.

4.1.3. Hopping Conduction

When the miniband conduction fails to form coherent conduction at higher electric fields, the incoherent conduction becomes the dominant mechanism which is referred to as the sequential resonant tunneling with a random wave phase. In the incoherent conduction, the states in the quantum wells (i.e., Kane state) become localized within the individual well and the carriers will transport via phonon-assisted tunneling (hopping) with a frequency of $q \varepsilon L/\hbar$. A better approach for analyzing the incoherent hopping conduction is to utilize carrier scattering mechanism. Carrier scattering tends to destroy the coherency of wave-functions, and hence the fully resonant threshold value will never be built-up. The mobility of hopping conduction is usually much lower than that of the miniband conduction. As the barrier layer thickness or the thickness fluctuation increases, the maximum velocity $v_{max}(=\Gamma L/2\hbar)$ and carrier mobility decrease. This is due to the fact that the relaxation time is nearly independent of superlattice period L. The mobility for the hopping conduction can be expressed as [49]

$$\mu_z \approx \frac{eL^2}{k_B T} A \exp\left[-\left(\frac{8m^*}{\hbar^2}(\Delta E - E_1)\right)^{1/2} L_b \right]. \qquad (29)$$

where L_b is the barrier layer thickness.

It is worth noting that the product of $v_{max}\tau$ is always greater than the mean free path λ in the miniband conduction. However, it will reduce to even smaller value than the superlattice period L in the hopping conduction limit. When the QWIPs are operating at cryogenic temperature, phonon-assisted tunneling is suppressed and other scattering sources such as ionized impurities, intersubband levels, interface roughness can also play important roles in the tunneling conduction process.

4.2. Dark Current

We first consider the theoretical calculations of dark current for a BTM QWIP, and compare the results with the measured values for the three BTM QWIPs listed in Table 1. Under dark conditions, electrons in a QWIP are transported out of the quantum wells and produce the observed dark current mainly via two mechanisms: One is attributed to the thermionic emission of electrons from the bound ground state to the continuum state above the barrier, which is the dominant dark current component for the conventional bound-to-continuum (BTC) QWIPs with bulk barrier. The other is the thermionic-assisted tunneling current transported via the global miniband states inside the quantum wells as in the case of a BTM QWIP.

Calculations of the dark current in a standard (BTC) QWIP have been reported by Levine et al. [6], and is briefly discussed as follows. The first step is to determine the effective number of electrons $n(V)$ excited out of the quantum well of a QWIP, which is given by

$$n(V) = \left(\frac{m^*}{\pi\hbar^2 L_b}\right) \int_0^{\infty} f(E)T*TdE, \qquad (30)$$

where L_b is the barrier width, $f(E)$ is the Fermi distribution function, $n(V)$ accounts for both the thermionic emission (for $E > E_b$; E_b is the barrier potential) and the thermionic-assisted tunneling ($E < E_b$) processes. The dark current is then calculated by using the expression

$$I_d(V) = n(V)qv(V)A_d, \qquad (31)$$

where $n(V)$ is given by Eq. (30), q is the electronic charge, A_d is the device area, and $v(V)$ is the average electron velocity which is field dependent [49]:

$$v(V) = \frac{\mu\varepsilon + v_s(\varepsilon/\varepsilon_o)^4}{1 + (\varepsilon/\varepsilon_o)^4}, \qquad (32)$$

where ε is the electric field, μ is the electron mobility, and ε_o is the critical field. For GaAs, $\mu = 8,000$ cm^2/V·s and $v_s = 7.7 \times 10^6$ cm/s at 300 K, and $\varepsilon_o = 4 \times 10^3$ V/cm.

For a GaAs/AlGaAs BTM QWIP, the dark current is dominated by the thermionic-assisted tunneling conduction via the global miniband formed inside the quantum well and superlattice barrier layers, whereas thermionic emission above the superlattice barrier is negligible because of the higher barrier potential (~ 324 meV) for this BTM QWIP. To find the net tunneling current, we apply the formulism [50, 51] of multibarrier tunneling to this QWIP. The current density J is calculated as the average product of the transmission probability $|T*T|$ and group velocity, $v(k) = \hbar^{-1}\nabla_k E$, which can be expressed as

$$J = \frac{q}{4\pi^3\hbar} \int_0^{\infty} dk_l \int_0^{\infty} dk_t \, [f(E) - f(E')]T*T\frac{\partial E}{\partial k_l}. \qquad (33)$$

The transverse components of J are equal to zero by symmetry requirement [50]. Using separation of variables, the product $|T*T|$ is only a function of the longitudinal energy [51], and the longitudinal (i.e., perpendicular to the well plane) component of J becomes

$$J_l = \int_0^{\chi} |T*T| g(E, V) dE_l$$

$$= \frac{4\pi q m^* k_B T}{h^3} \int_0^{\chi} |T*T| \ln\left(\frac{1 + \exp[(E_F - E_l)/k_B T]}{1 + \exp[(E_F - E_l - qV_a)/k_B T]}\right) dE_l, \quad (34)$$

where V_a is the bias voltage across one period of the superlattice barrier layer. E_l is the longitudinal energy of an electron along the superlattice barrier layer, and E_F is the Fermi level. Although transmission probability $|T*T|$ may change with temperature, it is far less sensitive to temperature when compared to the carrier energy distribution function, $g(E, V)$. The Fermi level E_F can be calculated using the expression

$$N_D = \frac{m^* k_B T}{\pi \hbar^2 L_a} \sum_i \ln(1 + e^{(E_F - E_i)/k_B T}) \simeq \frac{m^*}{\pi \hbar^2 L_a} (E_F - E_{W1}), \quad (35)$$

where L_a is the well width, E_i is the ith state, and E_{W1} is the ground state energy in the well. Equation (35) is valid for E_F well below the miniband and for $T < 100$ K. Note that a modified Fermi level should be used in Eq. (35) to take into account the exchange energy arised from the electron-electron interactions [36] in the heavily doped quantum wells. In the BTM QWIPs, high concentration of electrons is confined in the ground state of the quantum well at low temperatures. Therefore, the attractive force induced by the spin of electrons is significant, which results in an energy lowering effect (e.g., exchange energy). In the calculations of exchange energy, we neglected the interactions coming from the higher energy states where only few electrons exist at low temperatures. Also, the direct Coulomb force is negligible compared to the exchange energy. The calculated exchange energies for the BTM QWIPs were found to be about -20 meV. Obviously, this is an important correction for accurate calculations of the Fermi level and hence the dark current in the BTM QWIPs.

Since the tunneling process in n-type QWIPs involves states in the partially filled, quasi-continuous conduction band states, $g(E, V)$ inside the integrand of Eq. (34) may vary more strongly with temperature. This is one of the main effects which governs the temperature behaviour of the dark current in n-type QWIPs.

The measured dark current versus bias voltage at different temperatures for the three BTM QWIPs listed in Table. 1 are shown in Figs. 8

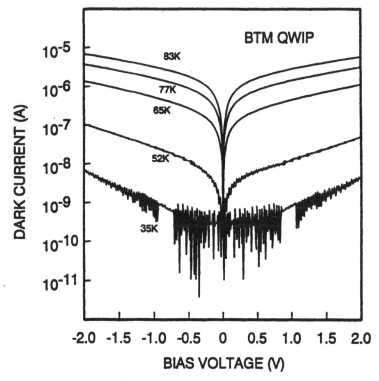

Fig. 8 The dark current versus bias voltage for the GaAs BTM QWIP (QWIP-A) measured at $T = 83, 77, 65, 52,$ and 35 K.

through 10. Figure 8 shows the measured dark current versus bias voltage for QWIP-A at different temperatures. The dark current for this GaAs BTM QWIP was lower than that of a conventional GaAs QWIP with bulk barrier and similar quantum well parameters. The reduction of dark current is a direct result of the barrier height enhancement and current transport via miniband states in this BTM QWIP. Similar results for the dark current versus bias voltage for QWIP-B and -C are also shown in Figs. 9 and 10, respectively. A comparison of the measured and calculated dark currents at low bias voltages for both QWIP-A and -B are shown in Fig. 11(a) and 11(b). As shown in this figure, the solid lines are the calculated values using the expressions of the dark current described above. Excellent agreement was obtained between the theoretical calculations and the measured values

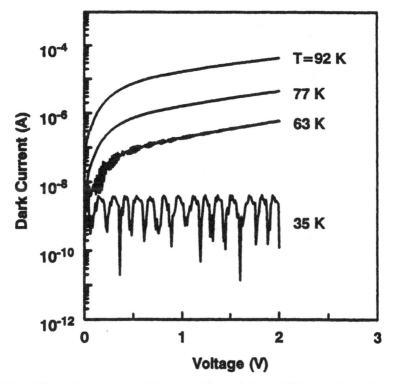

Fig. 9 The dark current versus bias voltage for the InGaAs SBTM QWIP (QWIP-B) measured at $T = 92, 77, 63$, and 35 K, resonant tunneling current dominates at 35 K.

at low bias voltages for both BTM QWIPs. The results clearly show that current conduction in the BTM QWIPs can be attributed to the thermionic-assisted tunneling via the global miniband states inside the quantum well.

5. SPECTRAL RESPONSIVITY

In this section we discuss the spectral response of the BTM QWIPs and present the results of the photoresponse measurements on the three BTM QWIPs listed in Table 1. As mentioned earlier that one of the features of a BTM QWIP is the tunability of its spectral response peak and photoresponse bandwidth. This can be realized by tuning

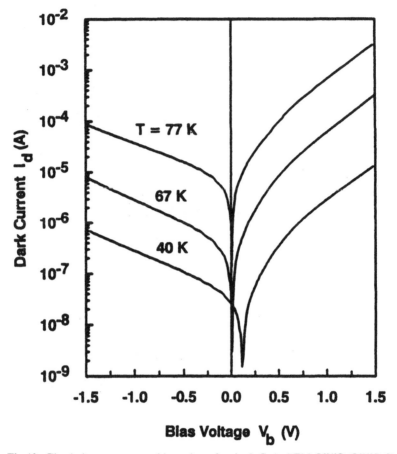

Fig. 10   The dark current versus bias voltage for the InGaAs BTM QWIP (QWIP-C) measured at $T = 83, 77, 67,$ and $40\,K$.

the relative position of the first bound excited state within the miniband states in the quantum well with bias voltage. For example, if the first bound excited state lies at the bottom edge of the miniband, then the spectral response will produce a red shift with narrow band-width. On the other hand, if the first excited state lies at the top of the miniband, then a blue shift results, and if the first excited state is located in the middle of the miniband, then a broader photoresponse curve is expected. This is illustrated in Fig. 12.

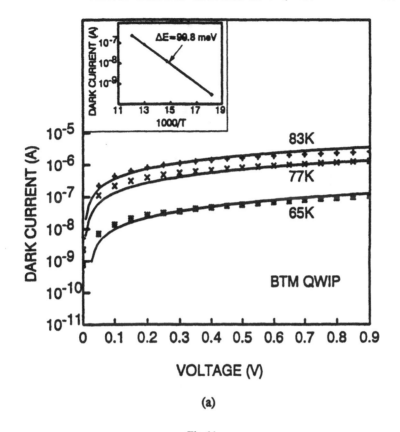

(a)

Fig. 11

Spectral responsivity R_A of a QWIP is defined by the photocurrent (A) produced per unit incident IR radiation power (W) at a specific wavelength. The responsivity depends on the detector quantum efficiency η and optical gain g, and is defined by

$$R_A = \frac{\lambda}{1.24}(\eta \cdot g)\ A/W,\qquad (36)$$

where

$$\eta = \frac{1}{2}(1 - e^{-m\alpha l}).\qquad (37)$$

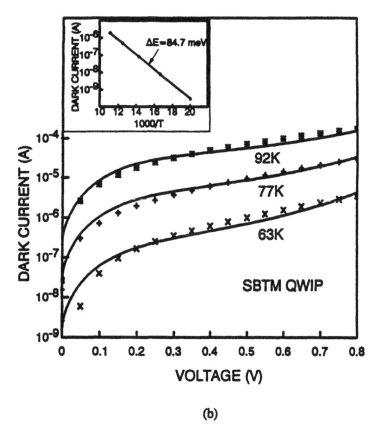

(b)

Fig. 11 A comparison of the calculated and measured dark current versus applied bias voltage for QWIP-A (*T* = 83, 77, 65 K) and QWIP-B (*T* = 92, 77, 63 K).

In Eq. (37), *m* is the number of absorption pass, and denotes the total quantum well thickness. The optical gain factor *g* is defined by the ratio of carrier lifetime τ and transit time τ_d of the QWIP.

Measurements of the spectral responsivity have been made for the three BTM QWIPs listed in Table 1 and the results are shown in Figs. 13(a), 13(b), and 13 (c). Figure 13(a) shows the responsivity versus wave-length for the GaAs BTM QWIP measured at $V_b = -0.2$V and $T = 77$ K, with peak response occured at $\lambda_p = 8.9\,\mu$m. The spectral responsivity for the InGaAs SBTM QWIP measured at $V_b = -2$V and $T = 77$ K is illustrated in Figure 13(b). The peak response for this QWIP is at $\lambda_p = 10\,\mu$m. Figure 13(c) shows the spectral responsivity

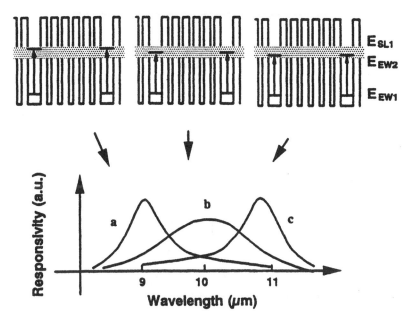

Fig. 12 The relative spectral responsivity versus wavelength for BTM QWIPs (a) E_{W2} lined up at the top of E_{SL1} miniband states (red-shift), (b) E_{W2} at the bottom of E_{SL1} (blue-shift), and (c) E_{W2} in the center of E_{SL1} miniband states.

versus wavelength for the lattice-matched InGaAs BTM QWIP (C) measured at $V_b = 0, -0.5V$ and $T = 67$ K. The peak response for the photovoltaic (PV) mode occurs at $\lambda_p = 10\,\mu\text{m}$ with a very narrow response bandwidth ($\Delta\lambda/\lambda_p = 7\%$), while for the photoconductive (PC) mode the peak response takes place at $\lambda_p = 10.3\,\mu\text{m}$ with a very broad response bandwidth ($\Delta\lambda/\lambda_p = 24\%$) covering wavelengths from 5 to 14 μm. The peak response and cutoff wavelengths, responsivities and detectivities for these three BTM QWIPs are listed Table 2.

6. NOISE CHARACTERIZATION

In this section we discuss the noise characterization for the BTM QWIPs. Noise characterization for the three BTM QWIPs listed in Table 1 has been made in the frequency range from 10 to 10^5 Hz. For frequencies between 10^2 and 10^4 Hz, the noise plateaus stemming from the trapping and detrapping of electrons in the quantum wells were observed in these BTM QWIPs. From the measured noise data,

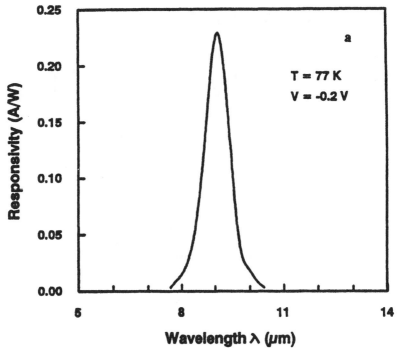

Fig. 13 *(Contd.)*

the bias dependent noise gain, and electron trapping probability were calculated for these QWIPs.

6.1. Noise Theory

The common sources of noise in a QWIP are the thermal noise, generation-recombination (*g-r*) noise, and the flicker noise. The thermal noise in a QWIP may be attributed to the random motion of the charge carriers. It produces a fluctuating emf across its terminals and is related directly to the detector temperature. In semiconductors, the number of carriers fluctuate because of generation and recombination via donors, traps, recombination centers, or band-to-band transitions. Therefore, the resistance, R, fluctuates. If $\delta R(t)$ represents this fluctuation and a dc current I flows through the sample, then a fluctuating emf $\delta V(t) = I\delta R(t)$ will be developed across the terminals. This noise is

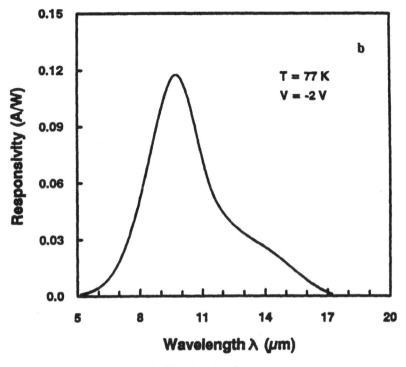

Fig. 13 *(Contd.)*

called the generation-recombination noise, or g-r noise. Most of the noise sources have a frequency-independent spectrum at low frequencies, whereas the spectral intensity decreases at higher frequencies. The only exception is the flicker noise, which usually has a $1/f$ spectrum throughout the entire frequency range. As a result, the flicker noise is often called the $1/f$ noise. We shall briefly discuss each of these noise sources as follows.

(i) Thermal Noise: Thermal noise theory was first reported by Nyquist in 1928 [52]. The thermal noise, also known as Johnson noise [53], is the dominant noise source at very low bias in a QWIP. From the Nyquist equation, the current spectral density for a BTM QWIP with resistance R and temperature T can be expressed by

$$S_i(f) = \frac{4k_B T}{R} \qquad (38)$$

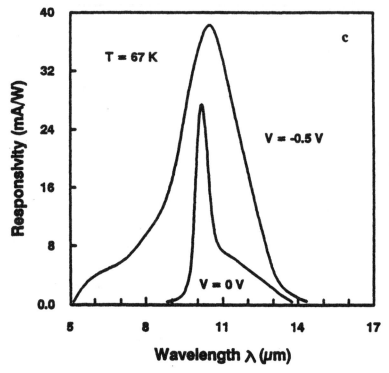

Fig. 13 The measured responsivity versus wavelength for QWIP-A, B, and C, with peak response occured at $\lambda_p = 8.9, 10$, and $10.3 \, \mu$m, respectively.

(ii) Generation-Recombination (g-r) Noise: The general expression for the fluctuation of free carrier density N in a QWIP can be expressed as

$$\frac{dN}{dt} = g(N) - r(N) + \Delta g(t) - \Delta r(t), \qquad (39)$$

where $g(N)$ and $r(N)$ denote the generation and recombination rates of carriers, respectively, and $\Delta g(t)$ and $\Delta r(t)$ describe the randomness in these rates. With $N = N_o + \Delta N, g(N_o) = r(N_o)$, and Poisson statistics, $\overline{\Delta N^2} = \overline{N}$, the g-r spectrum density can be written as

$$S_N(f) = 4\overline{\Delta N^2} \frac{\tau^2}{1 + \omega^2 \tau^2}. \qquad (40)$$

TABLE 2
Summary of the peak and cutoff wavelengths, responsivities, detectivities for the three BTM QWIPs listed in Table 1.

QWIP	$\lambda_p(\mu m)$	$\lambda_c(\mu m)$	R(A/W)	D* cm-Hz$^{1/2}$/W
	8.9(PC)	9.3	0.23	1.2×10^{10}
A	8.9(PV)	9.2	0.15	7.5×10^9
B	10(PC)	11.3	0.11	4.3×10^9
	10.3(PC)	11.7	0.038	5.8×10^9
C	10(PV)	10.4	0.027	5.7×10^9

This equation is valid for all the g-r processes described by fluctuations in carrier density N.

(iii) Flicker Noise: Flicker noise was discovered in vacuum tubes by Johnson [53] in 1925 and interpreted by Schottky in 1926 [54]. The main characteristic of the flicker noise is that it has a $1/f^\alpha$ spectrum with α close to unity. For that reason, the flicker noise is also referred to as the $1/f$ noise.

One popular explanation for the $1/f$ noise is the so-called McWhorter model [59] which interprets the $1/f$ spectrum as a superposition of a large number of Lorentzian spectra, $4A_i\tau_i/(1 + \omega^2\tau_i^2)$. Another possible explanation for the $1/f$ noise is given in terms of mobility fluctuations caused by lattice scattering and Brehm strahlung effect, which produce true $1/f$ noise. However, the observed excess noise in QWIPs at low frequencies does not exactly exhibit the $1/f$ frequency characteristic, and hence the McWhorter model appears to be more adequate for the QWIP devices.

6.2 Noise Gain and Electron Trapping Probability

(i) Noise Gain: In a BTM QWIP, the number of conduction electrons is equal to the total number of electrons N_c excited from the bound state to the miniband states. For an average current $\bar{I} = N_c i_e$, the noise current spectral density is given by

$$S_i(f) = i_e^2 S_N(f) = 4\bar{I}^2 \frac{\overline{\Delta N^2}}{N_c^2} \frac{\tau^2}{1 + \omega^2\tau^2}. \tag{41}$$

For a BTM QWIP that satisfies the Poisson satistics, the variance $\overline{\Delta N^2}$ is equal to N_c. Using $\bar{I} = qN_c/\tau_d$, Eq. (41) becomes

$$S_i(f) = 4q\bar{I}g \frac{1}{1 + \omega^2\tau^2} \tag{42}$$

where $g = \tau/\tau_d$ is defined as the noise current gain. For most QWIPs with $\omega\tau \ll 1$, the g-r noise current spectral density can be simplified to

$$S_i(f) \approx 4q\bar{I}g \qquad (43)$$

(ii) **Electron Trapping Probability:** To derive the electron trapping probability expression for a QWIP, three assumptions are made in the model: (a) the photon flux is independent of the position; (b) the dark current is limited by thermal effects and the interwell tunneling is negligible, and (c) perfect ohmic contacts.

Let us consider a BTM QWIP with N quantum wells. If p denotes the electron trapping probability for the excited carriers in the miniband states to be trapped into the bound ground state of the quantum well, then $(1 - p)$ is the probability that electrons will escape from the well region. Applying the current continuity condition, the photocurrent flows through the N quantum wells of the QWIP can be expressed by [56]

$$i_{\text{photo}} = q\phi\eta'\frac{(1-p)}{p} = q\phi\eta\frac{(1-p)}{Np} \qquad (44)$$

where the quantum efficiency $\eta \approx N\eta'$, and η' is the quantum efficiency of the single quantum well. The photocurrent gain g is related to the capture probability p by

$$g = \frac{(1-p)}{Np} \qquad (45)$$

Since the current gain of a QWIP is independent of the excitation methods, the photocurrent gain given in Eq. (45) should be the same as the noise gain.

6.3. Noise Measurements

The device parameters for the BTM QWIPs used in the noise measurements are listed in Table 1. Note that L_B is the total thickness of the superlattice barrier layer between the two adjacent quantum wells, and N_D is the doping concentration of the quantum well.

One important parameter entering in the noise expressions is the differential resistance of the QWIP. To determine this parameter, the dark currents of the BTM QWIPs were measured using a HP4145B semiconductor parameter analyzer at 77 K and the differential resistances were calculated from the measured current versus bias voltage data.

6.4. Results and Discussion

The noise current density spectra were measured under different bias conditions, and the results are discussed as follows. Figure 14 shows the noise current spectral density versus reverse bias voltage for these BTM QWIPs. Fig. 15 shows the noise gain versus reverse bias voltage for the three BTM QWIPs, using a value of $S_{i_{(thermal)}} = 2.0 \times 10^{-24}$ A²/ Hz, which shows that the noise gain tends to increase with increasing bias voltage. The trapping probability versus reverse bias voltage plots for these BTM QWIPs are displayed in Fig. 16.

(i) QWIP-A: Although the transition mechanism for QWIP-A and QWIP-B is identical, the doping concentration in the quantum well of QWIP-B is much lower than QWIP-A in order to achieve a lower dark current. Thus, the d.c. resistance and the differential resistance are quite high for this sample.

The measured noise current spectral density versus the applied bias voltage for QWIP-A is shown in Fig. 14. The arrow indicates the

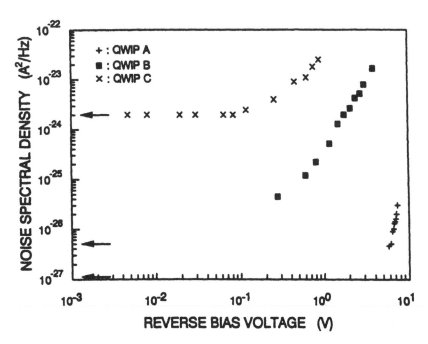

Fig. 14 The measured noise current spectral density versus reverse bias voltage for the three BTM QWIPs listed in Table 3. QWIP-A has the lowest noise and is dominant by *g-r* noise. Johnson noise dominates at very low biases for all three BTM QWIPs.

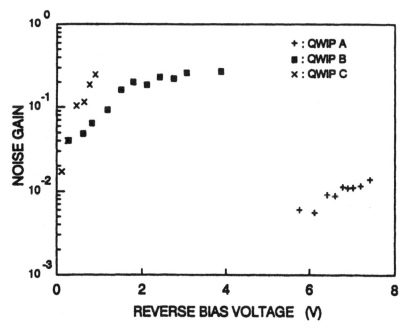

Fig. 15 The noise gain versus reverse bias voltage for the three BTM QWIPs shown in Fig. 14, using a value of $S_{i(\text{thermal})} = 2.0 \times 10^{-24} A^2/Hz$, which shows that the noise gain tends to incease with increasing bias voltage.

current noise level calculated from Eq. (43). From the Nyquist expression, the thermal noise of QWIP-B was found to be $1.0 \times 10^{-27} A^2/Hz$ for this GaAs BTM QWIP. Using this value, the noise gain was calculated from Eq. (47). The result shows that noise gain increases slightly with increasing applied voltage for this QWIP as can be seen in Fig. 15.

Using the number of quantum wells $N = 40$ and the noise gain from Fig. 15, we plot the trapping probability versus bias voltage for QWIP-A, and the result is shown in Fig. 16. It is noted that the trapping probability decreases slowly with increasing bias voltage. The arrow indicates the current noise level calculated using the Nyquist expression

$$S_i = 4kT\frac{dI}{dV} \, . \tag{46}$$

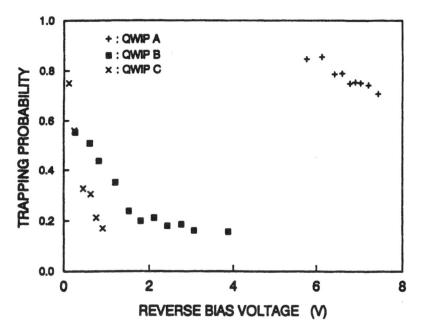

Fig. 16 The trapping probability versus reverse bias voltage plots for the BTM QWIPs shown in Fig. 14. Trapping probability decreases with increasing bias and reaches a plateau level at higher bias.

From the q-r noise expression given in Eq. (43), we obtain the noise gain

$$g = \frac{S_{i_{(g-r)}}}{4q\bar{I}}, \tag{47}$$

where $S_{i_{(g-r)}} = S_i - S_{i_{(\text{thermal})}}$ is the g-r current spectral density.

Since the electron trapping probability p can be expressed in terms of the noise gain, Eq. (45) becomes

$$p = \frac{1}{1 + Ng}. \tag{48}$$

With the help of Eq. (48) and using the number of quantum wells $N = 20$, we calculate p as a function of bias voltage for the three BTM QWIPs, and the results are displayed in Fig. 16, which shows that the electron trapping probability p decreases with increasing bias voltage.

(ii) **QWIP-B:** The lightly strained $In_{0.07}Ga_{0.93}As/GaAs/AlGaAs$ BTM QWIP was designed to bring down the conduction band edge of the quantum well below that of the superlattice barriers in order to lower the dark current of the device. Although the differential resistance for this QWIP is not as high as that of QWIP-A, the noise signal for this QWIP is below the detection limit of the measurement system for $V_b \leqslant 0.1$ V. From the noise measurements, it is shown that QWIP-B displays the g-r noise characteristics for $-4 < V_b < 0.9$ V, and the $1/f$ noise becomes dominant outside this bias range. The result of the noise spectral density, S_i, for this QWIP is also shown in Fig. 14.

(iii) **QWIP-C:** The noise current spectral density, S_i, for QWIP-C (InGaAs BTM QWIP) is plotted as a function of reverse bias voltage, and the result is shown in Fig. 14. It is noted that the thermal noise dominates for $V_b \leqslant 0.2$ V, while the g-r noise becomes dominant for $V_b \geqslant 0.2$ V.

The estimated thermal noise of QWIP-B (GaAs BTM QWIP) is about 5.3×10^{-27} A^2/Hz, and is roughly an order of magnitude smaller than the measured value, S_i. Therefore, it is reasonable to assume $S_{i_{(g-r)}} \approx S_i$ in this case. Using this approximation and quantum well number $N = 40$, the noise gain and trapping probability are calculated as a function of applied bias voltage, and results are shown in Figs. 15 and 16, respectively. It is noted that the noise gain increases with increasing bias voltage and saturation sets in for $V_b \geqslant 2$V. The trapping probability decreases with increasing bias and reaches a minimum value of 0.15 for $V_b \geqslant 2$V as shown in Fig. 16.

Based on the noise characterization of the three BTM QWIPs discussed above we conclude that:

- From our noise studies, the thermal noise is the dominant noise source in the very low bias region. As the applied voltage increases, the g-r noise becomes dominant. Upon further increasing the bias voltage, the $1/f$ noise becomes dominant again. QWIP-A has the lowest current spectral density among the three QWIPs studied in this work.

- The trapping probability p will decrease with increasing bias voltage. The barrier lowering due to the applied bias will increase electron emission from the quantum wells with increasing applied voltage. As a result, $(1 - p)$ will also be increased. This is consistent with our observations.

Finally, it is worth noting that our noise measurements made on a standard GaAs QWIP with bulk barrier reveal that the noise in such a QWIP is generally higher than that of a comparable BTM QWIP (QWIP-A) reported here. The lower noise observed in the GaAs BTM QWIP as compared to a standard GaAs QWIP with bulk barrier layers may be attributed to a lower dark current, better interface quality and lower defect densities in the superlattice barrier layers of QWIP-A.

6.5. Spectral Detectivity

The spectral detectivity D^* is an important figure of merit for a QWIP, which measures the responsivity and the noise equivalent power (NEP) of a QWIP with respect to the detector area and noise bandwidth. The spectral detectivity can be defined by

$$D^*(\lambda) = \frac{R_A \sqrt{A_d \Delta f}}{i_n}, \tag{49}$$

where Δf is the noise spectral bandwidth, i_n is the root-mean-square noise current $(\mathrm{A/Hz}^{1/2})$ given by,

$$i_n = \begin{cases} \sqrt{4EI_D g \Delta f} & \text{for the } g\text{-}r \text{ noise} \\ \sqrt{\dfrac{4k_B T}{R_d}} & \text{for Johnson noise} \end{cases} \tag{50}$$

The g-r noise is the dominant noise source in the PC mode detection QWIPs, while Johnson noise is the dominant noise source for the PV mode detection QWIPs. In general, the detectivity of a BTM QWIP can be calculated by substituting the measured responsivity and noise current into Eq. (49) at a specific temperature. Values of the responsivities, detectivities, peak and cutoff wavelengths for the three BTM QWIPs are summarized in Table 2.

7. DESIGN OF A 2-D PLANAR METAL GRATING COUPLER

It is known that n-type QWIPs do not absorb normal incident IR radiation, since the electric field vector of the incident radiation must have a component perpendicular to the quantum well in order to induce intersubband transitions [1, 57–59]. As a result, the angle of incidence with respect to the quantum well layers must be different from zero in order to produce photo-signal. Although IR absorption can be achieved by using a 45° polished facet on the edge of the QWIP substrate [12], only one-dimensional (1-D) linear arrays can be realized with such an arrangement. For IR imaging applications, a metal grating or dielectric grating coupler must be employed to couple the normal incidence light into the QWIP focal plane arrays (FPAs). Using a 1-D line grating [60–61] on a QWIP may serve this purpose but suffers from low coupling efficiency due to its polarization selectivity. A double period 2-D metal grating [27, 62–64] or random patterned dielectric grating [15] formed on top of the QWIPs has been used to couple the normal incidence IR radiation into an absorbable angle independent of light polarization. The most detailed analysis and measurements on QWIP grating coupling efficiency and optimization were done by Andersson and Lundqvist [27, 62], who considered both linear and 2-D gratings with and without waveguide structure. In our previous works we have reported a planar 2-D square dot [63] and square mesh metal grating coupler [64] for n-type QWIPs. The unique feature of a 2-D metal grating coupler lies in the simple planar structure which is formed by using photolithography and metal deposition processes. This eliminates the possible nonuniformity created by the etching process on the detector surface and hence greatly reduces the processing cost. In this section we describe the theoretical and experimental studies of the optical absorption and coupling efficiency in a planar 2-D square mesh metal grating coupled GaAs BTM QWIP. The detector is illuminated either from the front side or the back side under normal incidence condition. However, the results showed that coupling quantum efficiency of the detector using back side (i.e., substrate side) illumination is 1.65 times higher than that of the front side illumination. For comparison, calculations of the coupling quantum efficiency for a conventional GaAs QWIP using 2-D square mesh metal grating coupler have also been made, and the results are depicted in Sec. 3.

7.1. Basic Theory

To design a 2-dimensional (2-D) square mesh metal grating coupler, we consider the electromagnetic (EM) waves (i.e., IR radiation) impinging on a BTM QWIP under normal incidence illumination. The EM waves scattered by the 2-D metal grating consists of the

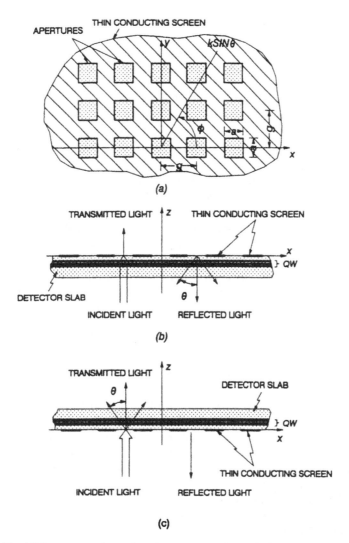

Fig. 17 A 2-D square mesh metal grating structure: (a) top view, (b) side view showing light impinging from the backside of the QWIP (transmission), and (c) side view showing normal incident light impinging from the front side of the QWIP (reflection). g is the grating period and a the aperture width of the square mesh metal grating.

transverse electric (TE) polarization with electric field parallel to the quantum well planes and the transverse magnetic (TM) polarization with one component of the electric field perpendicular to the quantum well planes. Only TM components of the IR radiation will lead to intersubband absorption in the quantum wells for type-I QWIPs. In a 2-D square mesh metal grating structure shown in Fig. 17, the total power of each higher order diffracted mode of the normal incident waves depends on the 'normalized wavelength' $s = \lambda/g$ (where g is the grating period and λ is the wavelength) and the 'strip factor' $h = a/g$ (where a is the aperture width of the square mesh metal grating). Furthermore, the angle between each higher order diffracted wave and the grating normal is determined solely by the parameter s. The intersubband absorption in the quantum well of a BTM QWIP is a function of the total power of the higher order TM waves and the diffracted angle. In this section we use the method of moments (modal expansion) to derive formula for calculating the total power of the higher order TM diffracted waves and the diffracted angle for a 2-D square mesh metal grating coupler formed on the BTM QWIP. From the present calculations, we can produce two universal plots, namely, the total power of the higher order diffracted waves and the cosine of the diffracted angle versus the normalized wavelength for different grating aperture widths and grating periods. Together with the calculated intersubband transition absorption coefficient, these two universal plots can be used to design the 2-D square mesh metal grating couplers for the QWIPs. The basic theory used in the derivation of unknown scattered waves of a 2-D square mesh metal grating coupled BTM QWIP is based on the modal expansion technique [65]. The square mesh metal grating consists of an infinitesimal thickness perfect conducting screen perforated with square apertures distributed periodically along two orthogonal coordinates $\hat{x}$ and $\hat{y}$ as shown in Fig. 17. In this figure we employed spherical coordinates, θ is the angle between the wave propagation vector $\vec{k}$ and the normal to the plane of the grating, and ϕ is the angle between the x-axis and the projection of $\vec{k}$ on the x-y plane. The mesh metal grating is formed on top of the QWIP. The distribution of the EM field near the grating array is in the form of Floquet modes according to wave theory in a periodic structure. Therefore, under normal incidence illumination the solution of the scalar time independent wave equation is given by [66]

$$\psi_{pq} = e^{-ju_p x} e^{-jv_q y} e^{-jw_{pq} z}, \tag{51}$$

where the wave vectors u_p, v_q, and w_{pq} are given respectively by

$$u_p = \frac{2\pi}{g} p = k \sin\theta_{pq} \cos\phi_{pq} \tag{52}$$

$$v_q = \frac{2\pi}{g} q = k \sin \theta_{pq} \sin \phi_{pq} \tag{53}$$

$$w_{pq} = \begin{cases} \sqrt{k^2 - t_{pq}^2} & \text{for} \quad k^2 \geqslant t_{pq}^2 \\ -j\sqrt{t_{pq}^2 - k^2} & \text{for} \quad k^2 \leqslant t_{pq}^2 \end{cases} \quad \text{where } p, q = 0, \pm 1, \pm 2, \ldots, \pm \infty$$

with

$$t_{pq}^2 = u_p^2 + v_q^2$$

The vector orthonormal mode functions for the TE and TM modes transverse with restpect to $\hat{z}$ can be expressed as

$$\vec{\Phi}_{pq}^{TE} = \frac{1}{g}\left(\frac{v_q}{t_{pq}}\hat{x} - \frac{u_p}{t_{pq}}\hat{y}\right)\psi_{pq} \quad \text{for TE modes}$$

$$\vec{\Phi}_{pq}^{TM} = \frac{1}{g}\left(\frac{u_p}{t_{pq}}\hat{x} - \frac{v_q}{t_{pq}}\hat{y}\right)\psi_{pq} \quad \text{for TM modes} \tag{54}$$

The wave admittances looking into the air in $+\hat{z}$ direction are

$$\zeta_{pq}^{TE} = \frac{w_{pq}}{k}\frac{1}{Z_0}$$

$$\zeta_{pq}^{TM} = \frac{k}{w_{pq}}\frac{1}{Z_0} \tag{55}$$

where $Z_0 = \sqrt{\mu_0/\varepsilon_0}$ is the free space characteristic impedance. Assuming that the impedance between the quantum well layers are perfectly matched and the intersubband absorption is complete so that no waves are reflected from the other side of the QWIP. Under this condition, the modal admittances for the TE and TM waves looking into the quantum well region from $z = 0^+$ plane are obtained by replacing Z_0 in Eq. (55) by $Z_d = \sqrt{\mu_0/\varepsilon_0 \varepsilon_d}$ of the QW medium, which yields

$$\zeta_{pq}^{TE} = \frac{w_{pq}}{k}\frac{1}{Z_d}$$

$$\zeta_{pq}^{TM} = \frac{k}{w_{pq}}\frac{1}{Z_d} \tag{56}$$

A plane wave with unit electric field intensity normal incidence in the ϕ plane can be expressed as the sum of TE and TM plane waves, that is,

$$\vec{E}^i = \sum_{r=1}^{2} A_{00r}\vec{\Phi}_{00r} \tag{57}$$

where A_{00r} is the magnitude of the incident field component which depends on the polarization direction. The third subscript $r = 1$ or 2 is used to designate the TE or TM Floquet modes, respectively. Similarly, the reflected waves and transmitted waves can also be expressed in terms of the Floquet modes with reflection coefficients R_{pqr} and transmission coefficients T_{pqr} given by

$$\vec{E}^t = \sum_{r=1}^{2} A_{00r} \vec{\Phi}_{00r} + \sum_{p} \sum_{q} \sum_{r=1}^{2} R_{pqr} \vec{\Phi}_{pqr}$$

$$= \sum_{p} \sum_{q} \sum_{r=1}^{2} T_{pqr} \vec{\Phi}_{pqr} . \tag{58}$$

Equation (58) includes the boundary condition that the tangential electric field in the aperture is continuous. The orthonormal waveguide modes $\vec{\Pi}_{mnl}$ of the square aperture itself other than the Floquet modes $\vec{\Phi}_{pqr}$ are used to expand the unknown electric field distribution in the aperture in order to satisfy the boundary condition, which has been shown to provide a faster convergence [66].

$$\vec{E}^t = \sum_{m} \sum_{n} \sum_{l=1}^{2} F_{mnl} \vec{\Pi}_{mnl} , \tag{59}$$

where the unknowns F_{mnl} are the coefficients given in the waveguide modes experssion. Finally, a matrix equation can be written as

$$[Y_{mnl}^{MNL}][F_{mnl}] = [I_{mnl}] \tag{60}$$

where

$$Y_{mnl}^{MNL} = \sum_{p} \sum_{q} \sum_{r=1}^{2} (\xi_{pqr} + \zeta_{pqr}^{d}) C_{pqr}^{MNL*} C_{pqr}^{mnl},$$

$$C_{pqr}^{mnl} = \iint_{\text{aperture}} \vec{\Phi}_{pqr}^{*} \cdot \vec{\Pi}_{mnl} da \tag{61}$$

and

$$I_{mnl} = 2 \sum_{r=1}^{2} A_{00r} \zeta_{00r}^{d} C_{00r}^{mnl*} \tag{62}$$

is the matrix of the incident waves. The higher order transmission coefficients and reflection coefficients are given respectively by

$$T_{pqr} = \sum_{m} \sum_{n} \sum_{l=1}^{2} F_{mnl} C_{pqr}^{mnl} \quad p, q \neq 0, 0 \tag{63}$$

and

$$R_{pqr} = \sum_m \sum_n \sum_{l=1}^{2} F_{mnl} C_{pqr}^{mnl} \quad p, q \neq 0, 0 \tag{64}$$

Since F_{mnl} is proportional to λ and C_{pqr}^{mnl} varies with $1/\lambda$ for a given g, the transmission coefficients T_{pqr} and the reflection coefficients R_{pqr} are a function of λ/g. The angle between the electric field of the higher order TM diffracted modes $T_{pq2}\,\vec{\Phi}_{pq2}$, $R_{pq2}\,\vec{\Phi}_{pq2}$ and the $\hat{z}$ direction is designated as γ_{pq2}, and the cosine of this angle depends on the diffracted order and the normalized wavelength $s(= \lambda/g)$, which can be expressed by

$$\cos\gamma_{pq2} = \frac{\lambda/n_r}{g}\sqrt{p^2 + q^2} \tag{65}$$

where n_r is the refractive index of the quantum well medium where the wave propagates.

The 2-D square mesh metal grating coupled GaAs BTM QWIP structure used in the present analysis is based on the transition from the bound ground state in the GaAs quantum well to the global miniband states formed in the superlattice barrier layer, as shown in Fig. 2. In a BTM QWIP, only electrons in the miniband are free to move in the direction perpendicular to the quantum well layers, which give rise to a current flow in the BTM QWIP. The intersubband absorption constant for a QWIP can be expressed as [66]

$$\alpha = \left(\frac{e^2\hbar^3}{m^* n_r \varepsilon_0 c}\right)\frac{n_e \cos^2\gamma}{\hbar\omega}\frac{TT^2\sqrt{(E - E_{\min})(E_{\max} - E)}}{(U + S_1)^2 - S_1(E_{\max} - E)} \tag{66}$$

where e is the electronic charge, m^* is the electron effective mass, c is the speed of light, n_e is the electron density in the wells, ω is the angular frequency of the EM waves, γ is the angle between the electric field vector of the normal incidence IR radiation and the motion vector of electrons, $E_{\max}$ and $E_{\min}$ are the top and bottom edges of the miniband. All other parameters, TT, U, and S_1 in Eq. (66) depend on the specific quantum well structure. By substituting Eq. (65) into (66), the absorption constant of a BTM QWIP can be calculated in the spectral range of interest.

7.2. Results and Discussion

The evanescent modes excited by the square mesh metal grating produce no photoresponse in the quantum well intersubband transition

[67], neither do the TE modes. Therefore, the following discussion will be focussed on the propagating TM modes in the quantum well region. It is convenient to use two normalized parameters, namely, the normalized wave-length '$s = \lambda/g$' and the strip factor '$h = a/g$' to illustrate the universal polts of the 2-D square mesh metal grating coupler for the BTM QWIPs. Rigorous calculations were carried out by using 40 waveguide modes, and adding more modes made no noticeable change in the results of the transmission and reflection coefficients. When analyzing a 2-D square mesh grating coupler with large strip factor ($a/g > 0.85$), a much larger number of waveguide modes may be required to simulate the wave distribution at the metal edge. However, large strip factor is not desirable for practical QWIPs, since it is more difficult to process such fine metal strips (e.g., $\leqslant 1 \, \mu m$) on the QWIP surface. In the case of back side illumination, we consider an EM wave with unit power density impinging on the square mesh metal grating at zero degree angle with respect to the grating normal as shown in Fig. 17. The coupling of IR radiation into the quantum wells is contributed only by the higher order diffracted TM Floquet modes. Fig. 18 shows a universal plot that illustrates the normalized total power of the first order TM diffracted waves $R_{012}\vec{\Phi}_{012}$, $R_{0-12}\vec{\Phi}_{0-12}$, $R_{-102}\vec{\Phi}_{-102}$ and $R_{102}\vec{\Phi}_{102}$ as a function of the normalized wavelength $s = \lambda/g$ for different values of h. We can also obtain the transmitted Floquet modes under front side illumination (see Fig. 17) by dividing the normalized power shown in Fig. 18 by a factor of 1.65 for the transmitted power, where the reflective effect due to the impedance mismatch between air ($n_r = 1$) and the GaAs quantum wells ($n_r = 3.25$ at 77 K) has also been eliminated. The first order diffracted waves emerge when the wavelength of the incident light in the GaAs quantum wells is smaller than the grating period, that is, $s = \lambda/g < 3.25$. For the same reason, within the spectral range shown in Fig. 18, only the zeroth order far field transmitted waves $T_{00i}\vec{\Phi}_{00i}(i = 1, 2)$ in the free space may be found, and all other higher order transmitted waves are being cut-off. This is due to the fact that free space wavelength λ is greater than the grating period g, and thus makes the grating coupler operating in the non-diffraction region. It is noted that the square mesh grating is indistinguishable between x and y directions. Thus, the total normalized power of the first order diffracted waves generated by the x and y components of the incident waves remains the same for different input polarizations (i.e., independent of polarization). As shown in Fig. 18, a relatively flat coupling curve with $h = a/g = 0.5$ may be considered as a better choice for

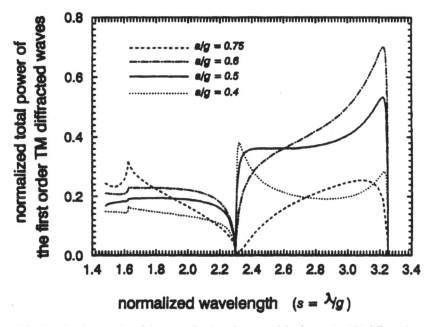

Fig. 18 A universal plot of the normalized total power of the first order TM diffracted waves $R_{012}\vec{\Phi}_{012}$, $R_{0-12}\vec{\Phi}_{0-12}$, $R_{-102}\vec{\Phi}_{-102}$ and $R_{102}\vec{\Phi}_{102}$ as a function of the normalized wavelength $s = \lambda/g$ for a GaAs BTM QWIP with different values of $h = a/g$.

coupling the IR radiation more effectively into the QWIP. Figure 19 illustrates the normalized total power of the second order TM diffracted waves $R_{\pm112}\vec{\Phi}_{\pm112}$ and $R_{-1\pm12}\vec{\Phi}_{-1\pm12}$ as a function of the normalized wavelength s for different values of h. Similarly, the second order transmitted TM components under front side normal incidence illumination (Fig. 17(c)) is obtained by dividing the coupling power shown in Fig. 19 by a factor of 1.65 if the reflection at the air/GaAs interface is modified. A comparison of Fig. 18 and Fig 19 reveals that the total power of the second order diffracted modes is directly proportional to the total power of the first order diffracted modes. The second order diffracted waves emerge for $s < 2.298$ with a total power about 50% smaller than that of the first order diffracted modes. Figure 20 shows a universal plot that relates $\cos\gamma_{pq2}$ to the normalized wavelength of the higher order diffracted waves for a GaAs BTM QWIP, where γ is the angle between the electric field vector and the grating normal. The relationship between $\cos\gamma_{pq2}$ and the normalized

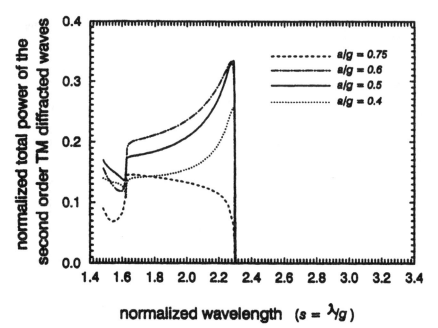

Fig. 19 The normalized total power of the second order TM diffracted waves $R_{\pm 1 1 2}\vec{\Phi}_{\pm 1 1 2}$ and $R_{-1\pm 1 2}\ \vec{\Phi}_{-1\pm 1 2}$ as a function of the normalized wavelength s for the GaAs BTM QWIP with different values of h.

wavelength λ/g for the first order diffracted waves (i.e., $|p| + |q| = 1$ in the notation of Floquet modes $\vec{\Phi}_{pqr}$) is given by

$$\cos\gamma_{pq2} = \frac{\lambda}{g} \qquad (67)$$

and for the second order diffracted waves with $|p| = |q| = 1$ is

$$\cos\gamma_{pq2} = \frac{\lambda}{g}\sqrt{2} \qquad (68)$$

We next calculate the absorption constant versus wavelength in a 2-D square mesh metal grating coupled GaAs BTM QWIP. The mesh grating period is chosen in the region where s falls between 0.7 and 1 for using the first order diffracted waves. In this case, $\cos^2\gamma$ is greater than 1/2, which corresponds to a 45° launching. If we select the curve with $h = 0.5$ in Fig. 18 by multiplying the grating period $g = 4\,\mu\text{m}$ to

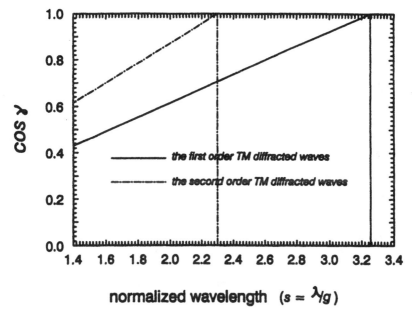

Fig. 20 A universal plot that relates $\cos\gamma_{pq2}$ to the normalized wavelength of the 1st. and 2nd order diffracted waves for a GaAs BTM QWIP, where γ is the angle between the electric field vector and the grating normal.

the coordinate, a relatively flat coupling quantum efficiency curve is obtained over a broad wavelength range (9–12 µm), as shown in Fig. 21 (curve (b)). By substituting the corresponding $\cos\gamma$ shown in Fig. 20 and using the QWIP parameters given in Table 1 into Eq. (71), we obtain the absorption coefficient versus wavelength for the GaAs BTM QWIP, and the result is shown in Fig. 22. Note that the dotted line represents the absorption constant of the QWIP with IR launched in a 45° polished facet. The aperture width of the square mesh grating used in the above design was $a = (4\,\mu m) \times 0.5 = 2\,\mu m$. The quantum efficiency η can be calculated using the expression

$$\eta = P_{\mathrm{eff}}(1 - e^{-\alpha l}) \tag{69}$$

where P_{eff} is the normalized coupling power of the incident IR radiation, l is the total length of the doped quantum wells, and α is obtained from Fig. 22. In the present case, $l = (88\,\text{Å/period}) \times (40\,\text{periods}) = 3520\,\text{Å}$, and η is shown (broken line) in Fig. 23.

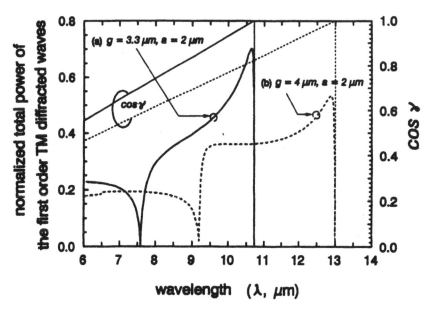

Fig. 21 The normalized total power of the first order TM diffracted waves for a 2-D square mesh grating under back side illumination, which shows a cut-off wavelength of the grating at about 10.7 μm. The solid line illustrates the optimum grating design for 10 μm IR coupling.

The optimal square mesh metal grating used in the above example for a GaAs BTM QWIP is $g = 3.3$ μm and $a = 2$ μm, corresponding to the curve with $h = 0.6$ shown in Fig. 18. The characteristic curve of the first order TM diffracted waves for a 2-D square mesh grating under back side illumination was plotted in Fig. 21, which shows that a cut-off wavelength of the grating is about 10.7 μm. The solid line in Fig. 21 illustrates the optimum grating design. Figure 23 shows the coupling quantum efficiency for a GaAs BTM QWIP using different light coupling schemes and grating parameters. The results show that coupling quantum efficiency of a square mesh metal grating coupler under the back side illumination is higher than that of the front side illumination. Furthermore, the coupling quantum efficiency for the optimal square mesh metal grating is higher than that of the 45° facet illumination.

7.3. Square Mesh Metal Grating Coupler for a GaAs QWIP

For comparison, the 2-D square mesh metal grating coupler discussed in the previous section has been applied to a conventional GaAs

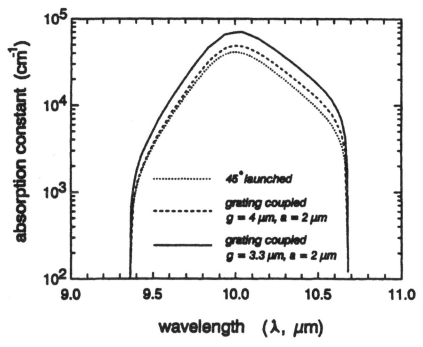

Fig. 22 The calculated absorption constant versus wavelength in a 2-D square mesh metal grating coupled GaAs BTM QWIP. The mesh grating period is chosen in the region where s falls between 0.7 and 1 for using the first order diffracted waves.

QWIP with bulk barriers. Calculations of the coupling quantum efficiency have been carried out for this QWIP under normal incidence backside illumination. A coupling quantum efficiency of about 25% was obtained for a standard GaAs QWIP using a 2-D square mesh metal grating coupler with $g = 3.3$ μm and $a = 2$ μm at $\lambda = 10$ μm. To design a 2-D square mesh metal grating coupler formed on a standard GaAs/AlGaAs QWIP with bulk barrier layers, the absorption constant versus wavelength in the wavelength of interest must be found first. From [67] and [68], the absorption coefficient for the GaAs QWIP were obtained by fitting the Lorentzian line shape according to the following form

$$\alpha(v,\gamma) \propto \frac{\Gamma(v)}{[(v - v_0)^2 + \Gamma^2(v)]} \cos^2\gamma , \tag{70}$$

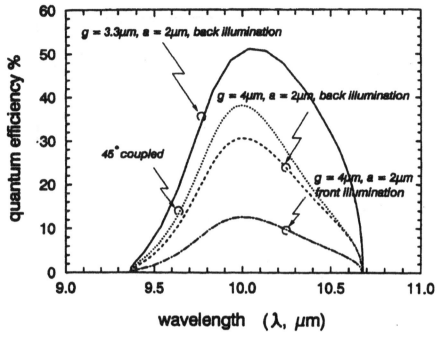

Fig. 23 The coupling quantum efficiency versus wavelength for a GaAs BTM QWIP using 2-D square mesh metal grating coupling scheme with different grating parameters as compared to a 45° facet illumination.

where $\Gamma(v)$ is the line width, v_0 is the peak energy position, and γ is the angle between the electric field vector and the normal of the quantum well plane. $\Gamma(v)$ and v_0 are used as fitting parameters.

The standard QWIP structure is the $GaAs/Al_xGa_{1-x}As$ QWIP used in the imaging camera reported by Bethea et al. [68] for $\lambda = 7 \sim 11$ μm IR detection. This GaAs QWIP consists of 50 periods of $L_w = 40$ Å quantum well (doped $n = 1.2 \times 10^{18}$ cm^{-3}) with an $Al_{0.25}Ga_{0.75}As$ barrier layer thickness $L_b = 500$ Å (i.e., total QW thickness $l = 2.7$ μm), sandwiched between 0.5 μm top and 1 μm bottom GaAs ohmic contact layers having the same doping density. By comparing the absorption constant given in [69], the Lorentzian fit is plotted by the dotted line shown in Fig. 24. In this figure, we also compared the absorption constant for the QWIP coupled by a 2-D square mesh metal grating, using grating period $g = 3.3$ μm and grating aperture width $a = 2$ μm (solid line). Note that absorption of all

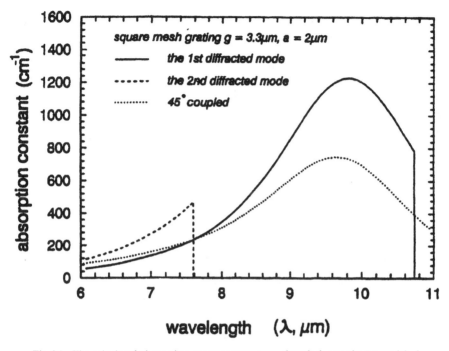

Fig. 24 The calculated absorption constant versus wavelength due to the 1st. and 2nd. order diffracted waves in a 2-D square mesh metal grating coupled GaAs QWIP with $g = 3.3$ μm and $a = 2$ μm, as compared to a 45° facet illumination case.

the evanescent TM modes was ignored [69] in the present calculations. When the wavelength of the incident IR radiation inside the QWIP is smaller than 7.58 μm, the second order diffracted waves will be excited but with a much smaller absorption constant (broken line in Fig. 24). The calculated coupling quantum efficiency corresponding to Fig. 24 is shown in Fig. 25. Obviously, the calculated results show that the 2-D square mesh grating coupled GaAs QWIP has a larger integrated quantum efficiency than that of the 45° polished facet QWIP.

8. CONCLUSIONS

We have presented in this chapter a detailed theoretical and experimental study of the intersubband absorption, dark current, spectral

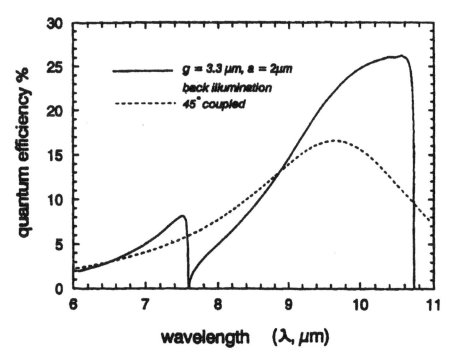

Fig. 25 The coupling quantum efficiency versus wavelength for a 2-D square mesh metal grating coupled GaAs QWIP with $g = 3.3\,\mu m$ and $a = 2\,\mu m$, as compared to a 45° facet illumination. The quantum efficiency for this standard QWIP is about 50% lower than that of the BTM QWIP shown in Fig. 23.

responsivity, noise figure, and detectivity in three 2-D metal grating coupled bound-to-miniband (BTM) transition QWIPs fabricated from the GaAs/AlGaAs and InGaAs/InAlAs material systems by MBE technique. The intersubband transition scheme for these BTM QWIPs is from the highly populated bound ground state in the enlarged quantum wells to the global miniband states formed by the superlattice barrier layers inside the quantum wells. By utilizing resonant tunneling and coherent transport along the superlattice miniband, a high performance long wavelength infrared (LWIR) photodetector has been demonstrated for 8 to 12 μm applications.

From the dark current measurements on these BTM QWIPs, we have found that the dark current in the BTM QWIPs is indeed lower than that of the conventional GaAs QWIPs with bulk barrier structure. From the noise characterization of the BTM QWIPs, it is shown that noise in the BTM QWIPs is inherently lower than that of a standard QWIP. The reasons for this may be attributed to the fact that introducing the superlattice barrier layer increases the barrier height and lowers the dark current and reducing interface defects, and hence reduces the excess frequency dependent noise in these devices.

The design of a 2-D square mesh metal grating coupler for efficient coupling of normal incident IR radiation in the BTM QWIPs and a standard GaAs QWIP has been described in details in this chapter. It is shown that for a given detector material the total power and diffracted angle of the higher (e.g., first and second) order TM diffracted waves depend on two normalized parameters $s = \lambda/g$ and $h = a/g$, which are a function of wavelength and grating size. By scaling the universal polts shown in Figs. 4.18. and 4.20, the optimal grating period and aperture width can be obtained for any infrared specturm. In addition, the absorption constant and coupling quantum efficiency for a 2-D square mesh metal grating coupled BTM QWIP can be calculated from these universal plots for any grating periods and aperture widths.

Acknowledgments

The author would like to thank several of his graduate students (Y. H. Wang, Daniel Wang and Y.C. Wang) for their dedications to the research work on quantum well infrared photodetectors and their assistance in preparing the materials used in this writing. He also like to acknowledge Dr. Pin Ho of Electronics Laboratory, Martin Marietta, Syracuse, for providing the MBE grown QWIP samples used in this study. This work was supported by the advanced Research Project Agency (ARPA) and monitored by the Office of Naval Research (ONR) under Navy grant No.N0014-91-J-1976.

References

1. L.C. West and S.J. Eglash, *Appl. Phys. Lett.*, **46**, 1156 (1985).
2. D.D. Coon and R.P.G. Karunasiri, *Appl. Phys. Lett.*, **33**, 495 (1984).
3. J.S. Smith, L.C. Chiu, S. Margalit, A. Ariv, and A.Y. Cho, *J. Vac. Sci. Techno.*, B, 1, 376 (1983).
4. B.F. Levine, K.K. Choi, C.G. Bethea, J. Walker, and R.J. Malik, *Appl. Phys. Lett.*, **57**, 383 (1990).
5. B. F. Levine, G. Hasnain, C.G. Bethea, and Naresh Chand, *Appl. Phys. Lett.*, **54**, 2704 (1990).
6. B.F. Levine, C.G. Bethea, G. Hasnain, V.O.. Shen, E. Pelve, R.R. Abbott, and S.J. Hsieh, *Appl. Phys. Lett.*, **56**, 851 (1990).

7. E.R. Brown and S.J. Eglash, *Phys. Rev.*, **41**, 7559 (1990).
8. L.S. Yu, S.S. Li, and Y.C. Kao, *Proc. Government Microcircuit Applications Conference*, Nov. 6–8, Las Vegas, 479 (1990).
9. L.S. Yu and S.S. Li, *Appl. Phys. Lett.*, **59**, 1332 (1991).
10. H.C. Liu, *J. Appl. Phys.*, **63**, 2856 (1988); 2749 (1991).
11. L.S. Yu, S.S. Li, and Pin Ho, *Appl. Phys. Lett.*, **59** (2712) (1991).
12. S.D. Gunapala, B.F. Levine, and Naresh Chand, *Appl. Phys. Lett.*, **70**, 305 (1991).
13. L.S. Yu, S.S. Li, Y.H. Wang, and P. Ho, *Appl. Phys. Lett.*, **60(8)**, 992 (1992).
14. Y.H. Wang, S.S. Li, P. Ho, and M.O. Manasreh, *J. Appl. Phys.*, **74(2)**, 1382 (1993).
15. B.F. Levine, *J. Appl. Phys.*, **74(8)**, R1–R85, Oct. 15 (1993).
16. L. Esaki and R. Tsu, *IBM J. Res. Develop.*, **14**, 65 (1970).
17. R.F. Kazarinov and R.A. Suris, *Fiz. Tekh. Poluprov.*, **5**, 797 (1971).
18. R. Tsu and L. Esaki, *Appl. Phy. Lett.*, **22**, 562 (1973).
19. L. Esaki and L.L. Chang, *Appl. Phy. Lett.*, **24**, 593 (1974).
20. F. Capasso, K. Mohammed, A.Y. Cho, R. Hull, and A.L. Hutchinson, *Appl. Phys. Lett.*, **47**, 420 (1985).
21. C. Rolland and P.B. Corkum, *J. Opt. Soc. Amer. B*, **3**, 1625 (1988).
22. K.J. Siemsen and H.D. Riccius, *Appl. Phys. A*, **35**, 177 (1984).
23. S.K. Chun, D.S. Pan, and K.L. Wang, *Phys. Rev. B*, **47(23)**, 15638 (1993).
24. A.K. Ghatak, K. Thyagarajan, and M.R. Shenoy, *IEEE J. Quantum Elec.*, **24(8)**, 1524 (1988).
25. M.O. Vassell, J. Lee, and H.F. Lockwood, *J. Appl. Phys.*, **54(9)**, 5206 (1983).
26. K.W. Goossen and S.A. Lyon, *Appl. Phys. Lett.*, **47**, 1257 (1985).
27. J.Y. Andersson, L. Lundqvist, and Z.F. Paska, *Appl. Phys. Lett.*, **59**, 857 (1991).
28. Y.C. Wang and S.S. Li, *J. Appl. Phys.*, **74(4)**, 2192 (1993).
29. D. Delagebeaudeuf, P. Deleckyse, P. Etienne, J. Massiers, M. Laviron, J. Chaplart, and T. Linh, *Electron. Lett.*, **18**, 85 (1982).
30. I. Hase, H. Kawai, K. Kaneko, and K. Watanabe, *Electron, Lett.*, **20**, 491 (1984).
31. P.J. Price, *IBM J. Res. Develop*, **17**, 39 (1973).
32. M. Artaki and K. Hess, *Superlattices and Microstructures*, **1**, 489 (1985).
33. A. Harwit, J.S. Harris, Jr., and A. Kapitulnik, *Appl. Phys., Lett.*, **59** 3211 (1986).
34. T.K. Gaylord, E.N. Glytsis, and K.F. Brennan, *J. Appl. Phys.*, **65**, 2535 (1989).
35. A. Raymond, J.L. Robert, and C. Bernard, *J. Phys. C* **12**, 2289 (1979).
36. J.W. Choe, Byungsung O, K.M.S.V. Bandara, and D.D. Coon, *Appl. Phys. Lett.*, **56**, 1679 (1990).
37. M. Ramsteiner, J.D. Ralston, P. Koidl, B. Dischler, H. Biebl, J. Wagner, and H. Ennen, *J. Appl. Phys.*, **67**, 3900 (1990).
38. K.M.S.V. Bandara and D.D. Coon, *Appl. Phys. Lett.*, **53**, 1865 (1988).
39. A. Harwit and J.S. Harris, Jr., *Appl. Phys., Lett.*, **50**, 685 (1987).
40. S.Y. Yuen, *Appl. Phys. Lett.*, **43**, 813 (1983).
41. A. Seilmeier, H.J. Hubner, G. Abstreiter, G. Weiman, and W. Schlapp, *Phys. Rev. Lett.*, **59**, 1345 (1987).
42. P. Yuh and K.L. Wang, *Phys. Rev.*, *B*, **37**, 1328 (1988).
43. Y.H. Wang, S.S. Li, and P. Ho, *Appl. Phys. Lett.*, **62**, 621 (1993).
44. J.J. Sakurai, *Advanced Quantum Mechanics*, Addison-Wesley, New York, (1967).
45. F. Bassani and G.P. Parravicini, *Electronic States and Opticall Transitions in Solids*, Pergamon, New York, (1975).
46. G. Hasnian, B.F. Levin, C.G. Bethea, R.A. Logan, J. Walker, and R.J. Malik, *Appl. Phys. Lett.*, **54**, 2515 (1989).
47. A. Madhukar, T.C. Lee, M.Y. Yen, P. Chen, J.Y. Kim, S.V. Ghaisas, and P.G. Newman, *Appl. Phys. Lett.*, **46**, 1148 (1985).
48. D. Calecki, J.F. Palmer, and A. Chomette, *J. Phys., C*, **17**, 5017 (1984).
49. S.D. Gunapala, B.F. Levine, and K. West, *J. Appl. Phys.*, **69**, 6517 (1991).

50. G.D. Shen, D.X. Xu, M. Willander, and G.V. Hansson, *Appl. Phys. Lett.*, **58**, 738 (1991).
51. S.S. Li, M.Y. Chuang, and L.S. Yu, *Semicon. Sci. Technol.*, **8**, S406 (1993).
52. H. Nyquist, *Phys. Rev.*, **32**, 110 (1928).
53. J.B. Johnson, *Phys. Rev.*, **26**, 71 (1925).
54. W. Schottky, *Phys. Rev.*, **28**, 74 (1926).
55. A.L. McWhorter, 1/f Noise and Related Surface Effects in Germanium, Lincoln Lab. Rpt. No. 80, Boston (1955).
56. H.C. Liu, *Appl. Phys. Lett.*, **61**, 2703 (1992).
57. C.F. Gerald and P.O. Wheatley, *Applied Numerical Analysis*, 3rd. edition, Addison-Wesley, New York, 1984.
58. D.D. Coon and R.P.G. Karunasiri, *Appl. Phys. Lett.*, **45**, 649 (1984).
59. M.J. Kane, M.T. Emeny, N. Apsley, C.R. Whitehouse, and D. Lee, *Semicond. Sci. Tech.*, **3**, 722 (1988).
60. W.J. Li and B.D. McCombe, *J. Appl. Phys.*, **71**, 1038 (1992).
61. J.Y. Andersson and L. Lundqvist, *J. Appl. Phys.*, **71**, 3600 (1992).
62. Y.C. Wang and S.S. Li, *J. Appl. Phys.*, **74**, 2192 (1993).
63. Y.C. Wang and S.S. Li, *J. Appl. Phys.*, **76**, Jan. 7 (1994).
64. C.C. Chen, *IEEE Trans* **MTT-18**, 627 (1970).
65. D.D. Coon, R.P.G. Karunasiri, and L.Z. Liu, *Appl. Phys. Lett.*, **47**, 289 (1985).
66. D. Cui, Z. Chen, Y. Zhou, H. Lu, Y. Xie, and G. Yang, *Infrared Phys.*, **32**, 53 (1991).
67. B.F. Levine, C.G. Bethea, K.K. Choi, and R.J. Malik, *Appl. Phys. Lett.*, **53**, 231 (1988).
68. C.G. Bethea, B.F. Levine, V.O. Shen, R.R. Abbott, and S.J. Hsieh, *IEEE Trans.*, **ED-38**, 1118 (1991).
69. M.O. Manasreh, F. Szmulowicz, D.W. Fisher, K.R. Evans, and C.E. Stutz, *Appl. Phys. Lett.*, **57**, 1790 (1990).

CHAPTER 4

Grating Coupled Quantum Well Infrared Detectors

J.Y. ANDERSSON and L. LUNDQVIST

Industrial Microelectronics Center (IMC), P. O. Box 1084, S-164 21 Kista, Sweden

1. INTRODUCTION AND OUTLINE

Long wavelength (7.5–10 µm) quantum well infrared photodetectors (QWIPs) based on AlGaAs/GaAs quantum wells (QWs) have been shown to exhibit high peak detectivities $D^* = 1.10^{10}$–9.10^{10} cm $Hz^{1/2}W^{-1}$ at 80 K and close to background limited operation [1, 2]. Due to the well established GaAs material and processing technology QWIPs are viable candidates for large, low cost LWIR (8–12 µm) focal plane arrays (FPAs) [3–6]. The quantum well detector and detector arrays are reviewed in Ref. 1.

QWIPs operate on account of intersubband transitions in doped QWs which implies photoexcitation of charge carriers from a bound ground state to quasi-bound or extended excited states where the charge carriers are freely mobile perpendicularly to the QW planes, thus enabling photoconductive action. QWIPs with n-doped QWs offer large values of D^* mainly as a result of the low electron effective mass. However, a drawback is that for n-doped QWs the quantum mechanical selection rules forbid absorption of radiation with incidence normal to the QW-layer plane. Therefore it is necessary to find optical geometries that overcome this fact and enhance the quantum efficiency η of the detectors [7–15]. The *absorptance* or *absorption quantum efficiency* is defined as the ratio of absorbed optical power to total incident optical power and is here denoted η.

The absorptance for normally incident radiation can be enhanced and become polarisation independent by the use of a crossed (doubly periodic) grating with a cladding layer and reach values close to unity for large detector mesa sizes (500 μm diameter) [10]. The crossed grating is etched into the top of the detector mesa and the waveguide is defined by the grating, the active QW layer and a cladding layer [16], from top to bottom. The random gratings now under investigation and presented in Refs. 14, 15 have the potential of offering even higher quantum efficiencies than the crossed gratings with a cladding layer.

In this work we present a theoretical and experimental investigation of grating coupled QWIPs, and the dependence of absorptance on various factors like grating type, the inclusion of a cladding layer, detector mesa size and type of contact is dealt with.

In order to theoretically model the absorptance properties of a QW infrared detector it is necessary to find proper models of its main constituents, i.e., the grating and the QW structure. The grating is conveniently modeled by the computation of its scattering matrix which describes the relation between incident and diffracted optical power. The scattering matrix is here obtained by using the modal expansion method which is a near-exact theory of grating scattering and which takes the vector properties of electromagnetic radiation into account. This is particularly important for QW grating detectors since these are polarization sensitive, and due to that the gratings operate close to first order grating cutoff where the simple scalar theory of light fails completely. The anisotropic properties of infrared absorption in the QW structure are modeled by a transfer matrix method, which also takes multiple internal reflection between layers into account.

The outline of this work is that first the theory of the grating detector is described in Sec. 2, including a theoretical treatment of grating scattering based on the modal expansion method (MEM), as well as of the transfer matrix method used to model quantum well absorption. The simple theory of MEM assumes gratings of infinite extension and grating metal of infinite conductivity. However, the theory can be extended to take into account the finite size of the grating, as well as the finite conductivity of the grating metal. An approximate treatment of grating scattering is also presented based on the single channel or cavity mode approximation. The advantage is that analytical expressions can be derived in closed form for both lamellar and crossed gratings, which gives considerable insight into the operation and optimizations of gratings. In Sec. 2.4 the theoretical framework presented in Sec. 2 will be used for a numeric simulation of grating detectors in order to find out the optimum grating geometry. Sec. 3 provides experimental results on grating scattering as well as on grating detectors. A comparison is made between theory and experiments and is finally discussed. A summary of the work is presented in Sec. 4.

2. THEORY OF THE GRATING COUPLED DETECTOR

2.1. Description of the Computational Model

The problem of the calculation of absorptance consists of two parts: i) to model the scattering behavior of the diffraction grating which is here described by the scattering matrix, and ii) to model the infrared absorption in the multi-quantum well structure (MQW). The grating scattering behavior is modeled by using the *modal expansion method* (MEM) [17] which is a near-exact way of calculating diffracted radiation intensities, taking the vector nature of the electromagnetic (EM) field into consideration. However, it can be used for certain grating profiles only, e.g. the lamellar grating with rectangular profile, or the crossed (doubly-periodic) grating with box-shaped or cylindrical cavities. In this work mainly etched phase gratings will be dealt with where the phase, not the amplitude, of the EM field is modulated by the grating. Phase gratings offer large diffraction efficiencies since ideally no losses due to transmitted radiation or absorption are present. Another type of grating that can be conveniently tackled by MEM is the mesh metal grating where the grating is defined by a thin metal film pattern deposited onto the detector [11,12]. The MQW detector structure, on the other hand, is modeled by using a transfer matrix method which describes the optical properties of the detector structure including the IR absorbing MQW layers and the doped contact layers. When the grating scattering matrix and the transfer matrices of the QW structure are known, the EM field may be calculated at any level (y coordinate) in the structure. From this the absorptance η is found.

It is common practice to distinguish between TE (transverse electric) and TM (transverse magnetic) polarization of the incident and diffracted waves which interact with the grating. TE and TM are defined here as having the electric or magnetic field vector, respectively, in the x-z plane (see Fig. 1).

MEM assumes the grating to be of infinite extension and the grating metal of infinite conductivity. The theory can, however, be extended to take account of finite extension (Sec. 2.5). For gratings with more than about ten grating periods approximations can be made. Finite conductivity of the grating metal is taken into account by a perturbation approach which is described in Sec. 2.6.

2.2. Calculation of the Grating Scattering Matrix of Ideal Gratings

2.2.1. The Modal Expansion Method

The modal expansion method (MEM) is based on the space above the grating being separated into two regions, divided by a plane parallel to the grating surface (at $y = 0$) and intersecting the topmost part of its surface: i) the channel (groove) or cavity region ($y < 0$), and ii) the free half space above it ($y > 0$) (see Fig. 1). The next step is to solve Maxwells equations separately in each region. The solutions are expressed as a sum of modal functions taking the appropriate boundary conditions into account. The two sets of solutions are then matched to each other with respect to the E- and H-fields, respectively, at the plane of intersection, which gives a system of equations containing the diffraction amplitudes. This system is finally solved for the amplitudes. The step by step solution will be exemplified below for the scattering matrix of the lamellar grating. The case of the crossed gratings is solved in an analogous manner.

An approximate treatment based on the single mode assumption is presented in Sec. 2.2.5 below. With this method analytic expressions in closed form can be derived, which provides insight into the problem of grating scattering of a both qualitative and quantitative nature.

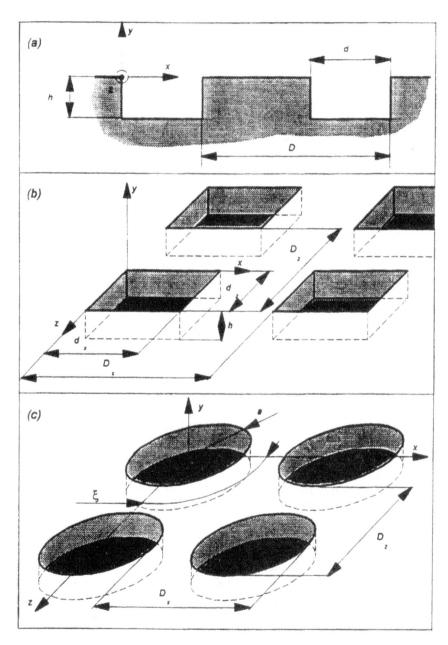

Fig. 1   Definition of the parameters describing the geometry of (a) a lamellar grating, (b) a crossed grating with box-shaped cavities, and (c) a crossed grating with cylindrical cavities.

2.2.2. Scattering Matrix of the Lamellar Grating

The geometry of the lamellar grating [18] with the coordinate system is shown in Fig. 1a. Here only the H polarization gives rise to QW absorption, since only in this case the E field possesses a component in the direction perpendicular to the QW plane, which is required by the QW selection rules. The scattering matrix can be written

$$\underline{S} = [S_p^r], \tag{1}$$

where the elements S_p^r denote the ratio of the pth-order scattered wave amplitude to the rth-order incident wave amplitude. The indexing is done in the following way: The radiation to be detected which enters from the GaAs substrate and the corresponding wave with the same propagation angle, but with opposite propagation direction, are labelled zeroth order. The latter type of wave arises from diffraction by the grating as well as from multiple reflections in the epilayer interfaces. Waves with the same propagation angle as the grating orders -1 and 1 are labelled minus first and plus first order, respectively, and so forth (see Fig. 9). In the free half-space above the grating (for $y > 0$) the TM EM plane waves can be written (expressed as the component of the E field in the x-z plane, or the *transverse* component)

$$R_p(x, y) = 1/\sqrt{D} \exp\left[i(\alpha_p x + \beta_p y)\right]\hat{x}, \tag{2}$$

with

$$\alpha_p = \alpha_0 + (2\pi p/D), \tag{3}$$

$$\alpha_0 = k \sin(\theta), \tag{4}$$

$$\beta_p = \sqrt{k^2 - \alpha_p^2}, \tag{5}$$

where D is the grating constant, θ the angle of incidence, and the wave vector $k = 2\pi/\lambda$, λ being the wavelength in the material. p is the grating order: $(= \ldots -1, 0, 1 \ldots)$ and $\hat{}$ signifies a unit vector. Eqs. (3–5) result from the fact that the boundary conditions enforces the total EM field pattern to possess the same periodicity as the grating in a plane parallel to the grating surface. The functions $R_p(x, 0)$ are orthonormal over the unit cell distance (grating constant $= D$).

The total transverse electric field for $y > 0$, and for unity amplitude of the rth-order incident wave, can be written as a series summation

$$E_t^{r,>} = R_r(x, -y) + \sum_{p=-\infty}^{\infty} S_p^r R_p(x, y), \tag{6}$$

where S_p^r are components of the scattering matrix.

The TM channel ($y < 0$) modal functions, on the other hand, can be written (also expressed as the transverse E field)

$$N_n(x, y) = M_n(x) \sin [u_n(y + h)], \tag{7}$$

with

$$M_n(x) = \sqrt{[(2 - \delta_{n0})/d]} \cos (n\pi x/d) \, \hat{x}, \tag{8}$$

$$u_n = \sqrt{k^2 - (n\pi/d)^2}, \tag{9}$$

where d is the channel width, h the channel depth, δ_{ij} the Kronecker delta, and the channel mode index $n = 0, 1 \ldots$. The functions $N_n(x, 0)$ or $M_n(x)$ are defined over the channel length d only, and constitutes an orthonormal set. The TM_0 channel mode is a TEM mode with both the E- and H-field in the x-z plane.

The total E-field for $y < 0$ can be written:

$$E_t^{r,<} = \sum_{n=0}^{x} c_n^r N_n(x, y) = \sum_{n=0}^{x} c_n^r M_n(x) \sin [u_n(y + h)], \tag{10}$$

where c_n^r are channel mode amplitudes.

In order to find S_p^r (and c_n^r) the boundary conditions of the E- and H-field for $y = 0$ is utilized.

The continuity of the electric field vector in the x-z-plane for $y = 0$ requires

$$E_t^{r,>} = E_t^{r,<}, \tag{11}$$

or:

$$R_r(x, 0) + \sum_{p=-x}^{x} S_p^r R_p(x, 0) = \sum_{n=0}^{x} c_n^r M_n(x) \sin (u_n h). \tag{12}$$

This expression is multiplied by $\overline{R_p(x, 0)}$ and integrated over the unit cell distance (the bar signifies the complex conjugate). Due to orthonormality of the functions $R_p(x, 0)$ one obtains

$$\delta_{rp} + S_p^r = \sum_{n=0}^{x} c_n^r I_n^p \sin (u_n h), \tag{13}$$

where

$$I_n^p = \int_0^d \overline{R_p(x, 0)} M_n(x) dx. \tag{14}$$

From Maxwell's equations the transverse H-field for $y > 0$ becomes

$$\hat{y} \times H_t^{r,>} = \frac{1}{Z_0} \left[\frac{k}{\beta_r} R_r(x, -y) - \sum_{p=-\infty}^{\infty} S_p^r \frac{k}{\beta_p} R_p(x, y) \right], \quad (15)$$

where Z_0 is the impedance of free space. For $y < 0$ the H-field becomes

$$\hat{y} \times H_t^{r,<} = \frac{i}{Z_0} \sum_{n=0}^{\infty} c_n^r \frac{k}{u_n} M_n(x) \cos [u_n(y + h)]. \quad (16)$$

Setting Eq. (15) and (16) equal for $y = 0$ gives

$$\frac{k}{\beta_r} R_r(x, 0) - \sum_{p=-\infty}^{\infty} S_p^r \frac{k}{\beta_p} R_p(x, 0) = i \sum_{N=0}^{\infty} c_N^r \frac{k}{u_N} M_N(x) \cos [u_N h]. \quad (17)$$

This expression is multiplied by $M_n(x)$ and integrated over the cavity length. Due to orthonormality of the functions $M_n(x)$ one obtains

$$\frac{k}{\beta_r} \overline{I_n^r} - \sum_{p=-\infty}^{\infty} S_p^r \frac{k}{\beta_p} \overline{I_n^p} = i c_n^r \frac{k}{u_n} \cos (u_n h), \quad (18)$$

Inserting the expression for S_p^r according to Eq. (13) one gets

$$\frac{k}{\beta_r} \overline{I_n^r} - \sum_{p=-\infty}^{\infty} \left\{ \left[\sum_{N=0}^{\infty} c_N^r I_N^p \sin (u_N h) - \delta_{rp} \right] \frac{k}{\beta_p} \overline{I_n^p} \right\} = i c_n^r \frac{k}{u_n} \cos (u_n h), \quad (19)$$

which can be simplified to

$$2 \frac{k}{\beta_r} \overline{I_n^r} = i c_n^r \frac{k}{u_n} \cos (u_n h) + \sum_{N=0}^{\infty} c_n^r \left\{ \sum_{p=-\infty}^{\infty} \frac{k}{\beta_p} I_N^p \overline{I_n^p} \right\}. \quad (20)$$

Eq. (20) is more conveniently written as

$$\sum_{N=0}^{\infty} C_{Nn} c_N^r = D_n^r, \quad (21)$$

or equivalently in the corresponding matrix form as

$$\underline{C} \cdot \underline{c} = \underline{D}, \quad (22)$$

where

$$C_{Nn} = \sin (u_N h) Q_{Nn} + i \frac{k}{u_N} \cos (u_N h) \delta_{Nn}, \quad (23)$$

$$D_n^r = 2 \frac{k}{\beta_r} \overline{I_n^r}, \quad (24)$$

$$Q_{Nn} = \sum_{p=-\infty}^{\infty} \frac{k}{\beta_p} I_N^p \overline{I_n^p}. \quad (25)$$

Eq. (21) or (22) is the starting point for a numerical calculation. In practice it is necessary to truncate the series to a finite number of terms. In the calculations below $p = -3$ through 3 are made, and the channel orders $n = 0-5$ are included throughout, if not otherwise stated. Eq. (22) is first solved for the unknown channel mode amplitudes c_n^r. The elements of the scattering matrix are then obtained from Eq. (13) or

$$S_p^r = \delta_{rp} + \sum_{n=0}^{\infty} c_n^r I_n^p \sin (u_n h). \tag{26}$$

One obtains for the scalar product

$$I_n^p = \left(\frac{2 - \delta_{n0}}{dD}\right)^{1/2} \frac{i\alpha_p}{\alpha_p^2 - (n\pi/d)^2} [(-1)^n \exp(-i\alpha_p d) - 1], \tag{27}$$

In order to make calculations more tractable only six waves (three incident and three scattered waves) are included in the scattering matrix, i.e., orders $-1, 0$ and 1. This is a reasonable approximation in view of the fact that for quantum well detection the grating operates close to first order cut-off (i.e., $\lambda \approx D$ and the grating orders -1 and 1 are diffracted at an angle close to $90°$).

2.2.3. Scattering Matrix of the Crossed Grating with Box-Shaped Cavities

In contrast to the case of lamellar gratings where one grating order index is sufficient, two indices (p, q) are necessary to denote reflections from a crossed grating [17, 19]. The grating constants are denoted D_x and D_z in the x and z directions, respectively. The cavity widths are d_x and d_z, and h the cavity depth. (see Fig. 1b). The scattering matrix is found in a similar manner to that described for the lamellar grating above. The major difference is that there are several indices to take care of for the crossed grating case. The free space EM modes valid for $y > 0$ are found in Appendix A. They constitute an orthonormal set over the grating unit cell area. The cavity modes for the case of box-shaped cavities are found in Appendix B. These modes are orthonormal over the area defined by the cavity opening. In order to realize the correspondence between the crossed and the lamellar grating case a similar notation has been used as far as possible. The scattering matrix is written

$$\underline{S} = [S_{pq,W2}^{rs,W1}], \tag{28}$$

where $(W1, W2 = \text{TE}, \text{TM mode})$, the superscripts $(rs, W1)$ denote the incident order and the subscripts $(pq, W2)$ the scattered order.

From the continuity of the total transverse E-field at $y = 0$ one obtains

$$S_{pq,\text{TE}}^{rs,W1} = -\delta_{rp}\delta_{sq}\delta_{W1,\text{TE}} + \sum_{m,n=0}^{\infty} (a_{nm}^{rs,W1} IEE_{nm}^{pq}) \sin(u_{nm}h), \qquad (29)$$

$$S_{pq,\text{TM}}^{rs,W1} = -\delta_{rp}\delta_{sq}\delta_{W1,\text{TM}}$$

$$+ \sum_{m,n=0}^{\infty} (a_{nm}^{rs,W1} IME_{nm}^{pq} + c_{nm}^{rs,W1} IMM_{nm}^{pq}) \sin(u_{nm}h), \qquad (30)$$

where $\delta_{W1,W2}$ is defined to be 1 when $W1 = W2 = (\text{TE or TM})$ otherwise zero, and

$$u_{nm} = \sqrt{k^2 - \left(\frac{n\pi}{d_x}\right)^2 - \left(\frac{m\pi}{d_z}\right)^2}, \qquad (31)$$

$((n, m) \neq (0, 0)$ in the summations. For the TM modes n or $m \neq 0)$.

The $a_{nm}^{rs,W1}$ and $c_{nm}^{rs,W1}$ are the TE and TM cavity mode amplitudes, respectively, and correspond to c_n^r for the lamellar grating TM modes. The quantities IEE_{nm}^{pq}, IME_{nm}^{pq}, and IMM_{nm}^{pq} are scalar products of the free space and cavity modes and are explicitly defined in Appendix B. They correspond to I_n^p of the lamellar grating. Since $IEM_{nm}^{pq} = 0$ it does not enter in the equations.

From the continuity of the H-field for $y = 0$ within the cavity region, the $a_{nm}^{rs,W1}$ and $c_{nm}^{rs,W1}$ can be found. One ends up with the following system of equations:

$$\sum_{NM=0}^{\infty} A_{NMnm}^{\nu} a_{NM}^{rs,W1} + C_{NMnm}^{\nu} c_{NM}^{rs,W1} = D_{nm}^{\nu,rs,W1}. \qquad (32)$$

with $\nu = 1, 2$ and $n, m = 0$ to ∞, and where

$$A_{NMnm}^{\nu} = \sin(u_{NM}h) P_{NMnm}^{\nu} + i\frac{u_{NM}}{k} \cos(u_{NM}h) \delta_{Nn}\delta_{Mm}\delta_{\nu,1}, \qquad (33)$$

$$C_{NMnm}^{\nu} = \sin(u_{NM}h) Q_{NMnm}^{\nu} + i\frac{k}{u_{NM}} \cos(u_{NM}h) \delta_{Nn}\delta_{Mm}\delta_{\nu,2}, \qquad (34)$$

$$P_{NMnm}^{1} = \sum_{p,q=-\infty}^{\infty} \left(\frac{\beta_{pq}}{k} IEE_{NM}^{pq} \overline{IEE_{nm}^{pq}} + \frac{k}{\beta_{pq}} IME_{NM}^{pq} \overline{IME_{nm}^{pq}}\right), \qquad (35)$$

$$P^2_{NMnm} = \sum_{p,q=-\infty}^{\infty} \left(\frac{k}{\beta_{pq}} IME^{pq}_{NM} \overline{IMM^{pq}_{nm}} \right), \tag{36}$$

$$Q^1_{NMnm} = \sum_{p,q=-\infty}^{\infty} \left(\frac{k}{\beta_{pq}} IMM^{pq}_{NM} \overline{IME^{pq}_{nm}} \right), \tag{37}$$

$$Q^2_{NMnm} = \sum_{p,q=-\infty}^{\infty} \left(\frac{k}{\beta_{pq}} IMM^{pq}_{NM} \overline{IMM^{pq}_{nm}} \right), \tag{38}$$

$$D^{1,rs,TE}_{nm} = 2\frac{\beta_{rs}}{k} \overline{IEE^{rs}_{nm}}, \tag{39}$$

$$D^{2,rs,TE}_{nm} = 0, \tag{40}$$

$$D^{1,rs,TM}_{nm} = 2\frac{k}{\beta_{rs}} \overline{IME^{rs}_{nm}}, \tag{41}$$

$$D^{2,rs,TM}_{nm} = 2\frac{k}{\beta_{rs}} \overline{IMM^{rs}_{nm}}, \tag{42}$$

(see Eq. (A7) for the expression for β_{pq}). In the numerical calculations the following grating orders are included in the scattering matrix: $(p,q) = (0,0), (-1,0), (1,0), (0,-1), (0,1)$ for both TE and TM modes. This is a reasonable approximation when $\lambda \approx D$, D being the grating period.

2.2.4. Scattering Matrix of the Crossed Grating with Cylindrical Cavities

The scattering matrix for crossed gratings with cylindrical cavities is derived in the same fashion as for box-shaped cavities [17, 20, 21]. One obtains

$$S^{rs,W1}_{pq,TE} = -\delta_{rp}\delta_{sq}\delta_{W1,TE} + \sum_{m,n=0}^{\infty}\sum_{l=1}^{2} (a^{rs,W1}_{nml} IEE^{pq}_{nml}) \sin(w'_{nm}h), \tag{43}$$

$$S^{rs,W1}_{pq,TM} = -\delta_{rp}\delta_{sq}\delta_{W1,TM} + \sum_{m,n=0}^{\infty}\sum_{l=1}^{2} a^{rs,W1}_{nml} IME^{pq}_{nml}[\sin(w'_{nm}h)$$

$$+ c^{rs,W1}_{nml} IMM^{pq}_{nml} \sin(w_{nm}h)], \quad ((n,m) \neq (0,0)) \tag{44}$$

with

$$w_{nm} = \sqrt{k^2 - \left(\frac{\chi_{nm}}{a}\right)^2}, \tag{45}$$

$$w'_{nm} = \sqrt{k^2 - \left(\frac{\chi'_{nm}}{a}\right)^2}, \tag{46}$$

where χ_{nm} is the mth zero of the Bessel function of order $n, J_n(x)$, whereas χ'_{nm} is the mth zero of the first $(x-)$ derivative of the corresponding Bessel function $J'_n(x)$, and a the cavity radius.

The TE and TM cavity mode amplitudes, $a^{rs,W1}_{nml}$ and $c^{rs,W1}_{nml}$, respectively, can be found from the following system of equations:

$$\sum_{N,M=0}^{\infty} \sum_{L=1}^{2} (A^v_{NMLnml} a^{rs,W1}_{NML} + C^v_{NMLnml} c^{rs,W1}_{NML}) = D^{v,rs,W1}_{nml}, \tag{47}$$

with

$$(n, m, l, v) = (0, 1, 1, 1) - (\infty, \infty, 2, 2);$$

$$A^v_{NMLnml} = \sin(w'_{nm}h) P^v_{NMLnml} + i\frac{w'_{NM}}{k} \cos(w'_{NM}h)\delta_{Nn}\delta_{Mm}\delta_{Ll}\delta_{v,1}, \tag{48}$$

$$C^v_{NMLnml} = \sin(w_{nm}h) Q^v_{NMLnml} + i\frac{w_{NM}}{k} \cos(w_{NM}h)\delta_{Nn}\delta_{Mm}\delta_{Ll}\delta_{v,2}, \tag{49}$$

where the expressions for P^v_{NMLnml} and Q^v_{NMLnml} are analogous to Eqs. (35) − (38) and are given in Appendix C,

$$D^{1,rs,TE}_{nml} = 2\frac{\beta_{rs}}{k}\overline{IEE^{rs}_{nml}}, \tag{50}$$

$$D^{2,rs,TE}_{nml} = 0, \tag{51}$$

$$D^{1,rs,TM}_{nml} = 2\frac{k}{\beta_{rs}}\overline{IME^{rs}_{nml}}, \tag{52}$$

$$D^{2,rs,TM}_{nml} = 2\frac{k}{\beta_{rs}}\overline{IMM^{rs}_{nml}}. \tag{53}$$

Two different grating symmetries are considered in the numerical calculations below due to their high symmetry: square and hexagonal. The angle ξ between x and z axes (see Appendix A) is 90° and 60°, respectively, for these symmetries.

The number of grating orders included is the same for the case of cylindrical cavities as for box-shaped cavities. This is again a good approximation near first order grating cut-off.

2.2.5. Approximative Treatment of Grating Scattering: The Single Mode Approximation

2.2.5.1. Background   When operating the grating above second order cut-off, i.e., at $\lambda > D/2$, the lowest order channel mode TM_0 for the lamellar grating and the TE_{10} and TE_{01} modes for the crossed grating with box-shaped cavities dominate over higher order modes [22]. Consequently, the amplitudes of all modes except the fundamental mode or modes can be set to zero (single mode approximation). This considerably simplifies the theoretical treatment and expressions in closed form are possible, which may provide insight into the problem. The TM_0 fundamental mode of the lamellar grating, which is a TEM mode, is exceptional since it never becomes evanescent even for small channel widths (see Eq. (7–9)). The first excited mode is the TM_1 mode which is propagating for $\lambda < 2d$ or $\lambda/D < 2d/D$. The grating detector operates at about $\lambda \approx (0.7–0.9)\cdot D$ and $d/D \approx 0.5$. This implies that only the TEM mode is propagating. However for large values of d/D and λ/D the approximation is questionable. For crossed gratings, no TEM mode exists, and all cavity modes including the TE_{10} and TE_{01} lowest order modes are evanescent for $\lambda > 2d$. Complete square symmetry is assumed in the following treatment in which case both of these modes are degenerate. The TE_{11} and TM_{11} modes start to propagate when $\lambda < \sqrt{2}d$. The single mode approximation is consequently good in the region where all modes except possibly the TE_{10} and TE_{01} modes are propagating, or differently expressed when $\lambda > \sqrt{2}d$.

2.2.5.2. Lamellar Gratings   Starting with the lamellar grating all cavity mode amplitudes c_n are assumed to be zero except c_0. As a result Eq. (21) is converted into

$$C_{00}c_0 = D_0, \tag{54}$$

Using

$$C_{00} = \sin(k\,h)\,Q_{00} + i\cos(k\,h), \tag{55}$$

$$D_0^r = 2\frac{k}{\beta_r}\overline{I_0^r}, \tag{56}$$

with the assumption that the incident radiation enters from the rth direction, one obtains when solving for the cavity amplitude that

$$c_0 = 2\left(\frac{k}{\beta_r}\right)\frac{I_0^r}{\sin(k\,h)\,Q_{00} + i\cos(k\,h)}. \tag{57}$$

The scattering matrix becomes using Eq. (26)

$$S_p^r = \delta_{rp} + 2\left(\frac{k}{\beta_r}\right)\frac{I_0^r I_0^p}{Q_{00} + i\cot(k\,h)}, \tag{58}$$

where

$$Q_{00} = \sum_{p=-\infty}^{\infty} \frac{k}{\beta_p}|I_0^p|^2. \tag{59}$$

Considering the dominant scattering matrix element for grating detector operation, i.e., the 0th to the 1st (or $-$ 1st) order, S_1^0(or S_{-1}^0), one obtains using

$$I_0^0 = \sqrt{d/D}, \tag{60}$$

$$I_0^1 = \frac{i}{2\pi}\sqrt{\frac{D}{d}}\left[\exp\left(-i\frac{2\pi d}{D}\right) - 1\right], \tag{61}$$

that

$$S_1^0 = \frac{1}{\pi}\cdot\frac{\left[\exp\left(-i\dfrac{2\pi d}{D}\right) - 1\right]}{\cot\left(\dfrac{2\pi h}{\lambda}\right) - iQ_{00}}. \tag{62}$$

From this expression the corresponding diffraction intensity in the y direction I_y can be found. Since according to Sec. 2.3

$$I_y = \frac{k}{\beta_1}|S_1^0|^2, \tag{63}$$

one obtains

$$I_y = \frac{4}{\pi^2}\cdot\frac{1 - \cos\left(\dfrac{2\pi d}{D}\right)}{\left[\operatorname{Im} Q_{00} + \cot\left(\dfrac{2\pi h}{\lambda}\right)\right]^2 + (\operatorname{Re} Q_{00})^2}\cdot\frac{1}{\sqrt{1 - (\lambda^2/D^2)}}. \tag{64}$$

If diffracted waves of 2nd order and higher are evanescent one has

$$\text{Re}(Q_{00}) = \frac{d}{D} + \frac{D}{d\pi^2}\left[1 - \cos\left(\frac{2\pi d}{D}\right)\right]\cdot\frac{1}{\sqrt{1 - (\lambda^2/D^2)}}, \qquad (65)$$

$$\text{Im}(Q_{00}) = -\frac{D}{d\pi^2}\sum_{p=2}^{\infty}\frac{1}{p^2}\left[1 - \cos\left(\frac{2\pi pd}{D}\right)\right]\frac{1}{\sqrt{(p^2\lambda^2/D^2) - 1}}. \qquad (66)$$

It should be noted that the diffraction properties of the grating can be expressed in terms of the dimensionless parameters $\lambda/D, d/D$ and h/λ. This also holds true for the near-exact theory presented above.

Results from the approximate treatment and the near-exact theory coincide except when $\lambda/D < 2d/D$. Some interesting conclusions may be drawn from the equations. From Eq. (64) the diffracted intensity is maximized for

$$\cot\left(\frac{2\pi h}{\lambda}\right) = -\text{Im}\,Q_{00}, \qquad (67)$$

which shows that the intensity is periodic in h with a period of $\lambda/2$. The maximum scattering intensity is obtained when $d/D \approx 0.6$. One obtains from Eq. (67) that in a broad range of $\lambda/D, h/\lambda \approx 0.25$. However, the optimum values of d/D and h/λ for the grating detector presented in part D is about 0.5 and 0.12, respectively. The discrepancy is due to that in the latter case the intensity integrated with respect to wavelength or photon energy is important, not the peak value.

A comparison between the near-exact and approximative theories is presented for the lamellar grating in Figs. 2–4. Evidently the near-exact and approximate theories accord except for the case stated above.

2.2.5.3. *Crossed Gratings*   The theory of *crossed gratings* with square symmetry is significantly simplified if only the cavity modes TE_{10} and TE_{01} are included and all others set to zero. We assume here box-shaped cavities with square cross-section, in which case both modes are degenerate. Eq. (32) is transformed into

$$\begin{bmatrix} A^1_{1010} & A^1_{1001} \\ A^1_{0110} & A^1_{0101} \end{bmatrix}\cdot\begin{bmatrix} a^{rs,W1}_{10} \\ a^{rs,W1}_{01} \end{bmatrix} = \begin{bmatrix} D^{1,rs,W1}_{10} \\ D^{1,rs,W1}_{01} \end{bmatrix}, \qquad (68)$$

with $W1 = \text{TE}$ or TM as before.

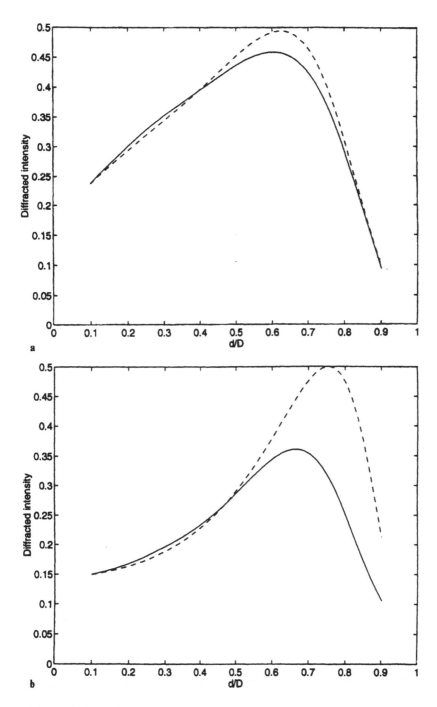

Fig. 2 The TM diffracted intensity in the -1st and $+1$st orders of a lamellar grating assuming *unpolarized* incident radiation in the 0th order direction, vs. the aspect ratio $d/D.h/\lambda = 0.27$. a) $\lambda/D = 0.9$, b) $\lambda/D = 0.985$. The full and dashed curves refer to the near-exact and approximate theory, respectively.

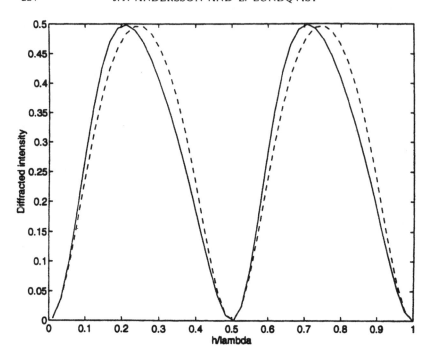

Fig. 3 The TM diffracted intensity in the −1st and +1st orders of a lamellar grating assuming *unpolarized* incident radiation in the 0th order direction, vs. h/λ. $\lambda/D = 0.9$ and $d/D = 0.6$. The full and dashed curves refer to the near-exact and approximate theory, respectively.

It turns out that

$$A^1_{1010} = A^1_{0101} = \sin(uh)P + i\frac{u}{k}\cos(uh), \tag{69}$$

$$A^1_{1001} = A^1_{0110} = 0, \tag{70}$$

where

$$u = u_{10} = u_{01} = \sqrt{k^2 - \left(\frac{\pi}{d}\right)^2}, \tag{71}$$

and

$$P = P^1_{1010} = P^1_{0101} = \sum_{p,q=-\infty}^{\infty}\left\{\frac{\beta_{pq}}{k}|IEE^{pq}_{10}|^2 + \frac{k}{\beta_{pq}}|IME^{pq}_{10}|^2\right\}. \tag{72}$$

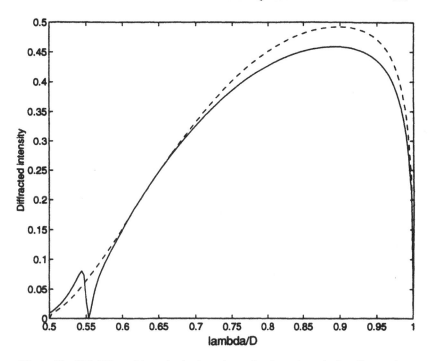

Fig. 4 The TM diffracted intensity in the -1st and $+1$st orders of a lamellar grating assuming unpolarized incident radiation in the 0th order direction, vs. λ/D. $d/D = 0.6$ and $h/\lambda = 0.27$. The full and dotted curves refer to the near-exact and approximate theory, respectively.

The cavity modes become

$$
\begin{bmatrix} a_{10}^{rs,W1} \\ a_{01}^{rs,W1} \end{bmatrix} = \frac{1}{\sin(uh)P + i\dfrac{u}{k}\cos(uh)} \begin{bmatrix} D_{10}^{1,rs,W1} \\ D_{01}^{1,rs,W1} \end{bmatrix}.
\tag{73}
$$

From Eqs. (29–30) the scattering matrix elements become

$$
S_{pq,\mathrm{TE}}^{rs,W1} = -\delta_{rp}\delta_{sq}\delta_{W1,\mathrm{TE}} + (a_{10}^{rs,W1} IEE_{10}^{pq} + a_{01}^{rs,W1} IEE_{01}^{pq})\sin(uh),
\tag{74}
$$

$$
S_{pq,\mathrm{TM}}^{rs,W1} = -\delta_{rp}\delta_{sq}\delta_{W1,\mathrm{TM}} + (a_{10}^{rs,W1} IME_{10}^{pq} + a_{01}^{rs,W1} IME_{01}^{pq})\sin(uh),
\tag{75}
$$

which are transformed into

$$
S_{pq,\mathrm{TE}}^{rs,\mathrm{TE}} = -\delta_{rp}\delta_{sq} + 2\frac{\beta_{rs}}{k}\frac{(\overline{IEE_{10}^{rs}}\, IEE_{10}^{pq} + \overline{IEE_{01}^{rs}}\, IEE_{01}^{pq})}{P + i\dfrac{u}{k}\cot(uh)},
\tag{76}
$$

$$S_{pq,\text{TE}}^{rs,\text{TM}} = 2\frac{k}{\beta_{rs}}\frac{(\overline{IME_{10}^{rs}}\,IEE_{10}^{pq} + \overline{IME_{01}^{rs}}\,IEE_{01}^{pq})}{P + i\frac{u}{k}\cot(uh)}, \qquad (77)$$

$$S_{pq,\text{TM}}^{rs,\text{TE}} = 2\frac{\beta_{rs}}{k}\frac{(\overline{IEE_{10}^{rs}}\,IME_{10}^{pq} + \overline{IEE_{01}^{rs}}\,IME_{01}^{pq})}{P + i\frac{u}{k}\cot(uh)}, \qquad (78)$$

$$S_{pq,\text{TM}}^{rs,\text{TM}} = -\delta_{rp}\delta_{sq} + 2\frac{k}{\beta_{rs}}\frac{(\overline{IME_{10}^{rs}}\,IME_{10}^{pq} + \overline{IME_{01}^{rs}}\,IME_{01}^{pq})}{P + i\frac{u}{k}\cot(uh)}. \qquad (79)$$

Considering the scattering matrix elements from $(r,s)=(0,0)$, TE to either $(p,q)=(0,1)$, TE or to $(p,q)=(1,0)$, TM, which are of major importance for grating detector operation one obtains

$$S_{10,\text{TE}}^{00,\text{TE}} = \frac{2\cdot\overline{IEE_{10}^{00}}}{P + i\frac{u}{k}\cot(uh)}\cdot IEE_{10}^{10}, \qquad (80)$$

$$S_{10,\text{TM}}^{00,\text{TE}} = \frac{2\cdot\overline{IEE_{10}^{00}}}{P + i\frac{u}{k}\cot(uh)}\cdot IME_{10}^{01}, \qquad (81)$$

$$S_{10,\text{TM}}^{00,\text{TE}} = S_{01,\text{TE}}^{00,\text{TE}} = 0 \qquad (82)$$

$((0,0)$, TE is defined to be polarized with the E field in the z direction). The scattering intensity is obtained from (see Sec. 2.3)

$$I_{y,\text{TE}} = \frac{\beta_{pq}}{k}|S_{10,\text{TE}}^{00,\text{TE}}|^2, \qquad (83)$$

$$I_{y,\text{TM}} = \frac{k}{\beta_{pq}}|S_{01,\text{TM}}^{00,\text{TE}}|^2. \qquad (84)$$

Since

$$|IEE_{10}^{00}|^2 = \frac{8d^2}{\pi^2 D^2}, \qquad (85)$$

$$|IEE_{10}^{10}|^2 = \frac{4}{\pi^2}\cdot\frac{1+\cos(2\pi d/D)}{(4d/D - D/d)^2}, \qquad (86)$$

$$|IME_{10}^{01}|^2 = \frac{4}{\pi^4}\cdot[1-\cos(2\pi d/D)], \qquad (87)$$

one ends up with

$$I_{y,\mathrm{TE}} = \frac{128}{\pi^4} \cdot \frac{1 + \cos(2\pi d/D)}{(4 - (D^2/d^2))^2} \cdot \frac{1}{R} \sqrt{1 - \frac{\lambda^2}{D^2}}, \tag{88}$$

$$I_{y,\mathrm{TM}} = \frac{128}{\pi^6} \cdot \frac{d^2}{D^2} \left(1 - \cos\left(\frac{2\pi d}{D}\right)\right) \cdot \frac{1}{R} \cdot \frac{1}{\sqrt{1 - (\lambda^2/D^2)}}, \tag{89}$$

where

$$R = \left| P + i \sqrt{1 - \frac{\lambda^2}{4d^2}} \cdot \cot\left(\frac{2\pi h}{\lambda} \sqrt{1 - \frac{\lambda^2}{4d^2}}\right) \right|^2. \tag{90}$$

The diffraction intensity is maximum if R is minimum. The optimum h is found from

$$\sqrt{1 - \frac{\lambda^2}{4d^2}} \cdot \cot\left(\frac{2\pi h}{\lambda} \sqrt{1 - \frac{\lambda^2}{4d^2}}\right) = -\operatorname{Im} P, \tag{91}$$

and when $\lambda < 2d$ the intensity is periodic in h with a period depending on the d/D and λ/D ratios. $d/D = 0.7$ and $\lambda/D = 0.9$ gives $\operatorname{Im} P = -3.5 \cdot 10^{-2}$ which in turn gives an optimum $h/\lambda = 0.31$ (the periodicity in h/λ is 0.65). This value is close to the value which optimizes the average absorptance of grating detectors. It should be noted that since the cavity mode wave vector u depends on λ/d and may even be zero this makes the grating scattering insensitive to variations in h or h/λ, in contrast to the lamellar grating. The latter may result in the diffracted intensity being close to optimum in a broad wavelength region, which may be utilized in grating detectors.

Eq. (91) is also valid for the case of evanescent TE_{10} and TE_{01} modes ($\lambda > 2d$). However, in this case no periodicity in h results.

Figure 5–7 present the diffracted intensity vs. the gratings parameters and wavelength. The accordance between the near-exact and approximate theories is satisfying.

An important feature of the crossed grating making coupling to QW layers very efficient is the preference for TM to TE polarized diffracted radiation. This preference can be expressed by the enhancement ratio $ER = I_{y,\mathrm{TM}}/I_{y,\mathrm{TE}}$. The approximate theory gives

$$\frac{I_{y,\mathrm{TM}}}{I_{y,\mathrm{TE}}} = \frac{d^2[1 - \cos(2\pi d/D)]}{\pi^2 D^2[1 + \cos(2\pi d/D)]} \cdot \frac{1}{1 - (\lambda^2/D^2)}. \tag{92}$$

Assuming again $d/D = 0.7$ and $\lambda/D = 0.9$ leads to $ER = 1.9$. For $\lambda/D = 0.95$ one obtains $ER = 3.7$. The near-exact theory gives $ER = 1.5$

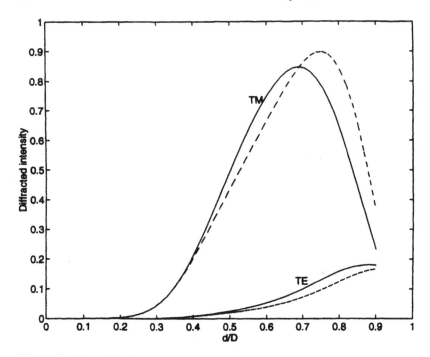

Fig. 5   The TE and TM diffracted intensity in the $(1, 0)$-orders of a crossed grating with box-shaped cavities assuming *unpolarized* incident radiation in the 0th order direction, vs. the aspect ratio d/D. $h/\lambda = 0.33$, $\lambda/D = 0.985$. The full and dashed curves refer to the near-exact and approximate theory, respectively.

in this case. When λ/D increases, ER increases to reach a value of about 40 near first order grating cut-off $(\lambda/D = 1)$ according to the near-exact theory (the approximate theory erroneously gives infinity in this limit).

Figure 8 depicts ER vs. the diffraction angle $= \arcsin(\lambda/D)$. Results are shown for the near-exact as well as for the approximate theory.

2.3. Calculation of Absorption Quantum Efficiency

2.3.1. Transfer Matrices and the Coupling of the Grating to the MQW Structure

Matrix techniques are suitable for modeling wave propagation in multilayer structures in which the optical properties of the medium

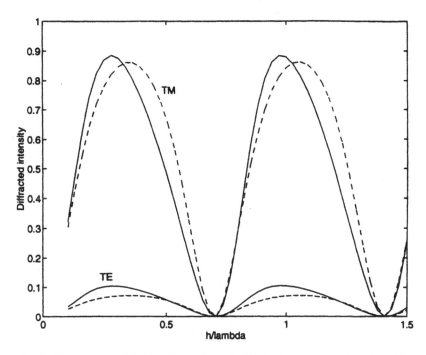

Fig. 6 The TE and TM diffracted intensity in the (1,0)-orders of a crossed grating with box-shaped cavities assuming *unpolarized* incident radiation in the 0th order direction, vs. h/λ. $\lambda/D = 0.985$ and $d/D = 0.7$. The full and dashed curves refer to the near-exact and approximate theory, respectively.

varies only in one direction (y). If the medium is linear coupling exists only between pairs of plane wave modes connected by reflection in the layer planes, as is evident from Fig. 9. Transfer matrices of the type dealt with here are of size 2×2 and act on two-wave vectors $(b^s_-, b^s_+)^t$ where b^s_- and b^s_+ are proportional to the optical field propagating in the $-y$ and $+y$ direction, respectively (t = transpose). The index s denotes all relevant parameters needed for the definition of the plane EM wave: angle of propagation, polarization (TE or TM mode) etc. One has

$$\begin{bmatrix} b^s_- \\ b^s_+ \end{bmatrix}_k = \begin{bmatrix} T^s_{11} & T^s_{12} \\ T^s_{21} & T^s_{22} \end{bmatrix} \cdot \begin{bmatrix} b^s_- \\ b^s_+ \end{bmatrix}_j, \tag{93}$$

where the index k denotes a location at a larger y than does j, as is also evident from Fig. 9. Transfer matrices of a layer sequence are

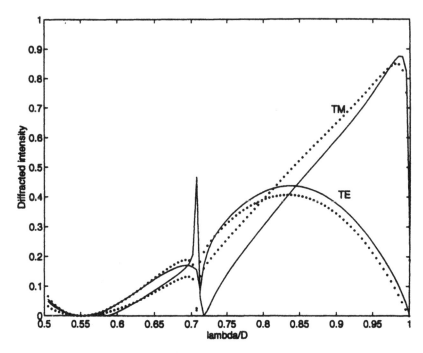

Fig. 7 The TE and TM diffracted intensity in the $(1, 0)$-orders of a crossed grating with box-shaped cavities assuming *unpolarized* incident radiation in the 0th order direction, vs. λ/D. $d/D = 0.7$ and $h/\lambda = 0.33$. The full and dotted curves refer to the near-exact and approximate theory, respectively.

obtained by a successive matrix multiplication of the matrices corresponding to single layers or interfaces. One distinguishes between reflection matrices that describes the transfer through an interface, and bulk transfer matrices that model the transfer through a single bulk layer of constant composition. Absorption is taken care of by using a complex refractive index. Since infrared absorption in the QW layers is anisotropic, different refractive indices for TE and TM mode are utilized.

A reflection matrix is written

$$R_{kj}^{W} = \frac{1}{t_{kj}^{W}} \begin{bmatrix} 1 & r_{kj}^{W} \\ r_{kj}^{W} & 1 \end{bmatrix}, \tag{94}$$

where $W = $ TE or TM.

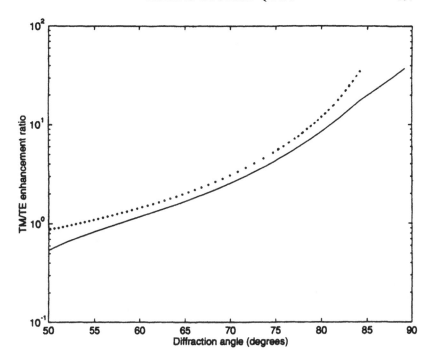

Fig. 8 TE/TM enhancement ratio vs. diffraction angle $= \arcsin(\lambda/D)$. $d/D = 0.7$ and $h/\lambda = 0.33$. The full and dotted curves refer to the near-exact and approximate theory, respectively.

The reflection and transmission coefficients are written

$$r_{kj}^{TE} = \frac{n_k \cos\theta_k - n_j \cos\theta_j}{n_k \cos\theta_k + n_j \cos\theta_j}, \tag{95}$$

$$t_{kj}^{TE} = 1 + r_{kj}^{TE}, \tag{96}$$

$$r_{kj}^{TM} = \frac{n_k \cos\theta_j - n_j \cos\theta_k}{n_j \cos\theta_k + n_k \cos\theta_j}, \tag{97}$$

$$t_{kj}^{TM} = 1 + r_{kj}^{TM}. \tag{98}$$

The bulk transfer matrix is written

$$B_j^W = \begin{bmatrix} \exp(-i\Phi_j) & 0 \\ 0 & \exp(i\Phi_j) \end{bmatrix}, \tag{99}$$

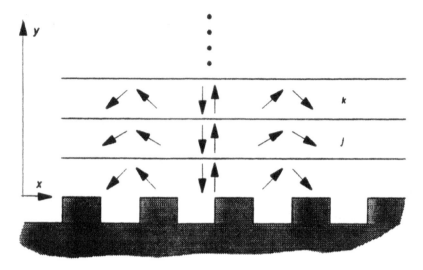

Fig. 9 The electromagnetic wave pattern of a detector consisting of the grating and multilayer epistructure. The arrows represent plane waves, their propagation direction indicated by the direction of the arrow. Only the waves of order -1, 0, and 1 are indicated. The orders correspond to the indices p and r of Eq. [1].

with

$$\Phi_j = n_j \frac{2\pi}{\lambda_0} \cdot b \cdot \cos\theta_j, \tag{100}$$

where n_j is the refractive index of layer number j, λ_0 the vacuum wavelength and b the layer thickness. The angle of propagation θ_j in layer j can be found from $\sin\theta_j = n_{\text{ref}}/n_j \sin\theta_{\text{ref}}$. The index ref denotes a reference layer, e.g., the semiinsulating GaAs substrate.

From the grating scattering matrix and the transfer matrices of the MQW structure the complete wave pattern in the structure may be calculated as follows (exemplifying with the lamellar grating): The incident and scattered wave field vector, respectively, close to a lamellar grating, become, if only orders -1, 0, and 1 are taken into account

$$\mathbf{b}_- = (b_-^{-1}, b_-^0, b_-^1)^t, \tag{101}$$

$$\mathbf{b}_+ = (b_+^{-1}, b_+^0, b_+^1)^t. \tag{102}$$

From the definition of the scattering matrix one obtains

$$\mathbf{b}_+ = S_- \cdot \mathbf{b}_-,\tag{103}$$

In a similar way as for $\mathbf{b}_-, \mathbf{b}_+$ denoting a location close to the grating, the vectors $\mathbf{d}_-, \mathbf{d}_+$ refer to a location in the gallium arsenide substrate. If the radiation enters the detector only through the zeroth order, which is normally the case, one has

$$\mathbf{d}_- = (0, 1, 0)^t.\tag{104}$$

From the definition of the transfer matrix of the complete detector structure one obtains

$$d_-^p = T_{11}^p b_-^p + T_{12}^p b_+^p.\tag{105}$$

From Eq. (103)

$$b_+^p = \sum_r S_p^r b_-^r,\tag{106}$$

which is inserted into Eq. (105) giving

$$d_-^p = \sum_r \{T_{11}^p \delta_{pr} + T_{12}^p S_p^r\} b_-^r = \sum_r M_{pr} b_-^r,\tag{107}$$

or in matrix form

$$\mathbf{d}_- = M \cdot \mathbf{b}_-.\tag{108}$$

Since $\mathbf{d}_-$ is known, $\mathbf{b}_-$ can be solved for by inverting (inv) the matrix $\underline{M}$ according to

$$\mathbf{b}_- = \mathrm{inv}(M) \cdot \mathbf{d}_-,\tag{109}$$

with

$$M_{pr} = T_{11}^p \delta_{pr} + T_{12}^p S_p^r.\tag{110}$$

This expression is easily generalized to the case of crossed gratings if the meaning of p(and similarly for r) is generalized to signify (p, q, W) where (p, q) denotes the two indices needed for modeling diffraction by crossed gratings, and $W = \mathrm{TE}$ or TM.

When $\mathbf{b}_-, \mathbf{b}_+$ are known by applying Eqs. (109) and (103), the optical field can easily be calculated at any point in the detector structure by letting transfer matrices T_p^s act on the two-wave vectors $(b_-^s, b_+^s)^t$ obtained from $\mathbf{b}_-, \mathbf{b}_+$. T_p^s is a partial transfer matrix, i.e., pertains only to the part of the detector structure from the grating surface to the level in question. From the values of the optical field the power density can be found at any point by the calculation of Poynting's vector, and from this the absorptance is easily extracted as discussed in the next section.

It is assumed in the calculations that the refractive indices of GaAs $n_{GaAs} = 3.3$, and of $Al_xGa_{1-x}As$, $n_{AlGaAs} = \sqrt{10.89 - 2.73x}$. A Drude model is used to model the doped GaAs contact layers.

The MQW stack is modeled in the following way taken into account the anisotropic infrared absorption. The complex dielectric constant of the QWs is taken as

$$\varepsilon_\perp = \varepsilon_x + \frac{f\omega_p^2 \varepsilon_0}{\omega_t^2 - \omega^2 - i\gamma\omega}, \tag{111}$$

where $\varepsilon_\perp$ is the perpendicular part of the dielectric function, ε_x the high-frequency dielectric constant ($= 10.89 \cdot \varepsilon_0$), f the oscillator strength, $\omega_p^2 = n_s q^2 / \varepsilon_0 m^* a_{QW}$, where ω_p is the plasma frequency, n_s the sheet carrier concentration, a_{QW} the QW width, m^* the effective mass, ε_0 the vacuum permittivity, ω_t the transition frequency and γ the relation rate. For the parallel part of the dielectric function $\varepsilon_\parallel$, the dielectric function of doped GaAs is assumed, implying that the Drude model is used.

The TM mode transmission matrices for QWs are due to anisotropy somewhat different than for isotropic materials which has been assumed in Eqs. (95)–(100). If

$$n_\perp = \sqrt{\varepsilon_\perp/\varepsilon_0} \quad \text{and} \quad n_\parallel = \sqrt{\varepsilon_\parallel/\varepsilon_0},$$

$$\Phi_{TE} = n_\parallel \frac{2\pi}{\lambda_0} \sqrt{1 - \left(\frac{n_{ref}}{n_\parallel} \sin\theta_{ref}\right)^2} \, a_{QW}, \tag{112}$$

$$\Phi_{TM} = n_\parallel \frac{2\pi}{\lambda_0} \sqrt{1 - \left(\frac{n_{ref}}{n_\perp} \sin\theta_{ref}\right)^2} \, a_{QW}, \tag{113}$$

for the phase angles Φ of the bulk transmission matrix.

The reflection and transmission coefficients needed for the reflection matrices become for TE mode if index k denotes the QW layer

$$r_{kj}^{TE} = \frac{n_k \cos\theta_k - n_\parallel \cos\theta_j}{n_k \cos\theta_k + n_\parallel \cos\theta_j}, \tag{114}$$

$$t_{kj}^{TE} = 1 + r_{kj}^{TE}, \tag{115}$$

where

$$\cos\theta_j = \sqrt{1 - \left(\frac{n_{\text{ref}}}{n_{\|}}\sin\theta_{\text{ref}}\right)^2}, \tag{116}$$

and for TM mode

$$r_{kj}^{\text{TM}} = \frac{n_k\cos\theta_k - n_{\|}\cos\theta_j}{n_k\cos\theta_k + n_{\|}\cos\theta_j}, \tag{117}$$

$$t_{kj}^{\text{TM}} = 1 + r_{kj}^{\text{TM}}, \tag{118}$$

where

$$\cos\theta_j = \sqrt{1 - \left(\frac{n_{\text{ref}}}{n_{\perp}}\sin\theta_{\text{ref}}\right)^2}. \tag{119}$$

2.3.2. Calculation of Quantum Efficiency

The absorptance is found from the Poynting vector [22]. The power density in the y direction becomes

$$I_y = \frac{1}{2}\text{Re}\{E \times \bar{H}\}_y, \tag{120}$$

which can be simplified according to

$$\{E \times \bar{H}\}_y = (E \times \bar{H})\cdot\hat{y} = -E_t\cdot\overline{(\hat{y} \times H_t)}. \tag{121}$$

Consider pairs of plane wave modes connected by reflection in the layer planes perpendicular to the y direction, as discussed above. Only TM polarized waves are taken into account since only these give rise to QW absorption. The transverse electric field is written

$$E_t(x, y, z) = b_+(y)\psi_+(x, z) + b_-(y)\psi_-(x, z). \tag{122}$$

Since for reflected pairs $\psi_+(x, z) = \psi_-(x, z) = \psi(x, z)$ one obtains

$$E_t(x, y, z) = [b_+(y) + b_-(y)]\psi(x, z). \tag{123}$$

The transverse magnetic field can be expressed

$$\hat{y} \times H_t(x, y, z) = -\frac{1}{Z_0}\left[\frac{k}{\beta_+}b_+(y) + \frac{k}{\beta_-}b_-(y)\right]\psi(x, z). \tag{124}$$

The quantities β_+ and β_- are the y component of the wave vector in the $+y$ and $-y$ direction, respectively, and consequently $\beta_+ = -\beta_- = \beta$. One thus obtains

$$\hat{y} \times H_t(x, y, z) = -\frac{k}{Z_0 \beta}[b_+(y) - b_-(y)]\psi(x, z), \tag{125}$$

$$E_t \cdot \overline{(\hat{y} \times H_t)} = -\frac{1}{Z_0}\overline{\left(\frac{k}{\beta}\right)}\{|b_+(y)|^2 + |b_-(y)|^2$$
$$+ 2i\,\mathrm{Im}[b_-(y)\overline{b_+(y)}]\} \cdot |\psi(x, z)|^2, \tag{126}$$

hence

$$I_y = \frac{1}{2Z_0}\left\{\mathrm{Re}\left(\frac{k}{\beta}\right)[|b_+(y)|^2 + |b_-(y)|^2]\right.$$
$$\left. + 2\,\mathrm{Im}\left(\frac{k}{\beta}\right)\mathrm{Im}[b_-(y)\overline{b_+(y)}]\right\} \cdot |\psi(x, z)|^2. \tag{127}$$

Finally, integrate over the unit cell of the grating. Due to orthonormality one obtains

$$I_{y,m} = \frac{1}{2Z_0}\left\{\mathrm{Re}\left(\frac{k}{\beta}\right)[|b_+(y)|^2 + |b_-(y)|^2]\right.$$
$$\left. + 2\,\mathrm{Im}\left(\frac{k}{\beta}\right)\mathrm{Im}[b_-(y)\overline{b_+(y)}]\right\}, \tag{128}$$

where $I_{y,m}$ denotes the power density averaged over the xz plane.

It is evident that when β is real, i.e., for propagating waves, only the first term in Eq. (128) is non-zero, and the power density is the sum of the power density for waves propagating in the $+y$ and the $-y$ direction. In contrast, for evanescent waves only the last term is non-zero. It is evident from Eq. (128) that a single evanescent wave in either the $+y$ or the $-y$ direction does not transport energy. However, if waves are present in both directions simultaneously the power density is non-zero due to interference. Finally, it can be noted that waves which do not constitute a reflected pair do not contribute to energy flow as a result of orthogonality between modes.

For the incident wave one obtains $I_{y,m,0} = 1/2Z_0$. The power density normalized with respect to incident power density thus becomes

$$\tilde{I}_{y,m} = \mathrm{Re}\left(\frac{k}{\beta}\right)[|b_+(y)|^2 + |b_-(y)|^2]$$

$$+ 2\,\mathrm{Im}\left(\frac{k}{\beta}\right)\mathrm{Im}[b_-(y)\overline{b_+(y)}]. \tag{129}$$

The total power density $\tilde{I}_{y,m}^{\mathrm{tot}}$ is obtained by adding the contributions from all two-wave vectors.

The absorptance which is the amount absorbed in a slice containing the QWs and confined between $y = y1$ and $y = y2$, is obtained from

$$\eta = \tilde{I}_{y2,m} - \tilde{I}_{y1,m}. \tag{130}$$

2.4. Grating Coupled Detectors–Numerical Results

2.4.1. General Assumptions

If not otherwise stated the quantum well structure used in the simulations is assumed to consist of 50 quantum wells of width 5.2 nm with 35.0 nm $Al_xGa_{1-x}As$ ($x = 0.295$) between. An upper contact layer (close to grating) of width 0.5 μm and a lower one of width 1.0 μm are assumed. The contact layers are n-doped to $1 \cdot 10^{18} \mathrm{cm}^{-3}$ whereas the QWs each are doped to $3.7 \cdot 10^{11} \mathrm{cm}^{-2}$ which is the sheet concentration that *optimizes detectivity* D^*. For detectors with a cladding layer (dielectric mirror) an $Al_xGa_{1-x}As$ layer with $x = 0.8$ and of width 3.0 μm is positioned below the bottom contact layer, close to the GaAs semiinsulating substrate. The thickness of the substrate is large compared to the wavelength or ca 0.5 mm. The grating constants D, D_x, and D_y are taken as 2.75 μm if not otherwise stated. The parameters of Eq. (111) are taken as: oscillator strength $f = 0.76$, transition wavelength $\lambda_t = 2\pi c/\omega_t = 8.95$ μm, and $\gamma = 3.7 \cdot 10^{13}$ Hz.

The parameter to be optimized is the absorptance (absorption quantum efficiency) averaged over the 8–12 μm window defined according to

$$\eta_m = \frac{\left[\int_{\lambda_1}^{\lambda_2} \eta(\lambda) \cdot d\left(\frac{1}{\lambda}\right)\right]}{\dfrac{1}{\lambda_1} - \dfrac{1}{\lambda_2}}, \tag{131}$$

with $(\lambda_1, \lambda_2) = (8, 12)$ μm.

The mean absorptance defined in this way is appropriate for thermal imaging applications and is nearly proportional to the photon current that results when exposing to a scene of temperature close to 300 K. The photon energy dependence of Planck's expression: photon flux density per photon energy interval, is neglected.

Values of absorptance are calculated with respect to unpolarized incident radiation throughout.

For reference purposes the 45° polished edge detector is investigated. The mean absorptance of such a detector becomes 4.9%. A summary of mean absorptances for detectors based on different type of couplers is presented in Table 1.

2.4.2. Absorptance of Grating Coupled Detectors

2.4.2.1. Lamellar Gratings Figure 10a is a contour plot displaying the mean absorptance versus the grating dimension parameters: channel width d and depth h for a lamellar grating detector without a cladding layer. It is found from the figure that an optimum exists at $(d, h) = (1.4, 0.35)\,\mu m$, where the mean absorptance η_m is 8.1%. The corresponding plot of a detector with a cladding layer (Fig. 10b) gives an optimum of $\eta_m = 10.2\%$ at $(d, h) = (1.3, 0.35)\,\mu m$. Figure 11 shows the spectral dependence of absorptance η for detectors with and without cladding layers. For the sake of comparison the results of a 45° polished edge detector are also included.

Evidently the inclusion of a cladding layer in the detector structure results in a 26% increase of η_m. However, the high η_m of the detector without cladding layer presumes large contributions from evanescent fields of the grating. There is experimental evidence that the evanescent part of the spectrum is suppressed in favor of the propagating wave part. The reason for this may be larger grating metal losses or stronger interaction with the lateral boundaries of the detector (i.e.,

TABLE 1

The mean absorptance η_m in the wavelength region 8–12 μm for different types of coupling configurations as obtained by numerical simulation. The detector is optimized for maximum detectivity D^*.

	Without cladding layer (%)	With cladding layer (%)
Lamellar	8.1	10.2
Crossed	14.7	19.3
Polished edge	4.9	–

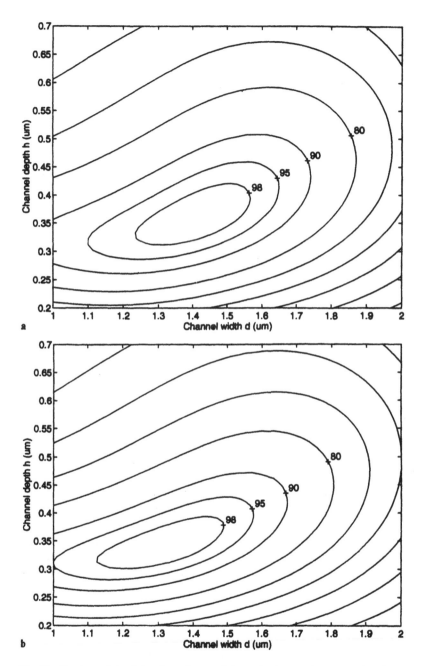

Fig. 10 Contour plot of the dependence of the ratio (mean absorptance/maximum mean absorptance in %) of a lamellar grating detector on the grating geometry parameters: (channel width d and depth h). (a) without cladding layer, and (b) with cladding layer. $D = 2.75\ \mu m$.

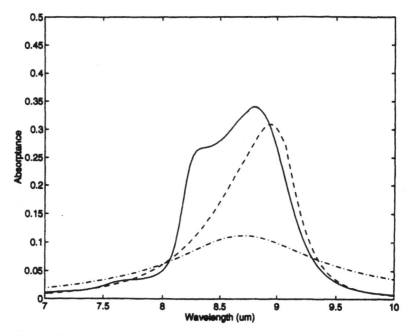

Fig. 11 The absorptance vs. wavelength for the case of optimum mean absorptance of a lamellar grating detector with a cladding layer (full curve), and without a cladding layer (dashed). The dash-dotted curve refers to a detector with a 45° polished edge. $D = 2.75 \, \mu\text{m}$.

the mesa edges). The absorptance of a cladding layer detector is mainly due to propagating waves and consequently the simulated and experimental results are more close.

The sensitivity of η_m to changes in d and h can be found from Fig. 10a–b. Considering the region within the 98% contour where the maximum variation of η_m is 2%, a value which is comparable to experimental pixel to pixel nonuniformities of detector arrays, one obtains: $\Delta d/d = \pm 10\%$ and $\Delta h/h = \pm 10\%$ for detectors both with and without cladding layer.

It should be added that only changes in d and h coherent across a large area (comparable to the detector mesa area) of the grating are relevant to this analysis. Random fluctuations within small areas of the grating cause a broadening of the η vs. λ curve, keeping η_m fairly constant.

2.4.2.2. Crossed Gratings   Figure 12a–b are contour plots of η_m vs. the area of the grating cavity openings d^2 and depth h for detectors with crossed gratings with box-shaped cavities, and with and without cladding layer, respectively. For detectors without cladding layer an optimum exists with $\eta_m = 14.7\%$ at $(d^2, h) = (2.4\,\mu m^2, 1.05\,\mu m)$. For detectors with cladding layer one obtains that $\eta_m = 19.3\%$ at $(d^2, h) = (2.6\,\mu m^2, 0.95\,\mu m)$. Plots displaying the spectral behaviour are presented in Fig. 13. It is evident that η_m increases by a factor of 31% if a cladding layer is included in the detector structure. However, the discussion above concerning the contribution from the evanescent field region to absorptance is also valid here, and in practice a larger relative increase is obtained. It is found that the optimum mean absorptance takes place for $d^2/D^2 = 0.33$. In contrast, when optimizing for maximum peak absorptance a value ≈ 0.5 is obtained.

The sensitivity of η_m to changes in d and h can be found from Fig. 12a–b. Considering the region within the 98% contour where the maximum variation of η_m is 2%, one obtains: $\Delta A_c/A_c = \pm 8\%$ and $\Delta h/h = \pm 10\%$ for both detectors with and without cladding layer (A_c = cavity area).

To find out the sensitivity to a general change in cavity shape a crossed grating with cylindrical cavities is considered. As a result of smearing during photolithography and subsequent etching a somewhat distorted cylindrical cavity is a good approximation to the experimentally obtained cavity shape. In order to find the grating geometry parameters that give the optimum η_m the cavity area $\pi a^2/4$ (a = cavity radius) and depth h are varied. It is found that the optimum parameters are the same as those found for box-shaped cavities, with an optimum $\eta_m = 14.6\%$ and 19.1% for detectors without and with cladding layer, respectively. The optimum absorptance η vs. λ is presented in Fig. 14.

In order to determine the influence of grating symmetry, gratings of hexagonal symmetry and cylindrical cavities are investigated. Even in this case the parameters cavity area and depth giving maximum η_m are the same as for box-shaped cavities. The optimum $\eta_m = 14.2\%$ and 18.3% for detectors without and with cladding layer, respectively. The absorptance η vs. λ at optimum is presented in Fig. 15.

The influence of the asymmetry of the cavity shape of a crossed grating being part of a detector is displayed in Fig. 16. In this case box-shaped cavities of rectangular cross-section are considered with d_x not necessarily equal to d_z, keeping the cavity area $d_x d_z$ constant. An

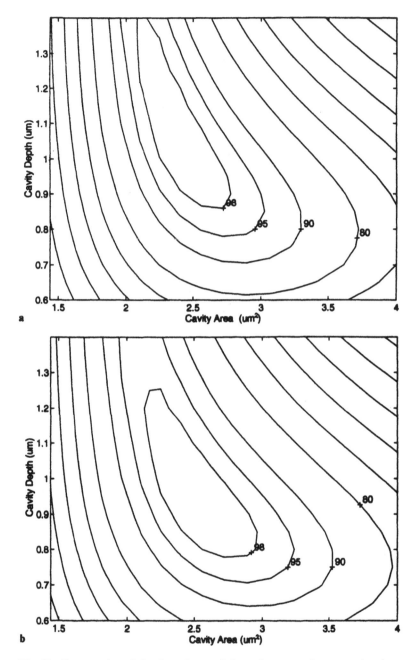

Fig. 12 Contour plot of the dependence of the ratio (mean absorptance/maximum mean absorptance in %) of a crossed grating detector with box-shaped cavities on the grating geometry parameters: (cavity area d^2 and depth h). (a) without cladding layer, and (b) with cladding layer. $D = 2.75 \ \mu m$.

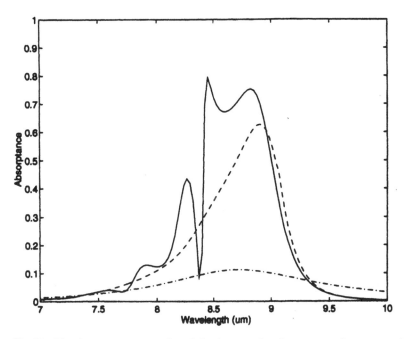

Fig. 13 The absorptance vs. wavelength for the case of optimum mean absorptance of a crossed grating detector with box-shaped cavities with a cladding layer (full curve), and without a cladding layer (dashed). The dash-dotted curved refers to a detector with a 45° polished edge. $D = 2.75$ μm.

asymmetry parameter (AP) is here defined as

$$AP = \frac{|d_x - \sqrt{d_x d_z}|}{\sqrt{d_x d_z}}. \qquad (132)$$

It is found that the AP resulting in 2% change in η_m is ≈ 0.03 for both detectors with as well as without cladding layer.

The dependence of η_m on the angle of incidence θ with respect to the normal of the grating plane has been calculated for both lameller as well as crossed grating detectors, and the results for crossed gratings are presented in Fig. 17 (the results for lamellar gratings are very similar). It is found that η_m is nearly constant up to at least $\theta = 15°$ (i.e., a 30° full cone angle of incidence) for both lamellar and crossed grating detectors.

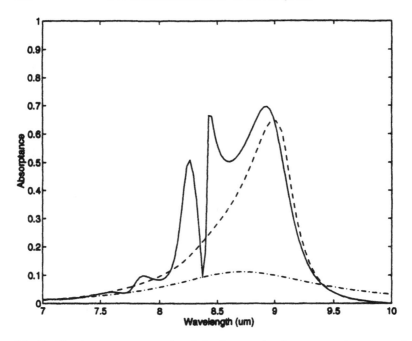

Fig. 14   The absorptance vs. wavelength for the case of optimum mean absorptance of a crossed grating detector with cylindrical cavities with a cladding layer (full curve), and without a cladding layer (dashed). The dash-dotted curve refers to a detector with a 45° polished edge. $D = 2.75$ μm.

2.5. Theory of Gratings of Finite Extension

2.5.1. General Background

The finite size of the grating-covered mesa has two major effects on the absorptance η vs. λ curve. The first one is connected with the finite grating area which results in a smearing of the angle of propagation of the different orders of diffraction. From the theory of gratings it is known that only gratings of infinite extension exhibit sharp discrete reflections at selected angles of propagation. In contrast, the reflections of a finite extension grating are diffuse, more so the smaller number of periods the grating possesses. This leads to a broadening of the η vs. λ curve but not necessarily to a lowering of the integrated quantum efficiency. The other effect is connected with the sharp mesa edges which tend to induce diffuse scattering of the radiation, which causes cross-talk and lower absorptance. It was earlier though that also direct penetration of radiation through the mesa edges was

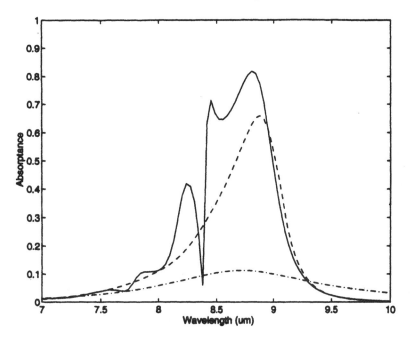

Fig. 15 The absorptance vs. wavelength for the case of optimum mean absorptance of a hexagonal symmetry grating detector with cylindrical cavities with a cladding layer (full curve), and without a cladding layer (dashed). The dash-dotted curve refers to a detector with a 45° polished edge. $D = 2/\sqrt{3} \cdot 2.75 = 3.18$ μm.

effective in lowering the absorptance. However, it has been experimentally shown that this effect is negligible.

2.5.2. Lamellar Gratings

For the case of infinite extension lamellar gratings the diffracted intensity is only obtained at discrete values of the wave vector components α_p (and thus of β_p), in contrast to the finite extension grating which may diffract at any value of α. The objective of this section is to derive expressions for the diffracted intensities or amplitude vs. α. The derivation will be performed according to MEM and analogously to the derivation of the scattering matrix of the lamellar grating above. The space above the grating is divided into two regions by the $y = 0$ plane, as before.

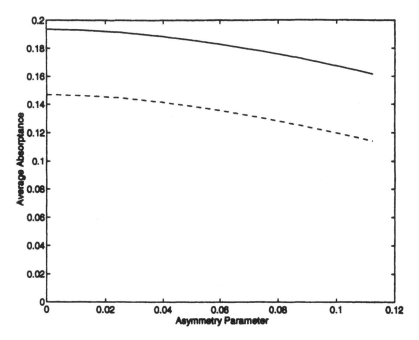

Fig. 16   The influence of asymmetry of the cavity shape on mean absorptance η_m for a detector with crossed gratings and box-shaped cavities. The asymmetry parameter is $|\Delta d_x|/\sqrt{d_x d_z}$, with the cavity area constant Unpolarized incident radiation is assumed. Without cladding layer (dashed curve), and with cladding layer (full).

Since in this case the incident and diffracted waves are continuous functions of α the scattering matrix also becomes continuous. The following relation applies

$$b_-(\alpha_1) = \int S(\alpha_1, \alpha_2)\, b_+(\alpha_2)\, d\alpha_2, \tag{133}$$

where $S(\alpha_1, \alpha_2)$ is the scattering matrix, and $b_+(\alpha_2)$ and $b_-(\alpha_1)$ are the incident and diffracted amplitudes.

For $y > 0$ one obtains for the transverse E field (TM polarization is assumed throughout)

$$R_\alpha(x, y) = \frac{1}{\sqrt{2\pi}} \exp\left[i(\alpha x + \sqrt{k^2 - \alpha^2}\, y)\right] \hat{x}. \tag{134}$$

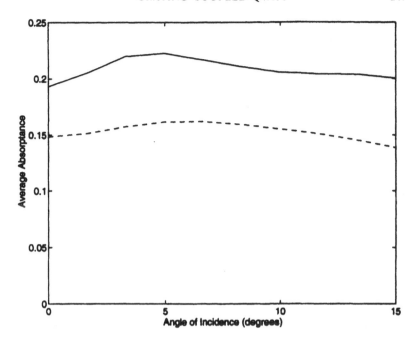

Fig. 17 The dependence of η_m on the angel of incidence (in air) for a crossed grating detector: without a cladding layer (dashed curve), and with cladding layer (full). The grating parameters are $D = 2.75$ μm, $d = 1.6$ μm, $h = 1.0$ μm.

The factor $1/2\pi$ is a normalization factor. The functions ψ are normalized over the complete $y = 0$ plane according to

$$\int R_{\alpha 1}(x,0) \cdot \overline{R_{\alpha 2}(x,0)} \, dA = \delta(\alpha 1 - \alpha 2), \qquad (135)$$

where $\delta(\alpha 1 - \alpha 2)$ is Dirac's delta function.

For $y < 0$ one has for one single channel the expressions according to Eq. (7)–(9). For a finite array of channels $M_n(x)$ and $N_n(x,y)$ are replaced by

$$M_n(x) = \sum_{r=-R}^{R} \exp(i\alpha_0 rD) M_n(x - rD) \qquad (136)$$

$$N_n(x, y) = \sum_{r=-R}^{R} \exp(i\alpha_0 rD) M_n(x - rD) \cdot \sin[\mu_n(y + h)], \qquad (137)$$

(Strictly speaking these expressions are the correct ones also for the infinite grating case if $R \to \infty$).

The scalar products according to Eq. (14) are replaced by

$$I'_n(\alpha) = \int_0^d \overline{R_\alpha(x,0)}\, m_n(x)\, dx = \left\{ \sum_{r=-R}^{R} \exp(-i\Delta\alpha r D) \right\} I_n(\alpha), \quad (138)$$

where

$$I_n(\alpha) = \int_0^d \overline{R_\alpha(x,0)}\, M_n(x)\, dx, \quad (139)$$

and $\Delta\alpha = \alpha - \alpha_0, (\alpha_0 = $ wave vector of incident radiation). But

$$\sum_{r=-R}^{R} \exp(-i\Delta\alpha r D) = \frac{\sin\left[\left(R + \frac{1}{2}\right)\Delta\alpha D\right]}{\sin[\Delta\alpha D/2]}$$

$$= \frac{\sin\left[\dfrac{N_G \Delta\alpha D}{2}\right]}{\sin[\Delta\alpha D/2]} = U(N_G, \Delta\alpha) \quad (140)$$

where N_G is the number of grating periods (and is here assumed to be odd).

One ends up with the following scattering matrix

$$S(\alpha_1, \alpha_2) = -\delta(\alpha_1 - \alpha_2)$$

$$+ U[N_G, (\alpha_1 - \alpha_2)] \sum_{n=0}^{\infty} c_n(\alpha_2) I_n(\alpha_1) \sin(\mu_n h), \quad (141)$$

where $\Delta\alpha = \alpha_1 - \alpha_2$, and α_1 and α_2 correspond to scattered and incident waves, respectively.

The channel mode amplitudes $c_n(\alpha_2)$ can be obtained from the system of equations

$$\sum_{N=0}^{\infty} C_{Nn} c_N(\alpha_2) = D_n(\alpha_2), \quad (142)$$

where

$$C_{Nn} = \sin(\mu_N h) Q_{Nn} + i\frac{k}{\mu_N}\cos(\mu_N h)\delta_{Nn}, \quad (143)$$

$$D_n(\alpha_2) = 2\frac{k}{\beta_2}\overline{I_n(\alpha_2)}, \quad (144)$$

$$Q_{Nn}(\alpha_2) = \int_{-\infty}^{\infty} U[N_G,(\alpha_1 - \alpha_2)] \frac{k}{\beta_1} I_N(\alpha_1) \overline{I_n(\alpha_1)} d\alpha_1. \qquad (145)$$

The function $U[N_G, \Delta\alpha]$ is periodic and for $N_G > \approx 11$ possesses sharp peaks at $\Delta\alpha = p \cdot 2\pi/D$. When $N_G \to \infty$, U approaches a sequence of delta functions, thus simulating the properties of a grating of infinite extension.

$$U[N_G, \Delta\alpha] \to \frac{2\pi}{D} \sum_{p=-\infty}^{\infty} \delta\left(\Delta\alpha - p \cdot \frac{2\pi}{D}\right). \qquad (146)$$

Assuming that N_G is sufficiently large Eq. (145) will transform into:

$$Q_{Nn}^{\alpha 2} = \frac{\cdot 2\pi}{D} \sum_{p=-\infty}^{\infty} \frac{k}{\beta_p} I_N^p \overline{I_n^p}, \qquad (147)$$

which is identical to Eq. (25) except for the factor $2\pi/D$ depending on a different normalization coefficient. However, this factor cancels out since the scalar products I_n^p contain the functions R_α which also includes the different normalization coefficient. The result is that Q_{Nn}^p according to Eqs. (147) and (25) are identical.

It is a good starting approximation to assume that the cavity mode amplitudes $c_n^{\alpha 2}$ do not change appreciably when changing from infinite extension mesas to finite extension ones. Using this assumption the diffracted amplitudes and intensities are calculated from Eq. (141). This equation is transformed into Eq. (26) when the grating area approaches infinity. The diffraction properties of a finite area grating can consequently be modeled in an approximate manner by first calculating the scattering matrix from Eq. (26) which gives the amplitudes integrated with respect to α. The spread in α around each α_p as given by $U[N_G, \Delta\alpha]$ is then taken into consideration. The spread in α or diffraction angle affects the QW absorption properties through the dependence of the transfer matrix elements on diffraction angle.

Based on this procedure, for mesa sizes $>$ ca 10 periods, the absorptance of the finite size detector can approximately be written

$$\eta_{\text{fin}}(\lambda, D_m) = \frac{1}{N_G \pi} \int_{-\infty}^{\infty} \eta_\infty(\lambda, D) \cdot \frac{\sin^2(N_G Q)}{Q^2} dQ, \qquad (148)$$

where η_{fin} is the absorptance of the finite size detector, η_∞ the absorptance for the case of infinite extension grating simulated according to Secs. 2.2 and 2.3, D_m the true grating constant of the finite area detector, D a grating constant implicitly used as an integration variable, and $Q(D, D_m) = \pi(D - D_m)/D_m$. In the approximate model of the finite mesa area detector, the more direct influence of mesa edges, e.g., transmission and reflection of radiation through the edges, as well as diffuse scattering down into the GaAs bulk induced by the presence of the mesa edge, are not taken care of.

2.5.3. Crossed Gratings

The theory presented above for lamellar gratings may easily be generalized to the case of crossed gratings exactly in the same way as was done in Secs. 2.2.2 and 2.2.3 for the case of infinite extension gratings. Especially Eq. (148) can be used for crossed grating detectors as well.

2.6. Theory of Grating Metal of Finite Conductivity

2.6.1. General

The theoretical approach presented above assumes a grating metal of infinite conductivity. High-conductivity metals such as pure gold, silver or aluminum are almost ideal in this respect. In order to calculate absorption losses in the metal a perturbation approach is conveniently used, where the basic assumption is that the EM field pattern obtained for a grating metal of infinite conductivity is unchanged when switching to a finite conductivity metal coating [22]. The magnetic field H tangential to the metal surface produces surface currents in it and thus power dissipation for finite conductivity. Infrared absorption in the grating metal is estimated by integrating the loss power density across the surface area of the metal. If the metal absorption is small enough not to perturb the original field pattern the calculated value is a good measure of the actual absorption.

The loss power density in the metal is

$$P_f = \frac{1}{2} \int \text{Re}(Z_m H_t^2) \cdot dA = \frac{1}{2} R_m \int |H_t|^2 \cdot dA, \qquad (149)$$

where $Z_m = \sqrt{\mu_0/\hat{\varepsilon}}$ is the intrinsic impedance of the metal and R_m its real part and μ_0 and $\hat{\varepsilon}$ denote the vacuum permeability and the complex dielectric constant, respectively. H_t is the magnetic field tangential to the surface of the metal. For a metal $\hat{\varepsilon} \approx i\sigma/\omega$ where σ is the conductivity. One obtains

$$R_m = \text{Re}(\sqrt{\mu_0/\hat{\varepsilon}}) = \text{Re}(\sqrt{\mu_0\omega/i\sigma}) = \sqrt{\mu_0\omega/2\sigma}. \qquad (150)$$

In order to calculate the total metal absorption per unit power density of the incident radiation, the contributions to P_f from all different parts of the grating surface located within one unit cell are calculated and subsequently added. One then divides by the total incident power per unit cell of the grating. In the calculation unit amplitude of the incident radiation is assumed, which implies a power density $= 1/2Z_0$. Hence the normalized power density becomes

$$\tilde{P}_f = \frac{R_m Z_0}{A_d} \int |H_t|^2 \cdot dA, \tag{151}$$

where A_d is the projected unit cell area of the grating.

2.6.2. Lamellar Gratings

It is convenient to divide the grating area into three subregions: upper grating surface, channel walls, and channel bottoms.

2.6.2.1. Upper grating surface
For $y > 0$ the H field is given by Eq. (15). The squared modulus is then given by

$$|H_t|^2 = |\hat{y} \times H_t|^2 = \frac{1}{Z_0^2} \sum_p \left| \frac{k}{\beta_p} (b_{p,-} - b_{p,+}) \right|^2, \tag{152}$$

where the quantities b are the wave amplitudes of the free space modes at $y = 0$. The contribution to the loss power density becomes

$$P_f^l = \frac{1}{2} R_m |H_t|^2 \cdot (D - d), \tag{153}$$

since $|H_t|^2$ is constant at the upper grating surface.

2.6.2.2. Channel Walls
The magnetic field in the channel region is given by Eq. (16) or

$$\hat{y} \times H_t = \frac{i}{Z_0} \sum_n \left\{ c_{n,\text{tot}} \frac{k}{u_n} g_n \cos\left(\frac{n\pi x}{d}\right) \cos[u_n(y + h)] \right\}. \tag{154}$$

The total channel mode amplitude is obtained by summing over the contributions from all waves incident onto the grating

$$c_{n,\text{tot}} = \sum_r b_r c_n^r. \tag{155}$$

For $x = 0$ (and for $x = d$ not considering the sign change)

$$\hat{y} \times H_t = \frac{i}{Z_0} \sum_n \left\{ c^r_{n,\text{tot}} \frac{k}{u_n} g_n \cos \left[u_n(y + h) \right] \right\}. \qquad (156)$$

One ends up with

$$|H_t|^2 = |\hat{y} \times H_t|^2 = \frac{1}{Z_0^2} \sum_{nN} \left\{ c^r_{n,\text{tot}} \overline{c^r_{N,\text{tot}}} \frac{k^2}{u_n u_N} g_n g_N \right.$$

$$\left. \times \cos \left[u_n(y + h) \right] \overline{\cos \left[u_N(y + h) \right]} \right\}. \qquad (157)$$

Finally the power loss is obtained from Eq. (149).

2.6.2.3. Channel Bottoms

The H field for the channel bottom is given by Eq. (16). One obtains

$$|H_t|^2 = |\hat{y} \times H_t|^2 = \frac{1}{Z_0^2} \sum_n \frac{k^2}{|u_n|^2} |c^r_{n,\text{tot}}|^2. \qquad (158)$$

The total power loss for one grating period is obtained by adding the contributions from each part.

The calculated power loss for lamellar gratings coated with pure gold is depicted in Fig. 18a together with the absorptance curve due to QW absorption. It is evident that as much as 5% of the incident (unpolarized) radiation is absorbed in the metal. For the ohmic contact metal gold-germanium the absorption increases further by a factor of ca three due to a factor of ten lower conductivity. It should be noted from the figure that the metal absorption increases abruptly close to first grating order cut-off, where the wave field becomes evanescent. It was assumed in these calculations that the conductivity of gold attains the bulk value $\sigma = 45 \cdot 10^6 \Omega^{-1} \text{m}^{-1}$, however, since thin films are used this value is probably too high. It is thus probable that the EM field pattern is perturbed, especially in the evanescent field region, which may explain the lower absorptance in this region.

2.6.3. Crossed Gratings

Even for the case of crossed gratings, when calculating power loss the integration domain used in Eq. (149) is conveniently divided into three regions: the upper grating surface at $y = 0$, the cavity walls, and the cavity bottoms. The integration is performed over one grating period and the average absorption loss per projected area of the grating is finally calculated.

2.6.3.1. Upper Grating Surface   One has

$$\hat{y} \times H_t = \frac{1}{Z_0} \sum_{p,q} \left[\frac{\beta_{pq}}{k} (b^{\text{TE}}_{pq,-} - b^{\text{TE}}_{pq,+}) \, R\text{TE}(x,0,z) \right.$$

$$\left. + \frac{k}{\beta_{pq}} (b^{\text{TM}}_{pq,-} - b^{\text{TM}}_{pq,+}) \, R\text{TM}(x,0,z) \right], \qquad (159)$$

where the quantities b are the wave amplitudes at $y = 0$ for the free space modes $R\text{TE}$ and $R\text{TM}$ as defined in Appendix A. The $+$ and $-$ signs stand for scattered waves and incident waves onto the grating.

The squared modulus of the transverse H field becomes

$$|H_t|^2 = |\hat{y} \times H_t|^2 = \frac{1}{Z_0^2} \sum_{p,q} \left[\left| \frac{\beta_{pq}}{k} (b^{\text{TE}}_{pq,-} - b^{\text{TE}}_{pq,+}) \right|^2 \right.$$

$$\left. + \left| \frac{k}{\beta_{pq}} (b^{\text{TM}}_{pq,-} - b^{\text{TM}}_{pq,+}) \right|^2 \right]. \qquad (160)$$

Since $|H_t|^2$ is constant over the area the contribution to the power loss density becomes, if square shaped unit cells (side length D) and cavities (side length d) are assumed

$$P^I_f = \frac{1}{2} R_m |H_t|^2 \cdot (D^2 - d^2). \qquad (161)$$

2.6.3.2. Cavity Walls   In this case there is no simple expression in closed form. However, the expressions for the components of the H field which are found in Appendix B (Eqs. (B12) – (B14)) can be integrated directly according to Eq. (149). The contributions from each of the four walls are calculated and subsequently added.

2.6.3.3. Cavity Bottom   One has for $y = -h$

$$\hat{y} \times H_t = \frac{i}{Z_0} \sum_{nm} \left\{ \frac{u_{nm}}{k} a_{nm,\text{tot}} \, M\text{TE}_{nm}(x,z) \right.$$

$$\left. + \frac{k}{u_{nm}} c_{nm,\text{tot}} \, M\text{TM}_{nm}(x,z) \right\}, \qquad (162)$$

where the functions $M\text{TE}$ and $M\text{TM}$ are found in Appendix B.

The total cavity mode amplitudes are obtained by summing over the contributions from all waves incident to the grating

$$c_{nm,\text{tot}} = \sum_{rs} b_{rs} c_{nm}^r.$$

Due to the orthonormality of the cavity modes one obtains

$$|H_t|^2 = |\hat{y} \times H_t|^2 = \frac{1}{Z_0^2} \sum_{nm} \left\{ \frac{|u_{nm}|^2}{k^2} |a_{nm,\text{tot}}|^2 + \frac{k^2}{|u_{nm}|^2} |c_{nm,\text{tot}}|^2 \right\}. \qquad (163)$$

Finally the power loss density becomes

$$P_f^{III} - \frac{1}{2} R_m |H_t|^2 \cdot d^2. \qquad (164)$$

As described above the contributions from each region are added. Finally the amount absorbed per unit incident power is obtained by dividing the sum by the incident optical power per grating unit cell.

Figure 18b shows results from the calculation of power loss for a crossed grating detector without cladding layer. It is found that for a coating of pure gold ca 5% of the total incident infrared power is absorbed, as was also found for lamellar gratings. Following the same reasoning as for the case of lamellar gratings it is probable that the EM field is strongly perturbed especially in the evanescent field region, which may explain the experimentally observed lower grating coupling efficiencies in this region.

3. EXPERIMENTAL RESULTS

3.1. Materials and Experimental Equipment

The experiments where diffracted intensities are directly measured optically, makes use of lamellar gratings fabricated by sputtering of gold onto a grating pattern etched into a gallium arsenide substrate. In order to obtain gratings of rectangular profile, reactive ion etching was used. The measurements of diffracted power are carried out using FTIR equipped with a specular reflection accessory. Only the zeroth order reflection was measured, and in order to separate the incident and diffracted orders an angle of incidence of 17° referred to a plane parallel to the grooves was utilized. The angle of incidence θ of Eq. (4) is zero (this angle is referred to a plane perpendicular to the grooves). The optical medium is air. The grating dimensions are $D = 4.0\,\mu\text{m}$, $d = 1.88\,\mu\text{m}$, and $h = 0.62\,\mu\text{m}$.

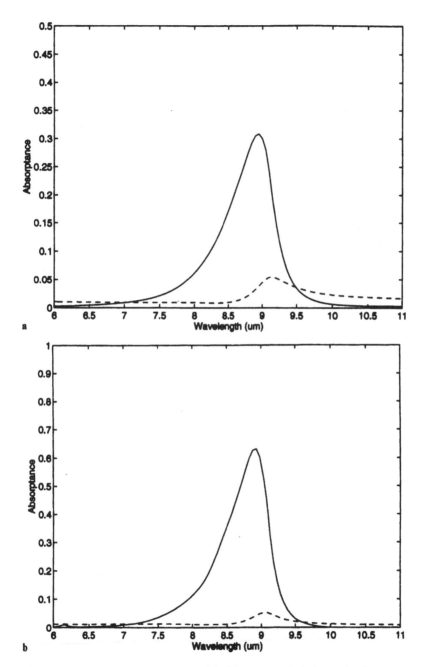

a

b

Fig. 18 Grating metal loss power per unit incident power (dashed curve) and quantum well absorptance for: (a) lamellar grating detector, and (b) crossed grating detector. The grating dimensions are chosen to be those that optimizes η_m. The grating metal is assumed to consist of pure gold ($\sigma = 45.10^6 \, \Omega^{-1} m^{-1}$).

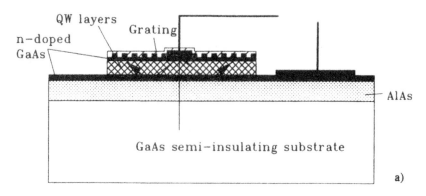

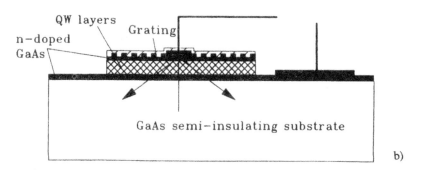

Fig. 19 Cross-section of a detector with an etched grating and (a) with cladding layer, and (b) without a cladding layer.

In order to experimentally compare the performance of QW infrared detectors with different types of gratings, and the influence of a cladding layer, two separate detector structures (sample A and B) are grown onto semi-insulating GaAs substrates. The epitaxial equipment consists of a low pressure metalorganic vapor phase epitaxy (MOVPE) reactor featuring wafer rotation. Both structures contain a 50 period $(GaAs/Al_xGa_{1-x}As)$ multi-QW stack sandwiched between a $1.0\,\mu m$ bottom and a $1.3\,\mu m$ top GaAs contact layer $(n = 7 \cdot 10^{17} cm^{-3})$ and one of them (sample A) an additional $3\,\mu m$ $Al_{0.78}Ga_{0.22}As$ cladding layer located between the substrate and bottom contact. The structure parameters are characterized by capacitance voltage etch profiling, x-ray diffraction and fourier transform infrared spectroscopy (FTIR) [23]. The values for sample (A, B) are: GaAs QW doping concentration $n = (3.7 \cdot 10^{17}, 3.7 \cdot 10^{17}) cm^{-3}$, GaAs QW width

$a = (5.2, 4.9)$ nm, AlGaAs barrier width $a_{barr} = (34.8, 37.7)$ nm and Al-GaAs barrier Al–content $x = (0.290, 0.286)$. The slightly different QW structure parameters lead to the type A structure possessing quasibound upper states, whereas the B type excited states are extended.

The detectors are fabricated as follows (see Fig. 19 for a scheme of the detector): gratings (linear and crossed with square symmetry) are etched by reactive ion etching (RIE) into the GaAs top contact layer. The processed grating dimensions viewed from the GaAs side are: grating depth $h = 0.75$ µm, grating constant $D = 2.8$ µm, crossed grating box shape cavity width $d = 1.6$ µm and linear grating cavity width $d = 1.4$ µm. Square shaped mesas of various sizes (from 25 to 500 µm side length) are defined by etching (RIE) down to the bottom contact layer, and finally AuGe/Ni ohmic contacts are deposited onto the top and bottom contact layer. Since the top contact also serves as a grating metal coating it is important that this contact possesses high reflectivity as well as good ohmic behaviour. Therefore two different types of top contacts are used, one (labelled N-type) where the whole mesa area is convered with AuGe/Ni and another (labelled S-type) with AuGe/Ni covering only the central 10% of the mesa area. After alloying, all the mesas are finally covered with a sputtered high reflective Au-layer (below designated "mirror"). In order to apply bond wires to the detectors, gold metallization connecting detector mesas with bond pads is utilized. Silicon nitride is used for insulation to the GaAs substrate. Finally, for characterization of the detectors, a grating monochromator, a 1000 K glowbar source and a pyroelectric reference detector are used. A 1000 K black-body source is exploited for determining black-body responsivities and detectivities as well as for calibration purposes.

3.2. Results and Discussion of Measurements – Comparison with Theory

In order to compare the theory of grating scattering based on MEM with experimental data, FTIR equipped with a specular reflection accessory is used. The results showing the ratio of diffracted optical power in the zeroth direction to incident optical power, are presented in Fig. 20 together with simulated curves obtained from the near-exact theory based on MEM. Evidently the accordance between theory and experiment is good. One feature present in the

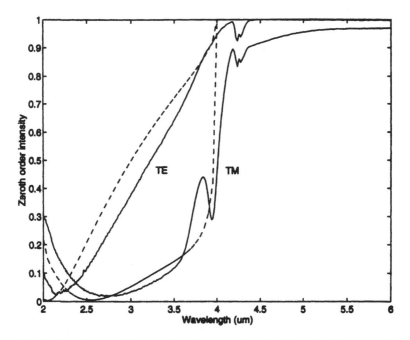

Fig. 20 Experimental (full curves) and simulated results (dashed) of the amount of diffracted optical power in the zeroth order vs. wavelength. Angle of incidence = 17°. The grating is of the lamellar type with dimensions: $D = 4.0\,\mu m$, $d = 1.88\,\mu m$, and $h = 0.62\,\mu m$. The measurements are performed in air.

experimental data but absent in the simulated ones is noteworthy. For TM polarization a local intensity minimum occurs at a wavelength somewhat smaller than $4\,\mu m$. The cut-off wavelength of the grating is $4\,\mu m$ so the minimum is probably connected with the transition from the propagating to evanescent region. It has been shown under Sec. 2.6 above that absorption losses due to the metal coating is especially strong in this region, and it is probable that the oscillation is connected with this fact.

Response measurements of detectors with different types of gratings as well as with or without cladding layer [24], are carried out and their absorptance calculated from

$$\eta = R_I hc/(\lambda qg), \tag{165}$$

where R_I is the current responsivity and g the photoconductive gain. g

is obtained from:

$$g = i_N^2/(4qI_d),\qquad(166)$$

where i_N is the generation-recombination (gr) noise current spectral density and I_d the DC detector current. In order to obtain the gr noise spectral density contribution, the Johnson noise should be subtracted from the total noise. Strictly speaking, the measured value of quantum efficiency is not the absorptance but equal to $\eta \cdot p$, where p is the escape probability of electrons from the upper state of the well to the continuum. p approaches unity when the bias voltage increases, mainly due to facilitated tunneling. Consequently, in order to obtain the absorptance, the measurements should be conducted at a large enough bias voltage to allow $p = 1$. All the results in this work are compensated for the 29% reflection loss at the air GaAs interface experienced by normal incidence radiation, thus simulating the presence of an ideal antireflective coating on the rear side of the GaAs detector wafer. Detector parameters for edge detectors will refer to unpolarized radiation. Finally, it should be mentioned that the detectors have been optimized with respect to detectivity D^* and not quantum efficiency, primarily by decreasing the QW doping concentration. Consequently this results in lower quantum efficiency than was previously presented in Ref. [10].

The respective 45° polished edge detectors (EDGE) serve as reference detectors within each of the groups A and B. This is necessary since the QW structures are not identical, and possess slightly different absorption properties. Table 2 collects results on the measurements of detector performance including absorptances and detectivities D^* for different detector types and geometries. Both the mesa dimension as well as the lateral dimension of the gold covered grating are indicated in it, the latter defining the optical area used in the calculations. In addition to the peak absorptance η, the integrated absorptance η_{int} is also included. η_{int} is here defined according to $\eta_{int} = \int_{\lambda_1}^{\lambda_2} \eta d\lambda$, with wavelengths $(\lambda_1, \lambda_2) = (7, 12)\,\mu m$, and is a useful measure when discussing LWIR (8–12 μm window) thermal imaging applications for QWIPs. η_{int} is nearly proportional to the mean η_m defined by Eq. (131). Evidently, compared to the EDGE detector with $\eta_{int}(A, B) = (2.5, 11)\%$, the quantum efficiency enhancement is substantial for detectors with crossed gratings (CG) and even more so for detectors with a crossed grating and waveguide (CGW). Considering large mesa areas of 500 μm, the enhancement ratio $r_{CG} = \eta_{int}\,(CG)/\eta_{int}(EDGE) = 2.1$, whereas the corresponding $r_{CGW} = 4.5$.

TABLE 2

Detector data for the different detector configurations utilized: CG = crossed grating, CGW = crossed grating with cladding layer, LG = lamellar grating, LGW = lamellar grating with cladding layer, EDGE = 45° polished edge detector, EDGEW = edge detector with a cladding layer. A and B refer to different QW structures. The measurements have been conducted at 77 K throughout.

Sample No.	Detector Type	Detector size [mirror size] (μm)	Bias voltage U_d (V)	Dark current density j_d (μA/cm²)	Photoconductive gain g	Peak current responsivity R_I (A/W)	Peak detectivity D^* (10^{10} cm Hz$^{1/2}$ W^{-1})	Peak quantum efficiency η (%)	Integrated quantum efficiency η_{int} (%)
A	EDGEW	150 [—]	4.0	200	0.19	0.10	1.7	7.4	9.5
A	CGW	500 [494]	4.0	210	0.19	0.71	14	53	43
A	CGW	150 [144]	4.0	200	0.18	0.63	13	49	46
A	CGW	100 [94]	4.0	210	0.18	0.55	10	42	41
A	CGW	45 [40.5]	4.0	190	0.18	0.42	8.0	32	36
A	CGW	36 [31.5]	4.0	210	0.18	0.39	6.9	31	36
A	CGW	25 [20.5]	4.0	220	0.17	0.31	5.2	26	33
A	LGW	150 [144]	4.0	210	0.18	0.32	6.2	25	20
B	EDGE	150 [—]	4.0	860	0.28	0.14	1.0	7.4	11
B	CG	500 [494]	2.5	400	0.38	0.76	7.6	29	23
B	CG	150 [144]	2.5	400	0.36	0.62	6.0	23	23
B	CG	100 [94]	2.5	390	0.37	0.55	5.4	22	23
B	CG	45 [40.5]	2.5	400	0.34	0.43	4.0	21	20
B	CG	36 [31.5]	2.5	400	0.36	0.42	3.8	16	22
B	CG	25 [20.5]	2.5	420	0.38	0.36	2.9	13	18
B	LG	150 [144]	2.5	400	0.34	0.17	1.8	7.1	8.6

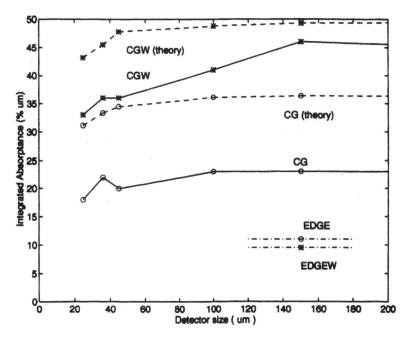

Fig. 21 Integrated absorptance η_m vs. detector mesa size for crossed grating detectors with a cladding layer (CGW), as well as without cladding layer (CG), according to experiment (full curves) and theory (dashed). The corresponding results of 45° polished edge detectors are included for the case of: with a cladding layer (EDGEW) and without (EDGE).

The ratio $r_{\mathrm{CGW}}/r_{\mathrm{CG}} = 2.1$, which clearly demonstrates the effect of the waveguide (cladding layer). The detectors with lamellar gratings with and without waveguide, respectively, (LGW and LG), are evidently inferior compared to the corresponding crossed grating detectors.

It is noteworthy that the quantum efficiency η and η_{int} decreases with decreasing mesa area, and more so than is predicted by Eq. (148). This is also evident from Fig. 21, which depicts the absorptance η_{int} vs. mesa dimension. It is, however, evident that even for smaller mesas the CGW detector type is superior as compared to the other types. Therefore, the described method of coupling radiation into QWIPs may advantageously be exploited in the fabrication of large focal plane arrays, where pixel dimensions $< 50\,\mu\mathrm{m}$ are needed.

When comparing grating detectors with N and S type contacts, detectors with S type contacts generally give a factor of two higher

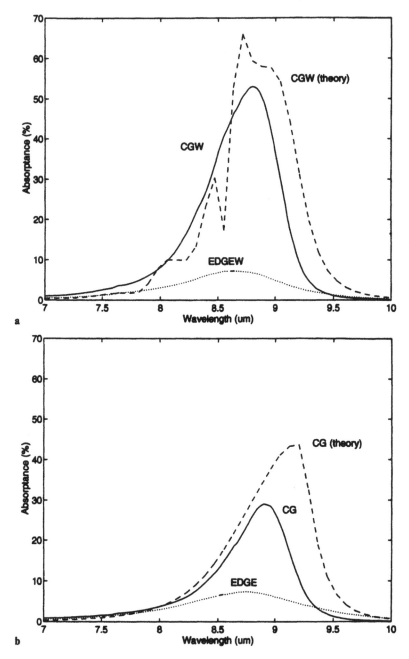

Fig. 22 Absorptance η vs. wavelength for detectors with a crossed grating, according to experiment (full curves), and theory (dashed). Results of 45° polished edge detectors are included for reference purposes. (a) detectors with a cladding layer, and (b) without cladding layer.

response compared to detectors with N type contacts, emphasizing the importance of a high reflectivity grating metal. The change from AlAs (used in our previous work[9, 10]) to $Al_{0.78}Ga_{0.22}As$ cladding layer material is justified by the better chemical stability of the latter. According to simulations, the change in cladding layer composition causes a decrease in η only in the short wavelength region of the response spectrum, leaving the remaining part of the spectrum unaffected.

Theoretical and experimental results of η vs. λ are presented in Fig. 22a and b. The simulation parameters were extracted by fitting to EDGE detector absorption and response curves. A comparison with the experimental curves shows that the main difference is the presence of a predominant evanescent response peak in the former which is lacking in the latter. We attribute this to broadening and absorption factors such as irregularities of the gratings and the finite conductivity of the metal which adds to the broadening due to mesa size included in the theory. It should be noted that broadening factors are taken account of for the smaller mesa sizes, but η_{int} is still too large compared to experimental data. Probably, the main reason for the low absorptance in the evanescent field region is considerable absorption losses in the metal as a result of a large electromagnetic field intensity close to the metal as has been discussed in Part 2.6.

As concerns peak D^*, very large values of $1.4 \cdot 10^{11} cm\, Hz^{1/2} W^{-1}$ are obtained for the $500\,\mu m$ mesa side length CGW detector at $\lambda = 8.7\,\mu m$ and 77K operation temperature, and the detector shielded from background radiation during noise measurement. To our knowledge this is the largest D^* reported to date for QWIPs at the wavelength and operation temperature specified.

Separate experiments were conducted in order to find out the dependence of absorptance on QW sheet carrier concentration. The QW parameters are close to those which are used in the experiments above for optimum D^*. When changing the sheet carrier concentration according to: $n_s = (1.6, 2.5, 4.1) 10^{11} cm^{-2}$ the peak absorptance varied as $\eta = (57, 77, 87)\%$.

To conclude, the absorptance of detectors with crossed gratings, with or without cladding layers has been shown to be substantial as compared to EDGE detectors. Even for mesa sizes as small as $45\,\mu m$, suitable for large area detector arrays, the enhancement factors are 3.8 and 1.8 for the CGW and CG detectors, respectively. Finally, the importance of a cladding layer (dielectric mirror) as well as of a high conductivity grating metal is demonstrated, each contributing a factor of two to η_{int}.

4. CONCLUSIONS AND SUMMARY

A theory of the reflection grating-coupled quantum well infrared detector is presented. The theory is based on the modal expansion method which is a near-exact way of calculating diffracted wave amplitudes of gratings of simple shapes. The theory in its basic form assumes gratings of infinite extension and grating metal of infinite conductivity. Both lamellar gratings of rectangular profile and crossed gratings with box-shaped or cylindrical cavities are analyzed. An approximative treatment of grating scattering is also presented, with the advantage of providing analytic expressions in closed form for lamellar as well as crossed gratings. The optical properties of the epilayers including the infrared-absorbing multi-QW stack are modeled by a transfer matrix approach. From the theory the absorptance η versus wavelength can be calculated, as well as the absorptance averaged over the $8-12$ µm atmospheric transmission window η_m. η_m is a suitable parameter of performance of a detector intended for thermal imaging. The theory is extended to include detector mesas of finite extension as well as grating metal coatings of finite conductivity.

The influence of grating shape and dimensions as well as the angle of incidence of the radiation on η_m are presented. Theoretical and experimental results on detectors with different types of gratings and with or without cladding layer are compared and discussed. The main results are collected in Tables 1 and 2. Table 1 shows data obtained by numerical simulation whereas Table 2 presents experimental results of grating coupled detectors. Both theory and experiments show that detectors with a crossed grating and cladding layer have a factor of about four higher absorptance than a 45° polished edge detector.

Acknowledgments

Thanks are due to J. Borglind, G. Landgren, N. Nordell, Z.F. Paska, and J. Wallin for providing the epitaxial structures fabricated by metal organic vapor phase epitaxy and for valuable discussions concerning growth technology.

APPENDIX A: FREE-SPACE ELECTROMAGNETIC MODES

In the free half-space above the grating the EM modes can be written

(expressing the E-field in the x-z plane)

$$R\,TE(x,y,z) = \frac{1}{\sqrt{D_x D_z \sin(\xi)\tau_{pq}}}(\gamma_{pq}\hat{x} - \alpha_p\hat{z})\,R_{pq}(x,y,z), \quad \text{(A1)}$$

$$R\,TM(x,y,z) = \frac{1}{\sqrt{D_x D_z \sin(\xi)\tau_{pq}}}(\alpha_p\hat{x} + \gamma_{pq}\hat{z})\,R_{pq}(x,y,z), \quad \text{(A2)}$$

for TE and TM modes, respectively, with

$$R_{pq}(x,y,z) = \exp\left[i(\alpha_p x + \beta_{pq} y + \gamma_{pq} z)\right], \quad \text{(A3)}$$

$$\tau_{pq} = \sqrt{\alpha_p^2 + \gamma_{pq}^2}, \quad \text{(A4)}$$

$$\alpha_p = \alpha_0 + \frac{2\pi p}{D_x}, \quad \text{(A5)}$$

$$\gamma_{pq} = \gamma_{00} + \frac{2\pi q}{D_z \sin\xi} - \frac{2\pi p \cot\xi}{D_x}, \quad \text{(A6)}$$

$$\beta_{pq} = \sqrt{k^2 - \alpha_p^2 - \gamma_{pq}^2}. \quad \text{(A7)}$$

The components of the incident wave along the coordinate axes are

$$\alpha_0 = k\sin\theta\cos\phi, \quad \text{(A8)}$$

$$\beta_0 = k\cos\theta, \quad \text{(A9)}$$

$$\gamma_{00} = k\sin\theta\sin\phi, \quad \text{(A10)}$$

where ϕ is the azimuth angle $(0 \to 2\pi)$ in the x-z plane measured from the z axis, and θ the polar angle $(0 \to \pi)$ measured from the y axis.

ξ is the angle between x and z axes. For square symmetry $\xi = 90°$ whereas for hexagonal symmetry $\xi = 60°$. For the case of box-shaped cavities only square symmetry with $\xi = 90°$ is treated.

APPENDIX B: ELECTROMAGNETIC CAVITY MODES OF BOX-SHAPED CAVITIES

The cavity modes for the case of box-shaped cavities can be written (expressing the transverse E field, i. e., the x-z plane), for $\xi = 90°$

$$N\,TE_{nm}(x,y,z) = M\,TE_{nm}(x,z)\sin\left[u_{nm}(y+h)\right], \quad \text{(B1)}$$

$$N\,TM_{nm}(x,y,z) = M\,TM_{nm}(x,z)\sin\left[u_{nm}(y+h)\right], \quad \text{(B2)}$$

for TE and TM modes, respectively, where

$$MTE_{nm}(x,z) = g_{nm}\left[\frac{m\pi}{d_z}\cos\left(\frac{n\pi x}{d_x}\right)\sin\left(\frac{m\pi z}{d_z}\right)\hat{x}\right.$$

$$\left. - \frac{n\pi}{d_x}\sin\left(\frac{n\pi x}{d_x}\right)\cos\left(\frac{m\pi z}{d_z}\right)\hat{z}\right], \qquad (B3)$$

$$MTM_{nm}(x,z) = g_{nm}\left[\frac{n\pi}{d_x}\cos\left(\frac{n\pi x}{d_x}\right)\sin\left(\frac{m\pi z}{d_z}\right)\hat{x}\right.$$

$$\left. + \frac{m\pi}{d_z}\sin\left(\frac{n\pi x}{d_x}\right)\cos\left(\frac{m\pi z}{d_z}\right)\hat{z}\right], \qquad (B4)$$

$$g_{nm} = \left(\frac{(2-\delta_{n0})(2-\delta_{m0})}{d_x d_z}\right)^{1/2}\left[\left(\frac{n\pi}{d_x}\right)^2 + \left(\frac{m\pi}{d_z}\right)^2\right]^{-1/2} \qquad (B5)$$

The quantities of the type IEM_{nm}^{pq} are scalar products of free space and cavity modes calculated at $y = 0$ within the cavity openings:

$$IEM_{nm}^{pq} = \int \overline{RTE_{pq}(x,z,0)}\, MTM_{nm}(x,z)\, dA, \qquad (B6)$$

and similarly for the other scalar products. dA is an area element $[= dx\, dz]$.

One obtains

$$IEE_{nm}^{pq} = \frac{g_{nm}\pi}{\sqrt{D_x D_z}\,\tau_{pq}}\left[\left(\frac{m\gamma_{pq}}{d_z}\right)IX_{nm}^{pq} + \left(\frac{n\alpha_p}{d_x}\right)IZ_{nm}^{pq}\right], \qquad (B7)$$

$$IEM_{nm}^{pq} = 0, \qquad (B8)$$

$$IME_{nm}^{pq} = \frac{g_{nm}\pi}{\sqrt{D_x D_z}\,\tau_{pq}}\left[\left(\frac{m\alpha_p}{d_z}\right)IX_{nm}^{pq} - \left(\frac{n\gamma_{pq}}{d_x}\right)IZ_{nm}^{pq}\right], \qquad (B9)$$

$$IMM_{nm}^{pq} = \frac{g_{nm}\pi}{\sqrt{D_x D_z}\,\tau_{pq}}\left[\left(\frac{n\alpha_p}{d_x}\right)IX_{nm}^{pq} + \left(\frac{m\gamma_{pq}}{d_z}\right)IZ_{nm}^{pq}\right], \qquad (B10)$$

where

$$IX_{nm}^{pq} =$$

$$\frac{i\alpha_p(m\pi/d_z)[(-1)^n\exp(-i\alpha_p d_x)-1][(-1)^m\exp(-i\gamma_{pq}d_z)-1]}{[\alpha_p^2 - (n\pi/d_x)^2][\gamma_{pq}^2 - (m\pi/d_z)^2]}, \qquad (B11)$$

$$IZ_{nm}^{pq} =$$

$$\frac{i\gamma_{pq}(n\pi/d_x)[(-1)^n \exp(-i\alpha_p d_x) - 1][(-1)^m \exp(-i\gamma_{pq} d_z) - 1]}{[\alpha_p^2 - (n\pi/d_x)^2][\gamma_{pq}^2 - (m\pi/d_z)^2]} \tag{B12}$$

In order to calculate the absorption losses in the grating metal the following explicit expression of the components of the H field is of use

$$H_x = \frac{i}{Z_0} \sum_{n,m} \left\{ \left[n\frac{u_{nm}}{k} \alpha_{mn}^{tot} - m\frac{k}{u_{nm}} c_{nm}^{tot} \right] \frac{\pi}{d} g_{nm} \right.$$

$$\left. \sin\left(\frac{n\pi x}{d}\right) \cos\left(\frac{m\pi z}{d}\right) \cos[u_{nm}(y+h)] \right\}, \tag{B13}$$

$$H_y = -\frac{i}{Z_0} \sum_{n,m} a_{nm}^{tot}(m^2 + n^2)\left(\frac{\pi}{d}\right)^2 g_{nm}$$

$$\cos\left(\frac{n\pi x}{d}\right) \cos\left(\frac{m\pi z}{d}\right) \sin[u_{nm}(y+h)] \right\}, \tag{B14}$$

$$H_z = \frac{i}{Z_0} \sum_{n,m} \left\{ \left[m\frac{u_{nm}}{k} a_{nm}^{tot} + n\frac{k}{u_{nm}} c_{nm}^{tot} \right] \frac{\pi}{d} g_{nm} \right.$$

$$\left. \cos\left(\frac{n\pi x}{d}\right) \sin\left(\frac{m\pi z}{d}\right) \cos[u_{nm}(y+h)] \right\}, \tag{B15}$$

if unit cells and cavity openings of square cross-section are assumed ($D_x = D_z = D$ and $d_x = d_z = d$). In this case the total amplitudes $a_{nm}^{tot} = \sum_{rs} a_{nm}^{rs}$ and $c_{nm}^{tot} = \sum_{rs} c_{nm}^{rs}$.

APPENDIX C: ELECTROMAGNETIC CAVITY MODES OF CYLINDRICAL CAVITIES

By using cylindrical coordinates (r, φ, y) and unit vectors $\hat{r}$ and $\hat{\varphi}$, the cavity modes for the case of cylindrical cavities can be written (expressing the transverse E field, i.e., the x-z plane)

$$NTE_{nml}(x, y, z) = MTE_{nml}(r, \varphi) \sin[w'_{nm}(y+h)], \tag{C1}$$

$$NTE_{nml}(x, y, z) = MTM_{nml}(r, \varphi) \sin[w_{nm}(y+h)], \tag{C2}$$

for TE and TM modes, respectively, where

$$MTE_{nml}(r, \varphi) =$$

$$g_{nm}\left[\hat{r}\frac{n}{r}J_n\left(\frac{\chi'_{nm}r}{a}\right) \times \begin{Bmatrix} -\cos(n\varphi) \\ \sin(n\varphi) \end{Bmatrix} + \hat{\varphi}\frac{\chi'_{nm}}{a}J'_n\left(\frac{\chi'_{nm}}{a}\right) \times \begin{Bmatrix} \sin(n\varphi) \\ \cos(n\varphi) \end{Bmatrix}\right], \quad \text{(C3)}$$

$$MTM_{nml}(r, \varphi) =$$

$$h_{nm}\left[\hat{r}\frac{\chi_{nm}}{a}J'_n\left(\frac{\chi_{nm}r}{a}\right) \times \begin{Bmatrix} \cos(n\varphi) \\ \sin(n\varphi) \end{Bmatrix} + \hat{\varphi}\frac{n}{r}J_n\left(\frac{\chi_{nm}r}{a}\right) \times \begin{Bmatrix} \sin(n\varphi) \\ -\cos(n\varphi) \end{Bmatrix}\right], \quad \text{(C4)}$$

where the lower expression in brackets is for $l = 1$ and the upper for $l = 2$.

The normalizing constants g_{nm} and h_{nm} can be written:

$$g_{nm} = \left(\frac{2 - \delta_{m0}}{\pi}\right)^{1/2} \frac{1}{J_n(\chi'_{nm})\sqrt{\chi'^2_{nm} - n^2}}. \quad \text{(C5)}$$

$$h_{nm} = \left(\frac{2 - \delta_{m0}}{\pi}\right)^{1/2} \frac{1}{J_{n-1}(\chi_{nm})\chi_{nm}}, \quad \text{(C6)}$$

$$P^1_{NMLnml} = \sum_{p,q = -\infty}^{\infty} \left(\frac{\beta_{pq}}{k}IEE^{pq}_{NML}\overline{IEE^{pq}_{nmi}} + \frac{k}{\beta_{pq}}IME^{pq}_{NML}\overline{IME^{pq}_{nmi}}\right), \quad \text{(C7)}$$

$$P^2_{NMLnml} = \sum_{p,q = -\infty}^{\infty} \left(\frac{\beta_{pq}}{k}IME^{pq}_{NML}\overline{IMM^{pq}_{nml}}\right), \quad \text{(C8)}$$

$$Q^1_{NMLnml} = \sum_{p,q = -\infty}^{\infty} \left(\frac{k}{\beta_{pq}}IMM^{pq}_{NML}\overline{IME^{pq}_{nml}}\right), \quad \text{(C9)}$$

$$Q^2_{NMLnml} = \sum_{p,q = -\infty}^{\infty} \left(\frac{k}{\beta_{pq}}IMM^{pq}_{NML}\overline{IMM^{pq}_{nml}}\right), \quad \text{(C10)}$$

The quantities of the type IEE^{pq}_{nml}, etc., are scalar products of free space and cavity modes at $y = 0$ within the cavity openings:

$$IEM^{pq}_{nml} = \int RTE_{pq}(x, z, 0)\, MTM_{nml}(r, \varphi)dA, \quad \text{(C11)}$$

analogous to the case of box-shaped cavities. They become

$$IEE^{pq}_{nml} = \frac{-g_{nm}2\pi a(-i)^{n-1}\chi'^2_{nm}}{\sqrt{D_xD_z}\sin(\xi)(a^2\tau^2_{pq} - \chi'^2_{nm})} \times J_n(\chi'_{nm})J'_n(\tau_{pq}a) \times \begin{Bmatrix} \sin(n\varphi') \\ \cos(n\varphi') \end{Bmatrix}, \quad \text{(C12)}$$

$$IEM^{pq}_{nml} = 0, \quad \text{(C13)}$$

$$IME_{nml}^{pq} = \frac{-g_{nm}\, 2\pi n(-i)^{n-1}}{\sqrt{D_x D_z} \sin(\xi)\, \tau_{pq}} \times J_n(\chi'_{nm})\, J_n(\tau_{pq} a)$$

$$\times \begin{cases} -\cos(n\varphi') \\ \sin(n\varphi') \end{cases}, \tag{C14}$$

$$IMM_{nml}^{pq} = \frac{-h_{nm}\, 2\pi a^2 \chi_{nm}(-i)^{n-1}\tau_{pq}}{\sqrt{D_x D_z}\sin(\xi)(a^2\tau_{pq}^2 - \chi_{nm}^2)} \times J'_n(\chi_{nm})\, J_n(\tau_{pq}a)$$

$$\times \begin{cases} \cos(n\varphi') \\ \sin(n\varphi') \end{cases}, \tag{C15}$$

$$\varphi' = \arctan(\alpha_p / \gamma_{pq}), \tag{C16}$$

where the lower expression in the brackets on the righthand side refers to $l = 1$ and the upper for $l = 2$.

References

1. B.F. Levine, *J. Appl. Phys.*, **74**, R1 (1993) and references therein.
2. J.Y. Andersson and L. Lundqvist, "*Intersubband Transitions in Quantum Wells*": Proceedings of the NATO Advanced Research Workshop, Cargese, Corsica, France Sept. 10–13, 1991. Eds. E. Rosencher, B. Vinter, B.F. Levine.
3. L.J. Kozlowsky, G.M. Williams, G.J. Sullivan, C.W. Farley, R.J. Anderson, J.K. Chen, D.T. Cheung, W.E. Tennant, and R.E. DeWames, *IEEE Trans Electron. Devices*, **38**, 1124 (1991).
4. B.F. Levine, C.G. Bethea, K.G. Glogovsky, J.W. Stayt, and R.E. Leibenguth, "*Narrow Band Gap Semiconductors*": Proceedings of the NATO Advanced Research Workshop, Oslo, Norway, June 25–27 (1991).
5. R.L. Whitney, F.W. Adams, and K.F. Cuff, "*Quantum Well Intersubband Transition Physics and Devices*", Proceeding of the NATO Advanced Research Workshop, Whistler, Canada, Sept. 7–10 1993. Eds.H.C. Liu, B.F. Levine, J.Y. Andersson.
6. J.Y. Andersson, L. Lundqvist, J. Borglind, and D. Haga, *Quantum Well Intersubband Transition Physics and Devices*", Proceeding of the NATO Advanced Research Workshop, Whistler, Canada, Sept. 7–10 1993, Eds. H.C. Liu, B.F. Levine, J.Y. Andersson.
7. K.W. Gossen, S.A. Lyon, and K. Alavi, *Appl. Phys. Lett.*, **53**, 1027 (1988).
8. G. Hasnain, B.F. Levine, C.G. Bethea, R.A. Logan, J. Walker, and R.J. Malik, *Appl. Phys. Lett.*, **54**, 2515 (1989).
9. J.Y. Andersson, L. Lundqvist, and Z.F. Paska, *Appl. Phys. Lett.*, **58**, 2264 (1991).
10. J.Y. Andersson and L. Lundqvist, *Appl. Phys. Lett.*, **59**, 857 (1991).
11. L.S. Yu, S.S. Li, and Y.C. Wang, *J. Appl. Phys.*, **72**, 2105 (1992).
12. Y.-C. Wang and S.S. Li, *J. Appl. Phys.*, **74**, 2192 (1993).
13. J.Y. Andersson and L. Lundqvist, *J. Appl. Phys.*, **71**, 3600 (1992).
14. B.F. Levine, G. Sarusi, S.J. Pearton, K.M.S. Bandara, and R.E. Leibenguth, *Quantum Well Intersubband Transition Physics and Devices*", Proceeding of the NATO Advanced Research Workshop, Whistler, Canada, Sept. 7–10 1993, Eds, H.C. Liu, B.F. Levine, J.Y. Andersson.
15. E. Yablonovitch and G.D. Cody, *IEEE Trans. Electron Devices*, **29**, 300 (1982).

16. M.J. Kane and N. Apsley, *J. Phys.*, (Paris) Colloq. **5,** C–545 (1987).
17. R. Petit, "*Electromagnetic Theory of Gratings*", Springer, Berlin, 1980, p. 26 and 227.
18. P. Sheng, *RCA Rev.* **39,** 513 (1978).
19. Y.-L. Kok, IEEE Proceedings of the 21st Southeastern Symposium on System Theory, Tallahassee, Florida, 1989, p. 493.
20. R.C. McPhedran and D. Maystre, *Appl. Phys.*, **14,** 1 (1977).
21. C.-C. Chen, IEEE *Trans. Microwave Theory Tech.*, **MTT-19,** 475 (1971).
22. See e.g.: D.K. Cheng, "*Field and Wave Electromagnetics*" 2nd ed., Addison-Wesley Publishing Co. 1989.
23. Z.F. Paska, J.Y. Andersson, L. Lundqvist, and C-O.A. Olsson, *J. Cryst. Growth*, **107,** 845 (1991).
24. L. Lundqvist, J.Y. Andersson, Z.F. Paska, J. Borglind, and D. Haga, *Appl. Phys. Lett.*, **63,** 3361 (1993).

CHAPTER 5

Normal Incidence Detection of Infrared Radiation in *P*-type GaAs/AlGaAs Quantum Well Structures

GAIL J. BROWN[1] and FRANK SZMULOWICZ[1,2]

[1]*Wright Laboratory, Materials Directorate, WL/MLPO,*
Wright-Patterson AFB, Ohio, 45433-7707 USA
[2]*University of Dayton Research Institute, Dayton, Ohio, 45469-0178 USA*

1. INTRODUCTION

Considerable progress has been made in the last ten years in the development of multiple quantum well heterostructures for long-wavelength (8–12 μm) infrared detection [1,2]. In these quantum well infrared photodetectors (QWIPs), intersubband transitions are used to photo-excite a carrier – an electron or a hole – from a filled ground state to a higher subband state near the top of the quantum well. These photo-excited carriers are then collected by applying an electric field perpendicular to the heterostructure layers. The III–V based QWIPs are of increasing interest because it is now possible to produce large area, high uniformity, two dimensional imaging arrays. Most previous work was focused on n-type (donor doped) GaAs/AlGaAs multiple quantum wells (MQWs) since the lower effective mass of the electron should lead to superior transport properties [1–3]. However, for these n-type QWIPs, intersubband transitions are limited by dipole selection rules to couple only to infrared radiation with a polarization component perpendicular to the quantum wells (i.e., normal incidence absorption is forbidden) [4,5]. This limitation leads to the use of gratings or 45° facets to bend the incident radiation so that it enters the quantum well stack with a polarization component in the growth direction (z polarization). Although excellent results have been obtained for n-type QWIPs with etched or metallized gratings, the use of gratings increases the number of process steps and adds the complication of achieving grating uniformity over a large two dimensional array [6–9]. An alternative approach is to use p-type (acceptor doped) QWIPs for which the intersubband selection rules intrinsically allow normal incidence absorption [10–12].

In p-type GaAs/AlGaAs multiple quantum well materials, there is strong mixing of the valence bands, especially with the conduction band, which imparts an additional s-like symmetry, which increases away from the zone center, to the normally p-like valence band wave functions. Transitions between a pair of subbands with s-like and p-like symmetry components, respectively, are dipole allowed by symmetry-related quantum mechanical selection rules, so that absorption of normal incidence radiation (x or y polarized) is allowed. In addition, the strong mixing of the light and heavy hole valence band states leads to the nonparabolicity and anisotropy (warping) of these bands. The presence of anisotropy of the constant energy surfaces also contributes to normal incidence absorption as in n-type QWs whenever the principal axes of their constant energy ellipsoids are oriented at an angle to the direction of the incident radiation. In *n*-type struc-tures, for proper orientations of the incident radiation with respect to the crystallographic axes, anisotropy leads to off-diagonal effective mass tensor components. In turn, the presence of nondiagonal effec-tive mass tensor elements allows electric field in the *x-y* plane to produce charge separation in the *z*-direction, i.e., to produce an oscil-lating electric dipole. A similar effect related to valence band aniso-tropy is also present in *p*-type structures. This desirable normal inci-dence photo-excitation of *p*-type quantum well infrared photodetec-tors was studied only recently [13–15]. In the samples studied, the photoresponse for light polarized in the *x-y* plane was stronger than for light polarized in the *x-z* plane (where *z* is the growth direction, see Fig. 1). This is in contrast to *n*-type QWIPs where only radiation with a *z*-component of polarization can couple and transfer its energy to the electrons in the system (i.e. photo-excite the electrons). The pola-rization dependence of the *p*-type photoresponse confirms that *x* and *y* components of the polarization are major contributors to the inter-subband optical transition as predicted by theory [12].

However, the exact nature of the optical transition involved in the *p*-type photoresponse has proven difficult to model. For instance,

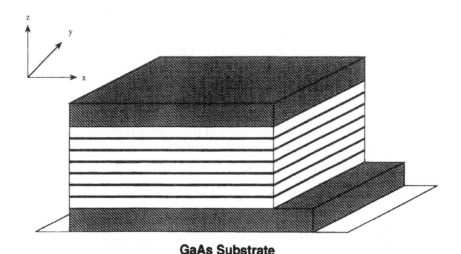

GaAs Substrate

Fig. 1 Drawing of a multi-quantum well mesa with the designated coordination sys-tem. The *z*-axis is in the growth direction. The *x-y* plane is the plane of the layers. The shaded layers are doped GaAs while the unshaded layers are AlGaAs.

most reports describe the photoresponse as due to a transition from the first heavy hole state to an extended subband state in the continuum. This extended state, proposed to be just above the quantum well barrier, has been attributed to either a heavy-hole subband (HH2) or an extended light hole (LH) subband [13, 15, 16]. While the HH1 to LH2 resonant subband transition would have a large optical matrix element [17], our calculations indicate that the LH2 resonance is broad, not much distinguished from ordinary continuum states, and at least 23 meV above the top of the well for the well widths used in those studies. A simple decoupled valence band model would inaccurately place the LH2 resonant position even higher in energy, nearly 100 meV above the top of the well. As a result, this resonance is not efficiently coupled to the localized HH1 state. In addition, the HH1 to HH2 transition would be a bound-to-bound state excitation, which does not match the experimental data. Therefore, the observed photoresponse is most likely due to a transition into the continuum of states above the barrier potential, similar to an extrinsic transition in a doped bulk semiconductor [18]. This coupling to the continuum states has been considered by previous theories which modeled bound-to-continuum n-type QWIPs [19, 20]. In order to obtain a better understanding of the types of intersubband transitions involved in these p-type QWIPs, we developed an intersubband absorption model which includes coupling to all states in the continuum, including resonant states. In addition, most previous models were based on a 4×4 envelope function approximation (EFA) model or a simplified 8×8 EFA model for calculating the energy band structures of the valence subbands [12, 16, 17]. Owing to the strong admixture of s-like states into valence band wave functions and the strong influence of valence band anisotropy, we have extended the theory to include a state-of-the-art 8×8 EFA calculation of the valence subbands, which more accurately reflects the coupling among the conduction, heavy-hole, light-hole, and spin-orbit split-off-hole bands.

2. WAVE FUNCTIONS AND BAND STRUCTURE

To calculate the absorption coefficient due to an intersubband transition in p-type GaAs/AlGaAs, we need to know the valence band structure $E_N(\mathbf{k}_\parallel)$ and the momentum matrix elements connecting the Nth and the Mth subbands $P_{MN}(\mathbf{k}_\parallel)$. The model for optical transitions between valence band subband levels is much more complex than that required for similar conduction band transitions. In the case of optical transitions between conduction subbands, a simple one band effective mass model is adequate in most cases because the conduction band in question is usually far removed from other bands [21, 22]. However, the valence band structure, being very anisotropic and nonparabolic, is most complex due to mixing between energetically close subbands. This band mixing is the consequence of the interaction between the degenerate three top valence bands and the proximity of the conduction band in the constituent bulk materials of the MQW. Thus, a multiple-band effective mass theory (and, consequently, a multiple band envelope function approximation) [21, 22] is required. To include the coupling among all four pairs of bands (conduction, heavy hole, light hole, and spin-orbit split-off) an eight band EFA model is used [23]. In particular, it is important to incorporate the coupling of the light- and heavy-hole states to the conduction band because it imparts an *s*-like character to the nominally *p*-like valence band states; this coupling is one mechanism which makes dipole transitions between valence band subbands possible at normal incidence. In the following, the electronic structure, wave functions, and optical matrix elements are obtained from an 8×8 envelope-function approximation (EFA) calculation.

Consider the quantum well formed by growing material *B* (GaAs) between two thick layers of material *A* and *C* (AlGaAs), where *A* and *C* has a band gap larger than *B* and where the band discontinuities are such that both types of carriers are confined in the *B* material (Fig. 2). In a quantum well, the total wave function for the *N*th subband is an eigenstate of parallel momentum $\mathbf{k}_\parallel$ in the plane of the well and, in principle, involves an infinite sum over all bands of the bulk material. To first order, the total wave function, in each layer of a heterostructure, can be written as [22]

$$
\Psi_N(\mathbf{k}_\parallel, \mathbf{r}) = e^{i\mathbf{k}_\parallel \cdot \mathbf{r}} \sum_\mu^A \left\{ F_\mu(N\mathbf{k}_\parallel, z) \left[u_\mu^0(\mathbf{r}) - \sum_\gamma^B \frac{\frac{\hbar}{m} \mathbf{k}_\parallel \cdot \mathbf{p}_{\gamma\mu}}{E_\gamma - E_\mu} u_\gamma^0(\mathbf{r}) \right] \right.
$$

$$
\left. + i \frac{\partial F_\mu(N\mathbf{k}_\parallel, z)}{\partial z} \sum_\gamma^B \frac{\frac{\hbar}{m} \mathbf{p}_{\gamma\mu}^z}{E_\gamma - E_\mu} u_\gamma^0(\mathbf{r}) \right\} \tag{1}
$$

**Conduction
Band Edge**

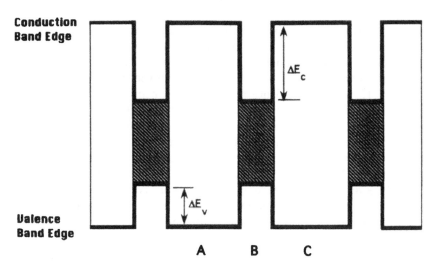

**Valence
Band Edge**

Fig. 2   Band edge diagram showing the band offsets which form the quantum wells in the conduction and valence bands. The confined states are in material *B*. For a GaAs/AlGaAs MQW, the GaAs layers would be material *B*.

where the first terms is the familiar product of envelope and Bloch wave functions. However, since the usual EFA formalism derives from the second-order degenerate $\mathbf{k} \cdot \mathbf{p}$ theory, the proper description of the total wave function requires a term of order $\mathbf{k}_\parallel$ and a second term which is the first derivative of the envelope function. These "extra" terms in Eq. (1) are important not only for satisfying boundary conditions but also for obtaining the correct dipole matrix elements between subbands. In practical application of the EFA method [12, 21 – 29], the infinite set of states at the center of the Brillouin zone is divided into two sets: the A-set of near states and the B-set of far states. In our case, set A contains eight states, while set B contains all other bands in the problem.

In the total wave function equation, Eq. (1),

$$F_\nu(N\mathbf{k}_\parallel, z) = \sum_{k_z} e^{ik_z z} F_\nu(N\mathbf{k}_\parallel, k_z), \qquad (2)$$

is the νth component of the envelope function, which is expanded in plane waves with amplitudes $F_\nu(N\mathbf{k}_\parallel, k_z)$. Function $u_\nu^0(\mathbf{r})$ is the νth Bloch wave function with energy E_ν at the center of the Brillouin zone of the bulk material (well or barrier). The momentum matrix elements between the zone center Bloch wave functions are denoted as $p_{\mu v}$. Exponents k_z are a set of numbers with the dimension of wave vectors which are the solutions of the $\mathbf{k} \cdot \mathbf{p}$ secular equation

$$|H(\mathbf{k}_{\parallel}, k_z) - E| = 0, \tag{3}$$

and H is the bulk $\mathbf{k} \cdot \mathbf{p}$ Hamiltonian of order K. Equation 4 yields $2K$ solutions for $k_z(\mathbf{k}_{\parallel}, E)$, which are wave vector- and energy-dependent and, in general, complex [27].

2.2. EFA Hamiltonian Equation

The standard EFA equation for the envelope function is obtained by replacing k_z by $-i(d/dz)$ in the bulk Hamiltonian equation [25]. The resulting multi-dimensional differential equation can be written as

$$\sum_{v}^{A} H_{\mu v}\left(\mathbf{k}_{\parallel}, -i\frac{d}{dz}\right) F_v(\mathbf{k}_{\parallel}, z) = E(\mathbf{k}_{\parallel}) F_{\mu}(\mathbf{k}_{\parallel}, z). \tag{4}$$

In second-order degenerate perturbation theory, the matrix elements of the bulk Hamiltonian are given by [24, 28]

$$H_{vv'}(\mathbf{k}) - E_v \delta_{vv'} = \frac{\hbar^2 k^2}{2m_0} \delta_{vv'} + \frac{\hbar}{m_0} \mathbf{k} \cdot \mathbf{p}_{vv}$$

$$+ \frac{1}{2}\left(\frac{\hbar}{m_0}\right)^2 \sum_{\gamma}^{B} \left[\frac{(\mathbf{k} \cdot \mathbf{p}_{v\gamma})(\mathbf{k} \cdot \mathbf{p}_{\gamma v'})}{E_v - E_\gamma} + \frac{(\mathbf{k} \cdot \mathbf{p}_{v\gamma})(\mathbf{k} \cdot \mathbf{p}_{\gamma v'})}{E_{v'} - E_\gamma}\right]$$

$$\equiv \sum_{ij} D_{vv'}^{ij} k_i k_j + \sum_{i} G_{vv'}^{i} k_i. \quad v, v' \in A \tag{5}$$

A complete 8×8 $\mathbf{k} \cdot \mathbf{p}$ Hamiltonian applicable to the quantum well problem can be found, for example, in the paper by Szmulowicz [26] or in Bastard's book [25]. In this Hamiltonian, hole confinement (the well potential) results from the discontinuity in the valence band profile across the heterostructure. This difference between the valence band edges of the well and barrier material, $E_v(\text{well}) - E_v(\text{barrier})$, appears only along the diagonal terms of Eq. (5).

For future reference, it is useful to expand the Hamiltonian in powers of k_z as [28]

$$H(\mathbf{k}) = H_2(\mathbf{k}_{\parallel}) k_z^2 + H_1(\mathbf{k}_{\parallel}) k_z + H_0(\mathbf{k}_{\parallel}). \tag{6}$$

For a given energy and parallel momentum, the bulk $\mathbf{k} \cdot \mathbf{p}$ Hamiltonians in the well and barrier materials can be solved for exponents k_z via equation 3 and for their associated right C and left L eigenvectors, which are defined through [28]

$$H(\mathbf{k}_{\|}, k_z) C(E, \mathbf{k}_{\|}, k_z) = EC(E, \mathbf{k}_{\|}, k_z)$$
$$L(E, \mathbf{k}_{\|}, k_z) H(\mathbf{k}_{\|}, k_z) = EL(E, \mathbf{k}_{\|}, k_z). \tag{7}$$

For the 8×8 case there are 16 linearly independent eigenvectors C (or L) corresponding to 8 pairs of doubly-degenerate wave vectors k_z.

2.3. Construction of Bound-State Wave Functions

For a symmetric quantum well, the total wave function, Eq. (1) can be chosen to be an eigenstate of parity, i.e., to be even or odd with respect to reflection in plane $z = 0$. When acted upon by the reflection operator, $O(IC_{2z})$ (consisting of 180° rotation about the z-axis, followed by inversion), each Bloch wave function transforms as [29]

$$O(IC_{2z}) u_v^0 = \sum_\mu \Gamma_{\mu v}(IC_{2z}) u_\mu^0, \tag{8}$$

which defines matrix $\Gamma(IC_{2z})$. This implies that the envelope functions must also have definite transformation properties upon reflection: for example, for the even state

$$\Gamma(IC_{2z}) F(\mathbf{k}_{\|}, -z) = F(\mathbf{k}_{\|}, z). \tag{9}$$

In general, the envelope function is a linear combination of sixteen eigenvectors $C(\mathbf{k}_{\|}, k_i)$ times a plane wave with the corresponding wave vector k_i. In the left-hand barrier, one selects eight of the exponents whose imaginary parts are negative (guaranteeing that the state decays under the barrier) and arranges them from 1 to 8. The resulting envelope function can be written as

$$F^A(\mathbf{k}_{\|}, z) = \sum_{i=1}^{8} A_i C^A(\mathbf{k}_{\|}, k_i) e^{ik_i^A(z+a)}, \quad z < -a.$$
$$\equiv C_A E_A(z+a) A. \tag{10}$$

The corresponding envelope function in the right-hand barrier for the even state is given by

$$F^C(\mathbf{k}_{\|}, z) = \sum_{i=1}^{8} A_i C^A(\mathbf{k}_{\|}, -k_i) e^{-ik_i^A(z-a)}, \quad z > a$$
$$\equiv (\Gamma C_A) E_A(a-z) A, \tag{11}$$

where we define that 8×8 matrices

$$\mathbf{\Gamma} \equiv \Gamma(IC_{2z}) \tag{12}$$

$$(C_A)_{ij} = C_j^A(\mathbf{k}_\|, k_i^A) \tag{13}$$

$$E_A(z)_{ij} = e^{ik_i^A z} \delta_{ij}. \tag{14}$$

In the well, all sixteen exponents are necessary to express the wave function. For the even state, the envelope function is given by

$$F^B(\mathbf{k}_\|, z) = \sum_{i=1}^{8} [C^B(\mathbf{k}_\|, k_i) e^{ik_i^B z} + C^B(\mathbf{k}_\|, -k_i) e^{-ik_i^B z}] B_i \quad -a < z < a$$

$$\equiv [C_B E_B(z) + \Gamma C_B E_B(-z)]B. \tag{15}$$

There are 16 unknowns: two 8×1 vectors with coefficients A_i and B_i. In order to solve for the wave function, sixteen boundary conditions are necessary, as will be shown next.

2.3.1. *Boundary Conditions*

Because the wave function has been chosen to have definite parity, we need to set up boundary conditions at one interface only. The continuity of the wave function at $z = -a$ gives eight equations

$$C_A A = [C_B E_B(-a) + \Gamma C_B E_B(a)]B \tag{16}$$

and the boundary condition for the derivative [22, 23] yields additional eight conditions

$$\left\{ H_2^A C_A K_A + \frac{1}{2} H_1^A C_A \right\} A = \left\{ H_2^B [C_B K_B E_B(-a) - \Gamma C_B K_B E_B(a)] \right.$$

$$\left. + \frac{1}{2} H_1^B [C_B E_B(-a) + \Gamma C_B E_B(a)] \right\} B, \tag{17}$$

where

$$K_{ij} = k_i \delta_{ij}. \tag{18}$$

These two boundary conditions can be combined to yield a homogeneous equation for energy eigenvalues and associated eigenvectors, in the form

$$\left\{ \begin{matrix} C_A & -M_B \\ H_2^A C_A K_A + H_1^A C_A & -(H_2^B N_B + \frac{1}{2} H_1^B M_B) \end{matrix} \right\} \begin{pmatrix} A \\ B \end{pmatrix} = 0,$$

where

$$M_B \equiv C_B E_B(-a) + \Gamma C_B E_B(a)$$

and

$$N_B \equiv C_B K_B E_B(-a) - \Gamma C_B K_B E_B(a).$$

The bound-state wave functions are then normalized and the matrix elements of momentum between each pair of bands is obtained. In particular, the intraband matrix element for each band is related to its energy gradient via

$$\frac{\hbar}{m_0} P_{NN}(\mathbf{k}_\parallel) = \frac{\partial E_N(\mathbf{k}_\parallel)}{\partial \mathbf{k}_\parallel}. \tag{19}$$

Using the normalized wave function, one can examine its band content by calculating the band-by-band decomposition of the total normalization

$$1 = \sum_n w_n, \tag{20}$$

where w_n gives the contribution of the nth band to the normalization, i.e., the band weight.

2.4. Continuum States

Continuum states are obtained in the second step of the calculation [26]. In our calculation, we eschew the use of an artificial large box to enclose the whole system in order to create two extra boundaries to confine the continuum wave functions, thus creating an extra quantization condition. By avoiding this stratagem, we are the first to treat this problem as a true continuum problem [30]. We do this as follows. First we select a parallel momentum and energy pair $(E, \mathbf{k}_\parallel)$. Next, we solve the bulk $\mathbf{k} \cdot \mathbf{p}$ Hamiltonians in the well and barrier materials in order to obtain exponents k_z and their associated eigenvectors C and L, Eq. (7). If any of the coefficients k_z in the barrier are real, we can proceed to construct a continuum state based on it.

For the energy range of interest to our study (below the spin-orbit energy of the well material), the continuum states derive from the heavy-hole or light-hole states in the barrier. Due to Kramer's degeneracy, there are two HH and two LH positive-going states in the barrier and the like number of negative-going states for a given energy and parallel momentum pair. For convenience, the positive- and negative-going states are superposed to be either even or odd with respect to the reflection in the plane through the middle of the well. Upon hitting the well, an incident state that is of either HH or LH character is scattered back and transmitted as a mixture of all eight states in the model. The exact admixture is determined by solving the appropriate boundary condition matrix for the scattering problem. This approach is capable of yielding not only regular continuum states but also resonant states in the continuum.

The detail of the procedure outlined above is as follows. Using propagating states in the barrier–i.e., those with $\mathrm{Re}\, k_v > 0$, corresponding to light or heavy-hole states–we construct a state incident from the left-hand side ($z < -a$) in material A, that is,

$$F^A(\mathbf{k}_\parallel, z) = C_v^A(E, \mathbf{k}_\parallel) e^{ik_v^A z}. \tag{21}$$

There are two linearly independent eigenvectors C for the given k_v, which we call "spin-up" and "spin-down" states. (Even though each state is a mixture of up and down spins, one state is more "up" than the other.) The above state derives from the vth Bloch state in the barrier with wave vector $\mathbf{k} = (\mathbf{k}_\parallel, k_v)$. Correspondingly, there is a back-scattered state in the barrier, which has contributions from eight negative-going states (or those that decay to the left). Altogether, the total wave function in the left hand barrier can be written as

$$F^A(\mathbf{k}_\parallel, z) = C_v^A(\mathbf{k}_\parallel) e^{ik_v^A z} + \sum_{i=1}^{8} c_i^A \Gamma(IC_{2z}) C_i^A(E, \mathbf{k}_\parallel) e^{-ik_i^A z},$$

$$\mathrm{Re}(k_i^A) > 0, \quad \mathrm{Im}(k_i^A) < 0. \tag{22}$$

The condition on the imaginary part guarantees that the reflected states do not blow up at negative infinity. The transmitted state in the right hand barrier is given by

$$F^C(\mathbf{k}_\parallel, z) = \sum_{i=1}^{8} c_i^C C_i^A(E, \mathbf{k}_\parallel) e^{ik_i^A z}, \quad \text{with } \mathrm{Re}(k_i^A) > 0, \quad \mathrm{Im}(k_i^A) < 0. \tag{23}$$

Correspondingly, there is a state incident from the right-band barrier.

The states incident from the right and the left are then superposed to yield even and odd parity stationary states for both up and down solutions. In shorthand notation, the even parity wave function can be written as

$$F^A(\mathbf{k}_\parallel, z) = C_A E_A(z+a)A + \Gamma C_A E_A(-a-z)C, \quad z < -a, \quad (24)$$

where for the νth incident channel only the νth component of column vector A is nonzero and is given by

$$A_i = \frac{\delta_{i\nu}}{\sqrt{2L}}, \quad (25)$$

which normalizes the incident wave function to unity over the length L of the structure. The scattered amplitudes are given by

$$C = E_A(a)(c^A + c^C). \quad (26)$$

For the transmitted state in the well, material $B, (-a < z < a)$, the even parity state can be written as before as

$$F(\mathbf{k}_\parallel, z) = \sum_{i=1}^{8} [C_i^B(E, \mathbf{k}_\parallel)e^{ik_i^B z} + \Gamma(IC_{2z})C_i^B(E, \mathbf{k}_\parallel)e^{-ik_i^B z}]B_i$$

$$\equiv [C_B E_B(z) + \Gamma C_B E_B(-z)]B. \quad (27)$$

Correspondingly, one can construct an odd state.

2.4.1. Boundary Conditions

Becasue the wave function has been chosen to have definite parity, we need to set up boundary conditions at one interface only. For the even state, proceeding similarly to the case of the bound-state wave function, the wave function and current continuity equations at $z = -a$ result in an inhomogenous equation for coefficients C and B in terms of the amplitude of the incident state A, Eq. (25), that is

$$\begin{pmatrix} C \\ B \end{pmatrix} = -\left[\begin{matrix} \Gamma C_A & -M_B \\ -\Gamma\left[H_2^A C_A K_A + \tfrac{1}{2}H_2^A C_A\right] & -\left(H_2^B N_B + \tfrac{1}{2}H_1^B M_B\right) \end{matrix} \right]^{-1}$$

$$\times \begin{bmatrix} C_A & 0 \\ 0 & H_{2A}C_A K_A + \tfrac{1}{2}H_{1A}C_A \end{bmatrix} \begin{pmatrix} A \\ A \end{pmatrix},$$

where the symbols have the same meaning as before.

The stationary solutions obtained by solving the secular equation can be shown to obey the zero-current condition, which is expressed through the relation

$$\sum_i [|A_i|^2 - |C_i|^2] v_i = 0,$$

where the velocity (which is proportional to the energy gradient) is given by

$$v_i \equiv \frac{1}{\hbar} \frac{\partial E}{\partial k_i} = \frac{1}{\hbar} L_i [2H_{2A} k_i + H_{1A}] C_i.$$

The zero-current condition is trivially satisfied by the well solution since the amplitudes for the positive and negative-going states are equal by construction.

We recall that for a simple band (single-band model) resonant states occur when an integral number of wave function half-wavelengths fit inside the well width, [25] i.e., $n\lambda/2 = 2a$ (or $ka = n\lambda/2$). For odd (even) parity states, the resonance occurs when n is odd (even). At resonance, the amplitude of the wave function inside the GaAs layer is maximum, which results in a quasi-bound continuum state. Indeed, since the energies of these quasi-bound states exceed the height of the potential well, the quasi-bound states are the continuation of the true bound states in the well and, at some larger well width, they enter the well as true bound states. The resonance condition in the valence band continuum is complicated by the coupled-band character of the problem.

2.5. Bound-State Energy Band Structure

Before dealing with continuum states, it is beneficial to list the results for the bound states, which are obtained using the input parameters listed in Appendix A. Figure 3 shows the calculated energy bands for a 50 Å GaAs/Al$_{0.3}$Ga$_{0.7}$As quantum well along the (11) direction in the Brillouin zone. Various subbands can be classified as either heavy-hole (HH), light-hole (LH), spin-orbit split-off hole (SH), or conduction electron (C) like according to their dominant character given by w_n (in the 8×8 model, the LH, SH, and C bands are coupled even at

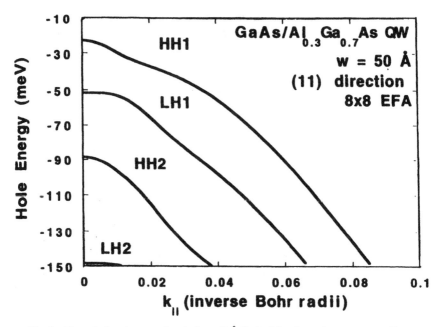

Fig. 3 The calculated energy bands for a 50Å GaAs/Al$_{0.3}$Ga$_{0.7}$As quantum well as a function of parallel momentum in the (11) direction of the Brillouin zone (Bohr radius = 0.529177Å). HH and LH refer to the heavy- and light-hole subbands. The zero of energy is at the top of the valence band of bulk GaAs. We use a 60/40 conduction-to-valence band offset ratio. The well depth is 149.6 meV.

the center of the Brillouin zone). The mixing of bands increases away from the Brillouin zone center as the band coupling is proportional to parallel momentum. This gives rise to avoided crossings and non-parabolicities of the resulting bands. As importantly, there is an admixture of the conduction band content into the valence band sub-bands. The ground state subband is labeled as HH1 and is followed by subbands LH1, HH2, and LH2. The latter band is located near the top of the well, and for well widths of about 47 Å and less it becomes a resonant state above the top of the well. Note that the bands are characterized by large nonparabolicities. In fact, the LH1 band first curves upward, resulting in an electron-like curvature.

Since the selection rules for bound-to-bound and bound-to-continuum transitions depend on band content of the final and initial states, it is of interest to present envelope function plots. In Fig. 4, the ground

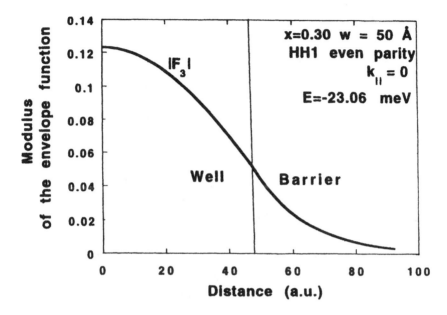

Fig. 4 Plot of the modulus of the normalized HH1 envelope function at $k_{\parallel} = 0$ for the case of a 50 Å GAs/Al$_{0.3}$Ga$_{0.7}$ As quantum well (a.u. = 0.529177 Å). This even parity HH1 state has a dominant $v = 3$ (heavy-hole) component.

state HH1 wave function at the center of the Brillouin zone is graphed. Because the heavy hole band is decoupled from the other bands at the center of the Brillouin zone, the wave function has only one heavy-hole component $v = 3$ (see Appendix B for the order of basis states), which has even parity. Similarly, the next heavy-hole state, HH2, has only the $v = 6$ component, whose parity itself is odd. The character of the light-hole states gets more interesting. The first light-hole state at the center of the Brillouin zone is dominated by the even-parity $v = 5$ component and has strong contributions from the conduction band component $v = 2$ (odd parity) and the split-off component $v = 8$ (even parity). The next light-hole state, LH2 in Fig. 5, is dominated by $v = 4$ (odd), but is also strongly coupled to a conduction band component $v = 1$ (even) and a split-off component $v = 7$ (odd). The knowledge of the parities of individual envelope components is of use in determining the selection rules for interband transitions for different light polarizations. Away from the Brillouin zone center, the

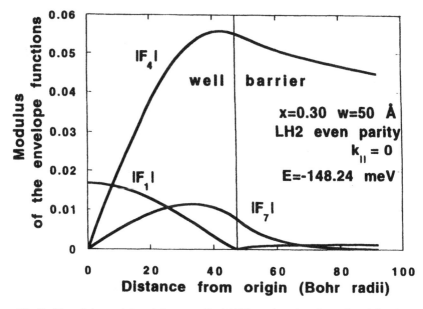

Fig. 5 Plot of the modulus of the normalized LH2 envelope function at $k_{\parallel} = 0$ for the case of a 50 Å GaAs/Al$_{0.3}$Ga$_{0.7}$As quantum well (a.u. = 0.529177 Å). This even parity LH2 state has a dominant $v = 4$ (light-hole) component and lesser $v = 1$ (conduction) and $v = 7$ (spin-orbit) components.

mixing of the bands increases since band coupling is proportional to parallel momentum.

In order to obtain a better understanding of the behavior of the valence subband wave functions over the entire Brillouin zone, it is best to provide their band-by-band decomposition as explained in the section on normalization. In Fig. 6, we plot the weights of various bands along the (11) ray in the Brillouin zone. Figure 6 for the HH1 shows that, at the origin of the Brillouin zone, it is entirely heavy-hole-like (component 3) as previously discussed. Away from the origin, the HH1 subband begins to acquire a significant light-hole character (component 5). As shown in Fig. 7, there is a small admixture of conduction band states as well. Similarly, the HH2 band starts out as completely heavy-hole-like (odd-parity component 6) and then acquires significant amounts of other valence-band and conduction-band components. The corresponding plots for the LH1 and LH2 bands show an even more mixed wave function character. For the LH2 band in Fig. 8, the dominant characteristic at the zone center is

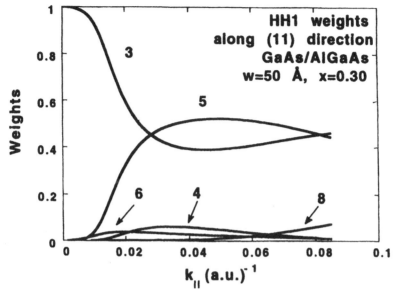

Fig. 6 Dominant component in the band decomposition of the HH1 band as a function of $\mathbf{k}_\parallel$ along the (11) direction for the case of a 50 Å GaAs/Al$_{0.3}$Ga$_{0.7}$As quantum well (a.u. = 0.529177 Å).

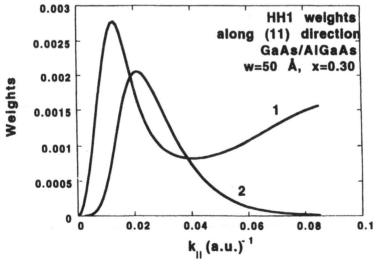

Fig. 7 Lesser components in the band decomposition of the HH1 band as a function of $\mathbf{k}_\parallel$ along the (11) direction for the case of a 50 Å GaAs/Al$_{0.3}$Ga$_{0.7}$As quantum well (a.u. = 0.529177 Å).

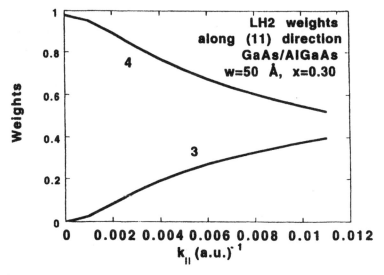

Fig. 8 Dominant component in the band decomposition of the LH2 band as a function of $k_{\parallel}$ along the (11) direction for the case of a 50 Å GaAs/Al$_{0.3}$Ga$_{0.7}$As quantum well (a.u. = 0.529177 Å).

light-hole-like (component 4) and, away from the center, some heavy-hole (component 3) character becomes significant. But more importantly, for optical transitions, there is a significant admixture of the conduction band and spin-orbit band states at the zone center, Fig. 9. In general, LH bands are more strongly hybridized with the conduction band states.

It is the presence of these conduction band components (especially in the light-hole subbands) which in part leads to strong normal incidence absorption in p-type MQWs. As the ground state is the predominantly p-like HH1 subband, the normal incidence absorption benefits from the s-like admixture of conduction band states in the LH1 and LH2 subbands, which is the case even at the center of the Brillouin zone. This admixture of conduction band states is a strong function of parallel momentum. Moreover, valence band anisotropy also contributes to normal incidence absorption, as it does in n-type MQWs, such as SiGe/Si heterostructures, for proper orientations of the constant energy ellipsoids with respect to incident radiation. The anisotropy of the valence subbands also increases away from the Brillouin zone center. For these reasons, it is important to determine

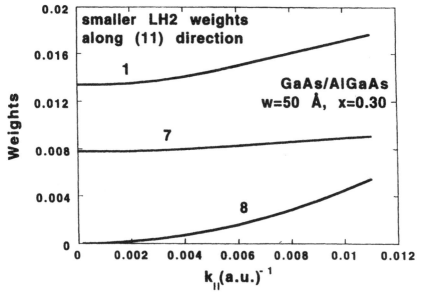

Fig. 9   Lesser components in the band decomposition of the LH2 band as a function of
$k_\parallel$ along the (11) direction for the case of a 50 Å GaAs/Al$_{0.3}$Ga$_{0.7}$As quantum well
(a.u. = 0.529177 Å).

the extent of the hole filling of the HH1 subband by calculating the
Fermi energy for a given level of doping.

2.6. Fermi Energy

For the range of doping used in our MQWs, only the lowest subband
(HH1) is occupied by holes. In order to calculate the extent of hole
filling of the ground-state subband, the density of states, Fig. 10, and
the integrated density of states, Fig. 11, were obtained based on the
energy bands of Fig. 3. The density of states gives the number of
energy states per unit energy interval and area available for hole
occupancy of the bands of Fig. 3. Figures. 10 and 11 were obtained by
integrating along rays in the two-dimensional Brillouin zones. Bands
HH1 and HH3 produce step-like onsets at about 23.06 and 88.45
meV, while the LH1 band produces the logarithmic singularity at
about 51.71 meV, which is the result of its electron-like curvature in
Fig. 3 (LH2 produces a step at 148.24 meV which is not shown in
Fig. 10). Integrating the density of states results in the integrated

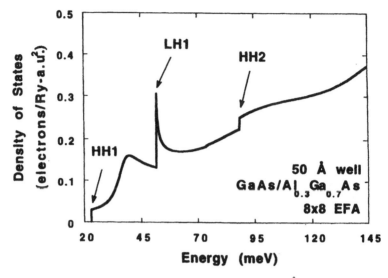

Fig. 10 The calculated density of states for the case of a 50 Å GaAs/Al$_{0.3}$Ga$_{0.7}$As quantum well as a function of energy (a.u. = Bohr radius = 0.529177 Å). Arrows point to the location of van Hove singularities in the density of states. HH and LH refer to the heavy- and light-hole subbands. The zero of energy is at the top of the valence band of bulk GaAs.

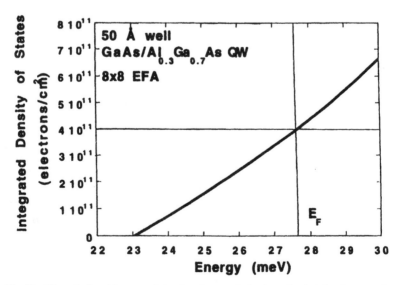

Fig. 11 The calculated integrated density of states (hole sheet density) for the case of a 50 Å GaAs/Al$_{0.3}$Ga$_{0.7}$As quantum well as a function of energy (a.u. = Bohr radius = 0.529177 Å). The arrow indicates the location of the Fermi level for the doping of 1×10^{18} cm^{-3} in the central 40 Å of the well.

density of states of Fig. 11, which gives the sheet density as a function of band filling. The Fermi level position indicated in the graph is for 1×10^{18} cm^{-3} doping with a 5 Å offset at each side of the well, which results in the hole sheet density of 4×10^{11} cm^{-2}.

The electronic structure calculations were performed for a set of five different *p*-type MQW structures for which the well width was varied from 30 Å to 50 Å in 5 Å increments. The doping concentration in the wells was the same for each MQW and was set at $N_A = 1 \times 10^{18}$ cm^{-3}. The barrier layers contained 30% aluminum and were 500 Å thick. The results for energy- and Fermi-level positions are given in Table 1, where the zero of energy is at the bottom of the well (i.e., the top of the valence band in GaAs). Because the heavy-hole effective mass is larger than the electron effective mass in *n*-type infrared detectors, for the same long-wavelength cutoff, the Fermi level in *p*-type detectors is lower than in *n*-type detectors with the same sheet carrier density. This gives *p*-type detectors the added advantages of a lower thermal noise as well as of a lower dark tunneling current. For example, consider a 40 Å well with an AlGaAs barrier composition of $x = 0.28$ for a *n*-type MQW and $x = 0.30$ for *p*-type. Both of these structures have a peak wavelength in the 8 micron region and similar cut-off wavelengths. For a well doping density of 1×10^{18} atoms/cm^3, the Fermi level in this *n*-type quantum well is 10.8 meV above the ground state subband while for the *p*-type well it is 3.47 meV above the heavy-hole ground state. Although this energy difference is small, the dark current due to thermionic emission varies exponentially with the Fermi energy [3], so this small difference would increase the dark current by a factor of three at 77 K.

Because of the relatively heavy doping of the wells, there could be several corrections to the positions of the energy levels. Using

TABLE 1
Subband Energies at $k_{\parallel} = 0$ and the Fermi Energies.

Well width	30 Å	35 Å	40 Å	45 Å	50 Å
HH1 m(eV)	44.89	37.25	31.36	26.74	23.06
LH1 (meV)	82.56	72.93	64.76	57.70	51.71
HH2 (meV)	146.09	131.78	115.61	101.02	88.45
ρ_s (cm^{-2})	2.0×10^{11}	2.5×10^{11}	3.0×10^{11}	3.5×10^{11}	4.0×10^{11}
E_F (meV)	47.16	40.12	34.83	30.79	27.66

The barrier widths were 500 Å and the composition was Al$_{0.3}$Ga$_{0.7}$As. The wells were doped to $N_A = 1 \times 10^{18}$ cm^{-3} in the center of the well, 5 Å from each interface.

Bandara et al.'s expressions for the exchange energy lowering [31], we estimate that the ground state, HH1 band for the 30 Å and 40 Å wells is about 12.4 and 15.2 meV, respectively, lower than predicted by the single electron theory above. Coulomb corrections to the calculated ground-state energies raise the energy bands on the order of 0.3 meV. These corrections are important to the calculation of the correct long-wavelength threshold and the expected magnitude of thermionic dark currents.

3. VALENCE BAND OFFSET

The accuracy of the model crucially depends on the quality of experimental input for a number of parameters used in the calculation, which are listed in Appendix A. The valence band offset determines the depth of the potential well experienced by the holes. As is well known, the ground-state band, lying relatively deep in the well, is largely unaffected by the height of the barrier. However, the electronic structure of the higher lying subbands is more sensitive to the magnitude of the valence band offset. Also, the magnitude of the valence band offset determines in part (along with the ground state band) the long-wavelength threshold for p-type infrared detectors. In order to accurately determine the actual valence band offset that should be used to model p-type intersubband transitions, admittance spectroscopy was used to measure the valence band offsets at the interfaces of p-type, Be-doped, $GaAs/Al_{0.3}Ga_{0.7}As$ multiple quantum wells [32].

In admittance spectroscopy, both the capacitance and the conductance of MQW mesas are recorded at several frequencies as a function of temperature. The peaks in the conductance curves (Fig. 12) indicate when the RC time constant of the specimen has become equal to the measurement frequency. From an Arrhenius plot of the natural log of the frequency lnf versus $1/kT$, the slope of the straight line yields the activation energy of the conductance over the barrier (Fig. 13). In this figure, the Arrhenius plots for three different MQW samples with well widths of 30 Å, 35 Å and 40 Å, respectively, are shown. The barrier layer thickness (500 Å) and composition ($x = 0.3$), and doping concentration ($N_A = 1 \times 10^{18}$ cm^{-3}) were kept the same for each of the MQW structures used. The conductance in the direction perpendicular to the layers is given by [33]

$$\sigma = \sigma_0 T^2 \exp\left[-(\Delta E_V - E_F + \delta)/kT\right], \qquad (28)$$

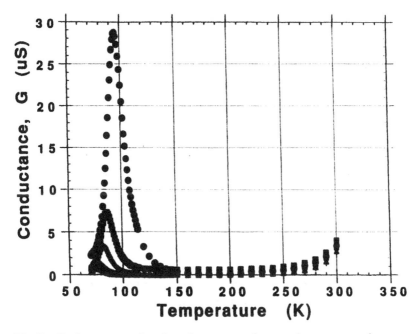

Fig. 12 Conductance as a function of temperature for several measurement frequencies. When the inverse of the RC time constant of the specimen equals the measurement frequency, a peak occurs in the conductance, measured at zero bias.

where σ_0 is a constant, ΔE_V the valence band discontinuity, E_F the Fermi level measured from the top of GaAs valence band in the well, and δ the lowering of the parabolic AlGaAs barrier due to space charge effects. In these samples, we take δ to be zero since the barriers were undoped and relatively narrow. The Boltzmann's constant is k, and T is the temperature. So, from the maximum of conductance in Eq. (28), the valence-band discontinuity is given by

$$\Delta E_V = E_A - 2kT + E_F, \qquad (29)$$

where E_A is the measured activation energy, and the quantity $-2kT$ arises as a result of the prefactor in Eq. (28) [33].

A calculation of the Fermi level E_F is all that is needed to yield a value for the valence band discontinuity. For this set of samples, the Fermi level positions as obtained from our calculations are listed in Table 1. The slopes of the curves shown in Fig. 13 yield an activation energy E_A of 0.123 eV for the 30 Å well, 0.137 eV for the 35 Å well, and 0.148 eV for the 40 Å well. On the basis of these numbers and the

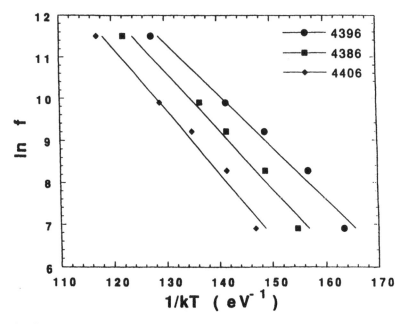

Fig. 13 Natural log of the measurement frequency vs. inverse kT. Plotting the ln of the measurement frequency f against the inverse of the temperature position of the peaks in the conductance yields a value for the activation energy of the band discontinuity.

calculated Fermi levels, the average valence-band discontinuity for the three samples is $\Delta E_V = 0.162 \pm 0.006$ eV. Notably, the valence-band offsets exhibit a perceptible shift with dopant sheet density, which could be attributed to the exchange lowering of the Fermi level. Using Bandara, Coon and O's [31] expression for exchange energy lowering at the Fermi energy, we estimate that the Fermi level for the 30 Å and 40 Å wells is about 7.5 and 8.8 meV, respectively, lower than predicted by singe-electron theory. Applying these exchange energy corrections, the average valence band offset reduces to 0.154 ± 0.006 eV. Thus, the standard 150 meV offset used in our calculations is within the experimental error of the measured band offsets.

4. VALENCE INTERSUBBAND ABSORPTION

The present 8×8 EFA band structure calculation employs a number of experimentally derived input parameters, including the conduction to valence band offset ratio, which are listed in Appendix A. As explained above, the first step is to obtain the energy dispersions and the associated wave functions as functions of the in-plane momentum. After calculating the wave functions for the bound states and continuum states, the next step is to obtain the matrix elements of momentum which govern the strength of optical transitions between the states. Also, the intraband matrix element gives the slope of the calculated subband with respect to the parallel momentum, which provides a check on the calculation. Energy gradients are also needed in calculating a variety of Brillouin zone integrals such as the density of states and the linear absorption coefficients.

4.1. Momentum Matrix Elements

The required matrix elements of momentum were obtained using an extension of the formalism of Chang and James to the case of finite barriers and of an 8×8 EFA (the extension is the subject of a separate paper) [34]. The matrix element is given by the following symmetrical form

$$\left\langle N\mathbf{k}_{\|} \left| \frac{\hbar}{m_0} \hat{\varepsilon} \cdot \mathbf{p} \right| M\mathbf{k}_{\|} \right\rangle =$$

$$\left(\frac{\hbar^2}{m_0} \right) \hat{\varepsilon} \cdot \sum_{vv'} \left\{ \int dz F_v(N\mathbf{k}_{\|}, z)^* P_{vv'}(\mathbf{k}_{\|}, z) F_{v'}(M\mathbf{k}_{\|}, z) \right.$$

$$+ \frac{1}{2} \int dz \left[-i\frac{d}{dz} F_v(N\mathbf{k}_{\|}, z) \right]^* Q_{vv'}(\mathbf{k}_{\|}, z) F_{v'}(M\mathbf{k}_{\|}, z)$$

$$\left. + \frac{1}{2} \int dz F_v(N\mathbf{k}_{\|}, z)^* Q_{vv'}(\mathbf{k}_{\|}, z) \left[-i\frac{d}{dz} F_{v'}(M\mathbf{k}_{\|}, z) \right] \right\}, \qquad (30)$$

where matrics $\mathbf{P}$ and $\mathbf{Q}$ are material dependent quantities and they are given in Ref. 34. The present form properly takes into account

a) valence band mixing,
b) polarization selection rules,
c) coupling to the conduction band,
d) hermiticity of the momentum operator,
e) wave function penetration into the barrier.

By taking wave function penetration into account, this equation has two, and not one, derivative terms multiplying the quantity $\mathbf{Q}$. The band parameters contained in $\mathbf{P}$ and $\mathbf{Q}$ are position dependent and the integrals in the above equation must be performed separately in every layer. The first term in the momentum matrix element is the overlap integral for direct coupling of subbands with the same symmetry, such as HH-to-HH or LH-to-LH transitions. The second and third terms in the momentum matrix equation are dipole coupling terms for HH-to-LH transitions.

The equations for $\mathbf{P}$ and $\mathbf{Q}$ for each part of the quantum well structure can be expressed in the form [12, 34]

$$\left(\frac{\hbar^2}{m_0}\right)\hat{\varepsilon}\cdot\mathbf{P}_{vv'} =$$

$$\frac{\hbar^2}{m_0}(\hat{\varepsilon}\cdot\mathbf{k}_\parallel)\,\delta_{vv'} + \frac{\hbar}{m_0}\,\hat{\varepsilon}\cdot\mathbf{p}_{vv'} + \left(\frac{\hbar}{m_0}\right)^2\sum_{\gamma}^{B}\left[\frac{(\mathbf{k}_\parallel\cdot\mathbf{p}_{v\gamma})(\hat{\varepsilon}\cdot\mathbf{p}_{\gamma v'})}{E_v - E_\gamma} + \frac{(\hat{\varepsilon}\cdot\mathbf{p}_{v\gamma})(\mathbf{k}_\parallel\cdot\mathbf{p}_{\gamma v'})}{E_{v'} - E_\gamma}\right],$$

or

$$\left(\frac{\hbar^2}{m_0}\right)\hat{\varepsilon}\cdot\mathbf{P}_{vv'} = \sum_i \varepsilon_i [(D^{xi}_{vv'} + D^{ix}_{vv'})k_x + (D^{yi}_{vv'} + D^{iy}_{vv'})k_y + G^i_{vv'}], \quad (31)$$

and

$$\left(\frac{\hbar^2}{m_0}\right)\hat{\varepsilon}\cdot\mathbf{Q}_{vv'} = \left(\frac{\hbar^2}{m_0}\right)\varepsilon_z\delta_{vv'} + \left(\frac{\hbar}{m_0}\right)^2\sum_{\gamma}^{B}\left[\frac{(p^z_{v\gamma})(\hat{\varepsilon}\cdot\mathbf{p}_{\gamma v'})}{E_v - E_\gamma} + \frac{(\hat{\varepsilon}\cdot\mathbf{p}_{v\gamma})(p^z_{\gamma v'})}{E_{v'} - E_\gamma}\right],$$

$$\left(\frac{\hbar^2}{m_0}\right)\hat{\varepsilon}\cdot\mathbf{Q}_{vv'} = \sum_i \varepsilon_i(D^{zi}_{vv'} + D^{iz}_{vv'}), \quad (32)$$

which, except for the extra G term, are exactly the expressions derived by Chang and James [12]. The coefficients D are the second derivatives of the $\mathbf{k}\cdot\mathbf{p}$ Hamiltonian at $\mathbf{k} = 0$, i.e.,

$$D^{ij}_{vv'} + D^{ji}_{vv'} = \left(\frac{\partial^2 H_{vv'}}{\partial k_i\partial k_j}\right)_0, \quad (33)$$

while the coefficients G are the first derivatives of the Hamiltonian

$$G^i_{vv'} = \left(\frac{\partial H_{vv'}}{\partial k_i}\right)_0. \quad (34)$$

where H is given by Eq. 5.

Expressions for the matrices $\mathbf{P}_{\mu v}$ and $\mathbf{Q}_{\mu v}$ (x, y and z polarizations), in terms of the modified Luttinger parameters and the modified electron effective mass, are provided in the paper by Szmulowicz [34].

These matrices are too large to be printed succintly on one page. However, a revised version of the Chang and James parameters A_1, A_2, B and C for the 4×4 case are listed in Table 2.

As pointed out by Chang and James [12], intersubband absorption with x-y polarization is possible in p-type quantum wells. Because of band mixing there are non-zero matrix elements for normal light incidence. For instance, in the $\boldsymbol{\varepsilon} \cdot \mathbf{P}$ case, heavy-heavy, heavy-light and light-light hole transitions are possible as long as the final and initial states have the same parity. The strength of these transitions depends upon the wave vector. In the $\boldsymbol{\varepsilon} \cdot \mathbf{Q}$ case, light-to-heavy and heavy-to-light transitions with a change of parity are possible for zero parallel wave vector. In addition, for structures with a large admixture of conduction band states into the hole wave functions, the x and y components of $\boldsymbol{\varepsilon} \cdot \mathbf{P}$ provide another channel for normal incidence absorption because of a term which is proportional to the valence-conduction band coupling parameter.

It is difficult to explain physically the various parts of the above formalism. In particular, there appears to be a dichotomy between the treatment of the momentum matrix elements for n- and p-type quantum wells. However, it is possible to unify the descriptions of light absorption in p and n-type quantum wells and to describe the momentum matrix element in a more physically-inspired fashion. Since $\mathbf{P}$ and $\mathbf{Q}$ ultimately derive from the $\mathbf{k} \cdot \mathbf{p}$ Hamiltonian, Eq. (30) can be

TABLE 2
Matrix Elements for Matrices **P** and **Q**.

$\hat{\boldsymbol{\varepsilon}} \cdot \mathbf{P}$	$\hat{\boldsymbol{\varepsilon}} \| \hat{\mathbf{x}}$	$\hat{\boldsymbol{\varepsilon}} \| \hat{\mathbf{y}}$	$\hat{\boldsymbol{\varepsilon}} \| \hat{\mathbf{z}}$
A_1	$-\gamma_1 q_x$	$-\gamma_1 q_y$	0
A_2	$-\gamma_2 q_x$	$-\gamma_2 q_y$	0
B	0	0	$\sqrt{3}\gamma_3(iq_x + q_y)$
C	$-\sqrt{3}(\gamma_2 q_x - i\gamma_3 q_y)$	$\sqrt{3}(i\gamma_3 q_x + \gamma_2 q_y)$	0
$\hat{\boldsymbol{\varepsilon}} \cdot \mathbf{Q}$	$\hat{\boldsymbol{\varepsilon}} \| \hat{\mathbf{x}}$	$\hat{\boldsymbol{\varepsilon}} \| \hat{\mathbf{y}}$	$\hat{\boldsymbol{\varepsilon}} \| \hat{\mathbf{z}}$
A_1	0	0	$-\gamma_1$
A_2	0	0	$2\gamma_2$
B	$i\sqrt{3}\gamma_3$	$\sqrt{3}\gamma_3$	0
C	0	0	0

These matrix elements are for the 4×4 case, in terms of the Chang and James [12] parameters A_1, A_2, B and C. These elements are the corrected version of the Table 1 in Chang and James [34].

rewritten in the following more transparent manner [34]:

$$\left\langle Nk_\parallel \left| \frac{\hbar}{m_0} \hat{\boldsymbol{\varepsilon}} \cdot \mathbf{p} \right| Mk_\parallel \right\rangle = \hat{\boldsymbol{\varepsilon}} \cdot \sum_{\mu\nu} \left\langle F_\mu(Nk_\parallel, z) \left| \frac{\partial^2 H_{\mu\nu}}{\partial \mathbf{k} \partial \mathbf{k}} \cdot \mathbf{k}_\parallel + \frac{\hbar}{m_0} \mathbf{p}_{\mu\nu} \right.\right.$$

$$\left.\left. + \frac{1}{2} \left[\left(\frac{1}{i} \frac{\overleftarrow{d}}{dz} \right)^* \frac{\partial^2 H_{\mu\nu}}{\partial \mathbf{k} \partial k_z} + \frac{\partial^2 H_{\mu\nu}}{\partial \mathbf{k} \partial k_z} \left(\frac{1}{i} \frac{\overrightarrow{d}}{dz} \right) \right] \right| F_\nu(Mk_\parallel, z) \right\rangle,$$

$$(35)$$

where the arrows indicate the direction in which the derivatives should be taken. The second term is the dominant term in valence-to-conduction band transitions. In the single band case ($\mu = \nu = 1$), the first two terms lead to on overlap integral between orthogonal final and initial envelope functions and, thus, yield zero. The last term is proportional to the inverse effective-mass tensor m_{iz}^{-1}, which can be exploited with a proper orientations of the incident light with respect to the crystal axes. For valence intersubband transitions in the 8×8 model, in normal incidence ($\varepsilon_z = 0$) and $\mathbf{k}_\parallel = 0$, the last term couples light and heavy-hole components of the initial and final states. This contribution is made possible by the presence of mixed $k_x k_z$ and $k_y k_z$ terms in the Hamiltonian, which points to the importance of valence band anisotropy in optical transitions. The second term is important only if the initial and final hole states have a significant admixture of conduction-band states. The first term couples same-parity components of the initial and final states away from the center of the Brillouin zone, which is important for heavier-doped MQWs. The above argument unifies the description of optical absorption in n- and p-type heterostructures and points to the dominant effect of anisotropy in both intervalence and interconduction subband transitions.

The results of applying these momentum matrix elements to a 50 Å wide p-type GaAs/Al$_{0.3}$Ga$_{0.7}$As quantum well are considered next. First, recall that the strength of an optical transition is proportional to the square of the momentum matrix element. In the following figures, the square of the momentum matrix elements for various intersubband transitions are plotted along the (11) ray in the two-dimensional Brillouin zone. The curves end whenever upper states in the transition become unbound. In Fig. 14, the allowed transitions from the HH1 ground state to the other valence subbands, for x polarization of the incident light, are shown. The HH1-to-LH2 (even-to-odd parity) transition is shown to be possible at the center of the Brillouin zone at

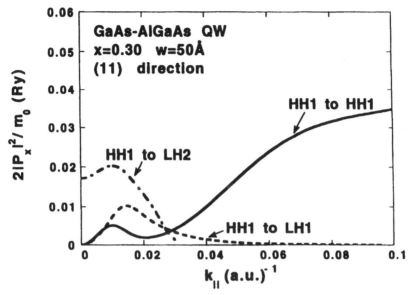

Fig. 14 Squared momentum matrix elements in the *x*-direction for intersubband transitions originating with the HH1 band as a function of $\mathbf{k}_\parallel$ along the (11) direction for the case of a 50 Å GaAs/Al$_{0.3}$Ga$_{0.7}$As quantum well (a.u. = 0.529177 Å).

normal incidence. In contrast, the HH1-to-LH1 (no change of parity) transition is allowed only for finite wave vectors. These selection rules are imbedded in the expression for the matrix element [12, 34]. Transitions within the HH1 subband also shown in this figure. By comparing Fig. 14 with Fig. 15, it can be seen that without the contribution of *s*-states, through coupling to the conduction band, the intervalence band transitions lose strength. In particular, the loss of the conduction band coupling significantly reduces the HH1-to-LH2 transition at the center of the zone. This verifies the importance of the conduction-to-valence band coupling for normal incidence absorption in these *p*-type quantum wells. However, for *z*-polarized light, the HH1-to-HH2 transition is allowed at the center of the Brillouin zone (see Fig. 16). Now, the HH1-to-LH2 transition only becomes significant at larger wave vectors (for *z*-polarization). Moreover, a neglect of the *s*-state coupling, Fig. 17, leads to a large underestimate of the strength of the transitions in Fig. 16. Without the *s*-state contributions, the HH1-to-LH2 transition for *z*-polarized light is reduced dramatically. However, at the center of the Brillouin zone, both HH1 and

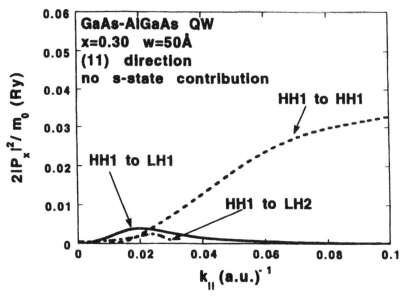

Fig. 15 Squared momentum matrix elements in the x-direction for intersubband transitions originating with the HH1 band as a function of $\mathbf{k}_\parallel$ along the (11) direction for the case of a 50 Å GaAs/Al$_{0.3}$Ga$_{0.7}$As quantum well (a.u. $= 0.529177$ Å). Here the contribution of conduction band components is neglected.

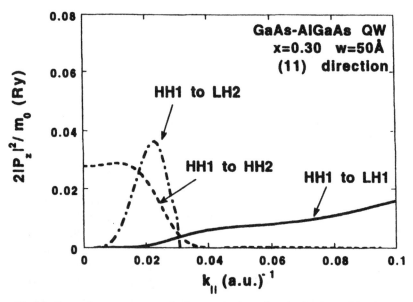

Fig. 16 Squared momentum matrix elements in the z-direction for intersubband transitions originating with the HH1 band as a function of $\mathbf{k}_\parallel$ along the (11) direction for the case of a 50 Å GaAs/Al$_{0.3}$Ga$_{0.7}$As quantum well (a.u. $= 0.529177$ Å).

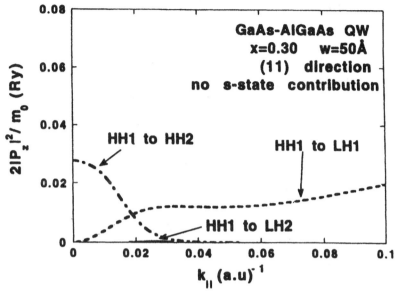

Fig. 17 Squared momentum matrix elements in the *z*-direction for intersubband transitions originating with the HH1 band as function of $k_{\parallel}$ along the (11) direction for the case of a 50 Å GaAs/Al$_{0.3}$Ga$_{0.7}$As quantum well (a.u. = 0.529177 Å). Here the contribution of conduction band components is neglected.

HH2 have no *s*-state admixture, so that there is no change in the HH1-to-HH2 between Figs. 16 and 17.

In addition, graphs were made of transitions originating at the LH1 subband. Figure 18 provides the graphs of the calculated matrix elements of momentum for normal-incidence transitions originating with the LH1 state. The LH1-to-HH2 transition is strongest at the zone center for *x*-polarized light, and the *s*-state contributions, Fig. 19, play a major role in this coupling to normal incident light. By neglecting *s*-states not even the slope of the LH1 band (which is proportional to the LH1-to-LH1 matrix element) is correctly predicted. On the other hand, coupling to the conduction band alone does not reflect the true strength of intersubband transitions either. For *z*-polarized light, the LH1-to-LH2 transition dominates and is very strong at the center of the Brillouin zone, as shown in Fig. 20.

So, for light polarized in the growth plane, it is found that the intersubband transitions arising from the dipole coupling terms (HH-LH transitions) dominate over the transitions from direct coupling

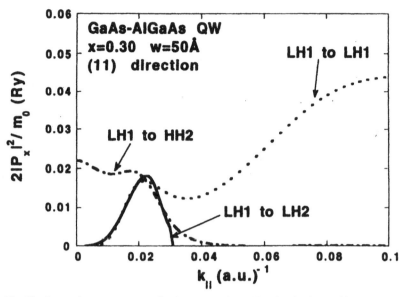

Fig. 18 Squared momentum matrix elements in the x-direction for intersubband transitions originating with the LH1 band as a function of $\mathbf{k}_\parallel$ along the (11) direction for the case of a 50 Å $GaAs/Al_{0.3}Ga_{0.7}As$ quantum well (a.u. = 0.529177 Å).

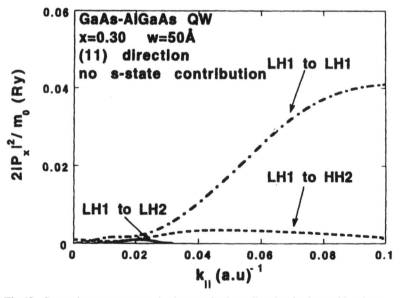

Fig. 19 Squared momentum matrix elements in the x-direction for intersubband transitions originating with the LH1 band as a function of $\mathbf{k}_\parallel$ along the (11) direction for the case of a 50 Å $GaAs/Al_{0.3}Ga_{0.7}As$ quantum well (a.u. = 0.529177 Å). Here the contribution of conduction band components is neglected.

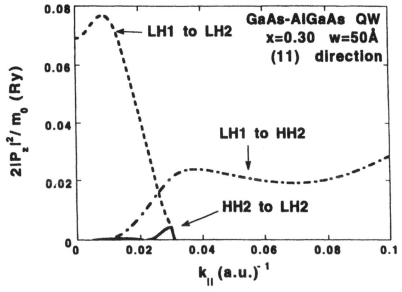

Fig. 20 Squared momentum matrix elements in the *z*-direction for intersubband transitions originating with the LH1 and HH2 band as a function of $\mathbf{k}_{\parallel}$ along the (11) direction for the case of a 50 Å GaAs/Al$_{0.3}$Ga$_{0.7}$As quantum well (a.u. = 0.529177 Å).

terms (HH-HH or LH-LH) for small $\mathbf{k}_{\parallel}$, i.e., near the bottom of the band. And, the strongest optical transition for normal incident light, at small $\mathbf{k}_{\parallel}$, is from the first heavy-hole subband to the second light-hole subband in the *p*-type MQWs under discussion. It is this transition that should be exploited in the design of a *p*-type GaAs/AlGaAs MQW for infrared detection.

4.2 Bound-to-Bound Absorption

For sake of completeness, we present our calculation of the bound-to-bound absorption. However, it must be borne in mind that a proper calculation should take account of the effect of depolarization and exciton shifts on the positions and shapes of calculated spectral features. In addition, the spectral shapes are influenced by thermal and nonthermal broadening mechanisms. As we are mainly interested in bound-to-continuum absorption, we will not dwell on these issues in the present section.

The linear absorption coefficient is given by [25]

$$\alpha(\omega) = \frac{4\pi^2 e^2}{ncm_0\omega\Omega} \sum_{if} \frac{1}{m_0} |\hat{\varepsilon}\cdot\vec{P}_{if}|^2 \delta(E_f - E_i - \hbar\omega)[f(E_i) - f(E_f)], \quad (36)$$

where n is the refractive index, c is the speed of light, ω is the frequency of radiation, Ω stands for the characteristic volume of the problem, ε is the polarization vector, f is the Fermi factor, i stands for the initial state, and f stands for the final state. For bound-to-bound absorption, the initial state are the occupied states, in our case, the doubly degenerate HH1 band. The final states are the excited state subbands, LH1, HH2, and LH2, depending on the width of the well.

Performing the indicated operations, we obtain

$$\alpha(\omega) = \frac{e^2}{ncm_0^2\omega L} \int d\mathbf{k}_\parallel \sum_{N,M} |\hat{\varepsilon}\cdot\vec{P}_{N,M}(\mathbf{k}_\parallel)|^2 \delta(E_N(\mathbf{k}_\parallel) - E_M(\mathbf{k}_\parallel) - \hbar\omega), \quad (37)$$

in which the initial states are filled and the final ones are empty.

For normally-incident, unpolarized light, the suitably averaged transition probability is given by

$$\langle|\hat{\varepsilon}\cdot\vec{P}(\mathbf{k}_\parallel)|^2\rangle = \frac{1}{2}(|P_x(\mathbf{k}_\parallel)|^2 + |P_y(\mathbf{k}_\parallel)|^2), \quad (38)$$

which is nearly isotropic in the Brillouin zone for small wave vectors; similarly, for z-polarized light

$$\langle|\hat{\varepsilon}\cdot\vec{P}(\mathbf{k}_\parallel)|^2\rangle = (|P_z(\mathbf{k}_\parallel)|^2, \quad (39)$$

which is also largely independent of the azimuthal angle in the Brillouin zone. By performing appropriate spatial averages, one can construct the appropriate expressions for light polarized in any plane of interest. Therefore, assuming the isotropy of the transition probability, the azimuthal integral gives 2π, which result in the final expression

$$\alpha(\omega) = \frac{2\pi e^2}{ncm_0^2\omega L} \sum_{N,M} \frac{k_0\langle|\hat{\varepsilon}\cdot\vec{P}_{NM}(k_0)|^2\rangle}{|\nabla(E_N - E_M)|}, \quad (40)$$

where the radial coordinate k_0 is evaluated from the condition

$$(E_N(\mathbf{k}_\parallel) - E_M(\mathbf{k}_\parallel) - \hbar\omega)|_{k_0} = 0. \quad (41)$$

The energy gradient in the denominator of Eq. (40) is easily evaluated using the intraband matrix elements of momentum

$$\frac{\hbar}{m_0} \vec{P}_{NN}(\mathbf{k}_\parallel) = \frac{\partial E_N(\mathbf{k}_\parallel)}{\partial \mathbf{k}_\parallel}. \qquad (42)$$

For numerical purposes, length L is taken to be the width of the quantum well, which is the only length in the problem. Therefore, when calculating quantum efficiencies, one should multiply the linear absorption coefficient by the total quantum well width (i.e., exclude barrier widths).

Fig. 21 shows the calculated linear absorption coefficient for a 50 Å GaAs/Al$_{0.30}$Ga$_{0.70}$As quantum well for xy-polarized light; the case of z-polarized light is shown in Fig. 22. In Fig. 21, there are prominent transitions HH1 to LH1 and HH1 to LH2. Because these pairs of bands have different curvatures, the resulting absorption lines acquire finite widths. Although, HH1 to LH1 is forbidden at zone center for xy-polarized light, the transition is allowed for finite wave vectors. The strength of the linear absorption coefficient for this transition, in

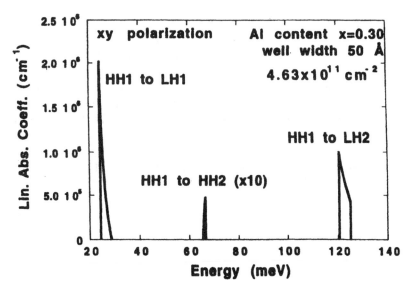

Fig. 21 The calculated bound-to-bound linear absorption coefficient for a 50 Å *p*-doped GaAs/Al$_{0.3}$Ga$_{0.7}$As quantum well at normal incidence as a function of photon wavelength for the sheet carrier density of 4.63×10^{11} cm^{-2}.

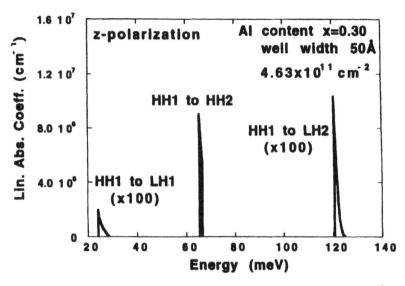

Fig. 22 The calculated bound-to-bound linear absorption coefficient for a 50 Å p-doped GaAs/Al$_{0.3}$Ga$_{0.7}$As quantum well and z-polarized light as a function of photon wavelength for the sheet carrier density of 4.63×10^{11} cm^{-2}.

part, derives from the $1/\omega$ factor in the equation for $\alpha(\omega)$. This brings up an interesting possibility of exploiting this transition in narrower MQWs, where LH1 can be made resonant with the top of the well. Among the advantages of such a transition is the fact that there would be no other excited bound states in the well to use up the available oscillator strength. For the mid-IR range relevant to this chapter, the HH1 to LH2 transition is most important. Once the LH2 state is pushed out of the well, the appreciable oscillator strength available for this transition becomes available to the resultant bound-to-continuum transition. In Fig. 22 for z-polarized light, HH1 to HH2 is a prominent transition whose strength derives from the fact these bands have opposite parities. In fact, this transition is reminiscent of the bound to first excited state transition in n-type quantum wells and is allowed for similar reasons [5].

4.3 Bound-to-Continuum Absorption

The expression for the linear absorption coefficient in the case of bound-to-continuum absorption is evaluated from the same starting expression as before, that is,

$$\alpha(\omega) = \frac{4\pi^2 e^2}{nc m_0 \omega \Omega} \sum_{if} \frac{1}{m_0} |\hat{\varepsilon} \cdot \vec{P}_{if}|^2 \delta(E_f - E_i - \hbar\omega)[f(E_i) - f(E_f)]. \quad (43)$$

However, the enumeration of states proceeds differently than in the case of the bound-to-bound absorption. For example, the initial states are enumerated by the sum

$$\sum_i = \frac{S}{(2\pi)^2} \int d\mathbf{k}_\parallel \sum_N, \quad (44)$$

where the sum proceeds over the bound occupied bands. Because of the conservation of parallel momentum in optical transitions, the enumeration of the final states in the continuum is restricted to states of the same parallel momentum as the initial state, which leaves only the sum over states which are characterized by index $v = H$ even parity, H odd parity, L even parity, L odd parity, and their conjugate partners.

$$\sum_f = \frac{L}{2\pi} \sum_v \int dk_v. \quad (45)$$

(For eigenstates of parity – which are sums and differences of right- and left-going states – the integral proceeds over positive momenta only.) Here, we must recall that the perpendicular momentum is a function of energy and parallel momentum, $k_v(E, \mathbf{k}_\parallel)$. Therefore, each energy in the continuum can be eightfold degenerate (for energies above the spin orbit energy in the barrier, each energy can be twelve- fold degenerate). The momentum matrix elements are calculated the same way as in the bound-to-bound case, using the expression from Eq. (30).

With these substitutions, the expression for the linear absorption coefficient becomes [26]

$$\alpha = \frac{e^2}{2\pi n c m_0^2 \omega} \sum_N^{\text{bound}} \int d\mathbf{k}_\parallel \sum_v^{\text{continuum}} \frac{|\langle N\mathbf{k}_\parallel |\hat{\varepsilon} \bullet \vec{p} | E, \mathbf{k}_\parallel, k_v \rangle|^2}{\left| \dfrac{\partial E}{\partial k_v} \right|_{E = E_N(k_\parallel) + \hbar\omega}}, \quad (46)$$

where it must be remembered that there is a factor of $1/L$ hidden in the normalization of continuum states. As in the case of bound-to- bound transitions, the momentum matrix element is largely isotropic, so that the azimuthal integral yields 2π, which reduces the Brillouin zone integral to a radial integral. In our case, we chose to integrate along the (11) ray in the Brillouin zone up to the Fermi momentum.

The last crucial step is to obtain the energy derivative in the denominator, which provides the density of states in the continuum. This derivative is most conveniently obtained from an extension of the Hellman-Feynman theorem [35, 36] (sometimes called the Ehrenfest theorem) to the case at hand

$$\frac{\partial E}{\partial k_z} = L\frac{\partial H(k_\parallel, k_z)}{\partial k_z} C. \tag{47}$$

A broadening factor was not included in the absorption coefficient equation. If included, broadening would soften the long-wavelength edge of the absorption spectrum. For the predicted scattering rates in these p-type materials, the broadening would be on the order of 10 meV [17]. Broadening can also affect the magnitude of the absorption coefficient for resonant states near the top of the well, as shown by Liu for n-type GaAs/AlGaAs quantum wells [19]. The transitions to nonresonant continuum states are much less affected.

The bound-to-continuum linear absorption coefficients for five different well widths of p-type GaAs/Al$_{0.3}$Ga$_{0.7}$As quantum wells were calculated. These linear absorption coefficients were calculated as a function wavelength, dopant concentration, and incident light polarization. Figures 23 and 24 display the calculated linear absorption coefficients for normally incident light in the case of 30 Å and 40 Å MQWs as a function of photon wavelength and two-dimensional doping in the well. Clearly, the higher the doping the stronger the absorption. The absorption for the 30 Å well is broad and featureless. Mathematically, higher doping necessitates integration to higher parallel momenta in the HH1 subband. The absorption for the 30 Å well is very broad. For the 40 Å well, the peak absorption moves toward the long-wavelength threshold and exhibits some structure.

This structure near the threshold needs some explaining. As is known, the onset of continuum (i.e., the presence of propagating states) is a function of the parallel momentum. At the center of the Brillouin zone, the onset of continuum is at the same energy for the heavy and light holes. However, the onset energy rises faster for the light than for the heavy holes as one moves away from the center of the Brillouin zone. Therefore, at higher k-parallel values (which come in at higher doping), above the absorption threshold for the heavy holes, the LH channel is not available as it takes greater photon energies to reach LH continuum states. As a result, absorption

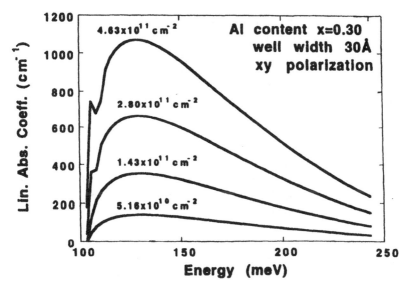

Fig. 23 The calculated bound-to-continuum linear absorption coefficient for a 30 Å *p*-doped GaAs/Al$_{0.3}$Ga$_{0.7}$As quantum well at normal incidence as a function of photon wavelength for several carrier densities.

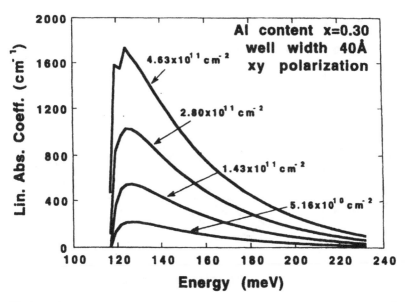

Fig. 24 The calculated bound-to-continuum linear absorption coefficient for a 40 Å *p*-doped GaAs/Al$_{0.3}$Ga$_{0.7}$As quantum well at normal incidence as a function of photon wavelength for the several carrier sheet densities.

decreases until the threshold for HH1-to-LH continuum absorption is reached at higher energies.

Figure 25 compares the calculated linear absorption coefficients for 30 Å through 50 Å well widths at a constant hole sheet density. The absorption spectrum for the 45 Å well is the strongest and sharpest, corresponding to the presence of the LH2 resonance just above the top of the well. Once the LH2 resonance becomes a bound state in the 50 Å well, the absorption decreases again. There is a distinct contrast between the spectra for 40, 45, and 50 Å wells, as absorption rises by a factor of two and then falls by a factor of three. In addition to changes in the magnitude of the absorption coefficient, there is also a change in cut-off wavelength. The cut-off wavelength shifts to higher energies (shorter wavelengths) as the first heavy hole subband moves lower in energy (toward the well bottom) with increasing well width. The peak absorption position for transitions to the continuum also gradually moves to lower energy for the 30 Å to 45 Å well widths, but shifts to higher energy as the LH2 subband becomes bound.

The absorption spectra in Figs. 23 through 25 are for unpolarized light in the the plane $(x$-$y)$ of the quantum well, i.e., for illumination

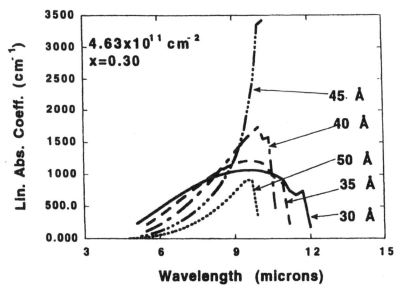

Fig. 25 The calculated linear absorption coefficient for five well widths of p-doped GaAs/Al$_{0.3}$Ga$_{0.7}$As wells at normal incidence as a function of photon wavelength for the equal sheet carrier density of 4.63×10^{11} cm^{-2}.

perpendicular to the growth direction (*z*-direction). Similar absorption spectra were calculated for *z*-polarized light, i.e., for edge illumination. Figs. 26 and 27 show the corresponding bound-to-continuum spectra for *z*-polarized light. For most of the well widths studied, the *z*-polarized light has a much smaller absorption coefficient. For *z*-polarization, it is the HH1-to-HH2 transition that dominates the spectra. However, even in the case of the narrowest well studied, 30 Å, the HH2 state is bound. Nevertheless, in this quantum well, the HH2 subband is near the top of the valence band well, which produces the largest admixture of HH2-like continuum states. This explains its greater absorption coefficient for *z*-polarized light than for *x*-*y* polarized light. The absorption of *z*-polarized light is at least an order of magnitude weaker for the other four well widths. As the well width is increased from 40 Å to 50 Å, the absorption coefficient temporarily increases by a factor of two and then decreases again. For the 45 Å quantum well, the presence of the LH2 resonance even enhances the weak *z*-polarization absorption as a result of the large density of states associated with the presence of the LH2 resonance.

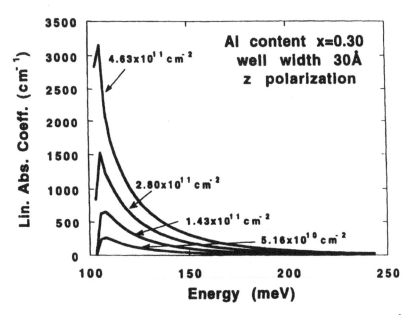

Fig. 26 The calculated bound-to-continuum linear absorption coefficient for a 30 Å *p*-doped GaAs/Al$_{0.3}$Ga$_{0.7}$As quantum well for *z*-polarized light as a function of photon wavelength for several carrier sheet densities.

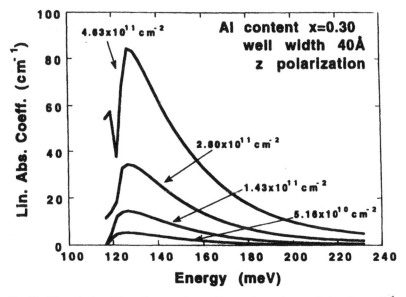

Fig. 27   The calculated bound-to-continuum linear absorption coefficient for a 40 Å
p-doped GaAs/Al$_{0.3}$Ga$_{0.7}$As quantum well for z-polarized light as a function of photon
wavelength for the several carrier sheet densities.

In Fig. 28, we give a comparison between the bound-to-bound and
bound-to-continuum absorption for the 50 Å well for which the LH2
resonance has just entered the well. Clearly, HH1-to-LH2 absorption
is much stronger and narrower than the bound-to-continuum absorption. This is similar to what has been observed experimentally in n-type
multi-quantum wells. The disadvantage of using bound-to-bound absorption is that the photo-excited charge carrier must tunnel through
the tip of the potential barrier in order to produce a photocurrent.

In Fig. 29, we put together the spectra for the three samples for
which photoresponse measurements were made. Here, we calculated
the expected exchange energy lowering of the HH1 ground state and
shifted each spectrum accordingly. These shifts amounted to about
12.4 and 15.2 meV for the 30 Å and 40 Å wells, respectively. This
shifted the long-wavelength cut-off from 11.8 to 10.6 μm for the 30 Å
well and from 10.5 to 9.3 μm for the 40 Å well. Because of the extended
nature of the continuum states (which gives rise to a small overlap
between the ground and continuum state wave functions), the depolarization and exciton shifts are expected to be small.

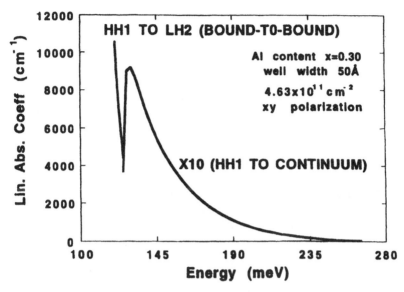

Fig. 28 The juxtaposition of the calculated bound-to-bound and bound-to-continuum linear absorption coefficients for a $50\,\text{Å}$ *p*-doped GaAs/Al$_{0.3}$Ga$_{0.7}$As quantum well at normal incidence as a function of photon wavelength for the sheet carrier density of $4.63 \times 10^{11}\,\text{cm}^{-2}$.

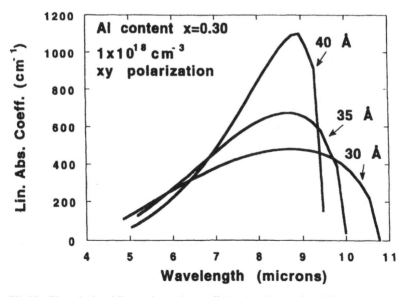

Fig. 29 The calculated linear absorption coefficient for three well widths, corresponding to the three experimentally measured *p*-doped GaAs/Al$_{0.3}$Ga$_{0.7}$As MQWs at normal incidence as a function of photon wavelength. Doping of $10^{18}\,\text{cm}^{-3}$. The exchange energy lowering of the HH1 subband was taken into account.

The absorption coefficients calculated above are based on the absorption in the optically active thickness of the multiple quantum wells. The thickness of the barrier layers was not included. So, for the 50 period multi-quantum well with 40 Å well widths, the total optical thickness is 2×10^{-5} cm. The definition of the optical thickness to use for the absorption coefficient in MQWs does vary between authors. Some prefer to include the barrier layers and use the total MQW thickness. Nevertheless, the experimentally relevant quantity, the quantum efficiency, is independent of the choice of length, since it involves the product of the linear absorption coefficient and the optical thickness.

There have been other calculations of the normal incidence absorption coefficient for a GaAs/Al$_{0.3}$Ga$_{0.7}$As p-type QWIP with a 40 Å well width. For instance, Man and Pan [16] reported a peak absorption coefficient of $\alpha_p = 7,500$ cm^{-1} for a dopant concentration of $N_A = 4 \times 10^{18}$ cm^{-3}. If we scale this to $N_A = 1 \times 10^{18}$ cm^{-3}, then the peak absorption coefficient becomes $\alpha_p = 1,875$ cm^{-1}. Similarly, the calculation by Teng [17] predicts $\alpha_p = 72$ cm^{-1} for a dopant concentration of $N_A = 2 \times 10^{16}$ cm^{-3}. Again with linear scaling to $N_A = 1 \times 10^{18}$ cm^{-3}, this peak absorption coefficient becomes $\alpha = 3,600$ cm^{-1} for the identical p-type MQW. However, Teng modelled continuum absorption by calculating a few above-barrier states as if they were bound in a larger external well. Our value for the peak absorption coefficient, for the same MQW heterostructure, is 1,100 cm^{-1} for $N_A = 1 \times 10^{18}$ cm^{-3}. Moreover, our calculated peak absorption coefficient for a 45 Å well width is about 3,400 cm^{-1} for nearly the same acceptor doping concentration.

There have been no reported direct measurements of the absorption coefficient spectrum for the p-type QWIPs that have been studied to date. However, there have been several reports of the quantum efficiencies in the p-type QWIPs, as determined from the optical gain and responsivity values [3,13,14]. For two samples with different dopant concentrations, but nominally the same well width ($L_W = 40$ Å) and barrier composition of $30 \pm 1\%$ aluminum, the reported quantum efficiencies are $\eta = 21.4\%$ for $N_A = 3 \times 10^{18}$ cm^{-3} and $\eta = 21.5\%$ for $N_A = 2 \times 10^{18}$ cm^{-3}, at a bias of 2 volts. To ascertain an approximate absorption quantum efficiency (η_a), the escape probability (p_e) must be taken into account for the applied bias. After normalizing these quantum efficiencies for p_e and N_A, the average absorption quantum efficiency is $10 \pm 2\%$ for $N_A = 1 \times 10^{18}$ cm^{-3}. Using the expression $\eta_a = 1 - e^{-2\alpha \ell}$, where ℓ is the total thickness of 50 quantum wells, and putting in the values of $\eta_a(77\text{K}) = 0.10$ and $\ell = 2.0 \times 10^{-5}$ cm, $\alpha_p(77\text{K})$ is determined to be about $2{,}630 \pm 560$ cm^{-1}. This is about a factor of two to three larger than our calculated peak absorption coefficient for the same MQW structure.

So, the estimates of the absorption coefficient range between $1{,}100$ cm^{-1} and $3{,}200$ cm^{-1} for a 40 Å well *p*-type QWIP with $N_A = 1 \times 10^{18}$ cm^{-3}, $\Delta\lambda/\lambda = 25\%$ and $\lambda_c \approx 9$ μm. These absorption coefficients are of about the same order of magnitude as those measured in bound-to-continuum *n*-type MQWs [3]. For instance, the measured absorption coefficients for an equivalent *n*-type QWIP $(L_W = 40 \text{ Å}, 26\%$ to 31% Al) – i.e., with comparable cut-off wavelengths and broadening – fall into the range of $4{,}700$ to $5{,}700$ cm^{-1}, when adjusted for absorption per well and a dopant concentration $N_D = 1 \times 10^{18}$ cm^{-3} [3, 37, 38]. We should note that these absorption coefficients in *n*-type QWIPs are for light coupled in through polished 45° facets. It appears then that the normal incidence absorption coefficient for a *p*-type QWIP is about a factor of 2 to 3 lower than the *z*-polarized absorption coefficient of a similar *n*-type QWIP. However, the *p*-type wells can be doped higher than 1×10^{18} cm^{-3} to increase the absorption coefficient and quantum efficiency without rapidly increasing the dark current. For instance, increasing the acceptor concentration to 5×10^{18} cm^{-3} would make the normal incidence absorption coefficient for *p*-type superior to that for *n*-type QWIPs.

For further comparison of *n*- and *p*-type structures, we quote Whitney [39], who gives the following relationships for comparing the peak absorption coefficients in *n*-type and *p*-type GaAs/AlGaAs QWIPs:

$$\alpha = 1.28 \times 10^{-16}(\rho_s/L_p)f_o(\lambda^2/\Delta\lambda)K_\alpha \text{ cm}^{-1} \qquad (48)$$

for *n*-type, where it is assumed that $m^* = 0.067\, m_0$, $n_r = 3.27$, and that 45° is the internal angle for the infrared radiation, while for *p*-type

$$\alpha = 5.4 \times 10^{-17}(\rho_s/L_p)f_o(\lambda^2/\Delta\lambda)K_\alpha \text{ cm}^{-1}, \qquad (49)$$

where $m^* = 0.34\, m_0$, $n_r = 3.27$ and the angle of incidence is 0°. In these expressions, ρ_s is the two-dimensional sheet charge density, L_p is the superlattice period, f_o is the oscillator strength, $\Delta\lambda$ is the absorption half linewidth, and λ is the peak wavelength. The correction factor K_α is used to correct the absorption coefficients calculated from this simplified expression to agree with experimentally determined results. These expressions would lead us to believe that the *n*-type QWIPs should have a higher absorption coefficient for the same peak wavelength, FWHM, and doping density. For the same oscillator strength and K_x, the n-type QWIP would have an absorpion coefficient about 2.5 times larger than the matching p-type. These estimates for the relative strength of n- and p-type QWIP peak absorption coefficients are perhaps surprisingly close to experimental results.

However, practical detector structures require backside illumination, and the proper comparison of n- and p-type detectors should be performed with this geometry in mind. The absorption quantum efficiency (η_a) is a parameter used to measure the fraction of incident light that is absorbed in the active region of an infrared detector. This parameter takes into account reflections from front and back surfaces and the absorption coefficient of the material, as well as the optical pathlength. Typical detector structures have metallized contacts (and/or gratings for n-type) on the top and are illuminated through the substrate side at normal incidence. For p-type devices, radiation is absorbed during its first pass through and after it is reflected off the top contact. The absorption quantum efficiency for two passes through the active region is given by

$$\eta_a = (1 - r)(1 - e^{-2\alpha \prime}), \qquad (50)$$

where r is the reflectivity of the polished substrate ($r = 0.28$ for GaAs), α is the absorption coefficient, and ℓ is the total thickness of the quantum wells. For a n-type device under the same conditions, the incident radiation will not be absorbed on its first pass through the structure. Only after reflecting back from the grating at some angle, will the radiation be absorbed. Thus, the n-type absorptance is given by

$$\eta_a = (1 - r)(1 - e^{-\alpha \prime})\beta, \qquad (51)$$

where β is a polarization correction factor. For a one dimensional diffraction grating, $\beta = 0.5$, so that only half of the radiation in an unpolarized beam will be diffracted in order to provide a component in the z-direction. Two-dimensional gratings are polarization independent, so that $\beta = 1$. It should be noted that these simplified expressions do not include the effective coupling power of the gratings. These expressions can also be modified to include additional passes due to internal reflection at a waveguide layer or at a thinned substrate [1]. So, let us compare the quantum efficiencies for a p-type and n-type structure under backside illumination, assuming no additional passes resulting from internal reflection at a back surface. Using the weak absorption approximation, $(1 - e^{-\alpha \prime}) \approx \alpha \ell$, and substituting $r = 0.28$, the expressions become

$$\eta_a = 1.44 \, \alpha \ell \quad (p\text{-type})$$

and   (52)

$$\eta_a = 0.72 \, \alpha \ell \, \beta \quad (n\text{-type}).$$

This implies that, for the same absorption coefficient and thickness, the *p*-type QWIP would have a higher quantum efficiency. And, if the absorption coefficient is about a factor of two higher for a more heavily doped *p*-type QWIP, the normal incidence absorption quantum efficiency will be 8 times higher than the *n*-type quantum efficiency with a linear grating.

The absorption quantum efficiency (η_a) is not the same as the detector quantum efficiency (η). Also, a high absorptance does not necessarily result in a high detector responsivity since the photoexcited carriers must escape the wells efficiently to give rise to a large photocurrent. The total net detector quantum efficiency is thus determined by $\eta = \eta_a p_e$, where p_e is the probability that a photoexcited carrier will escape from the quantum well and contribute to the photocurrent rather than be recaptured by the well. The escape probabilities for bound-to-continuum *p*-type and *n*-type QWIPs are similar [1]. The escape probability for holes is about 50% at zero bias and increases toward unity at high bias ($V_b > 2$ volts). The reported quantum efficiencies for *p*-type QWIPs are on the order of 25% at 77K [3, 13–15]. These quantum efficiencies are not determined from absorption coefficient measurements but are based on responsivity and optical gain results. The *p*-type quantum efficiencies compare favorably to similar *n*-type QWIPs, which have one-dimensional grating values of about 10% [40].

5. PHOTORESPONSE

Another important parameter used to characterize how well an infrared detector performs is the responsivity. The responsivity (R) is the ratio between the electrical output and the infrared radiant input. For MQW-based infrared detectors, it is customary to use the current responsivity, which is measured in amps per watt of incident radiant power. The spectral current responsivity is expressed as

$$R = (e\lambda/hc)\eta g \quad (A/W),$$ (53)

where η is the detector quantum efficiency and g is the photoconductive gain.

Normal incidence photoresponse was measured in the p-type MQW heterostructures discussed in the theory and admittance spectroscopy sections (see Table 3). The photoresponse spectra for three of the five well widths is shown in Fig. 30. The results for the 45 Å and 50 Å well widths are not included due to MBE growth errors in the composition of these materials. The position of the peak response remains nearly the same ($\lambda_p \approx 8.5$ µm) for each of these wells. This relative insensitivity of the response peak to well width shows that the peak position is not determined solely by the transition energy for a HH1 (bound) to LH2 (resonant) intersubband transition. The energy of the LH2 resonant state changes rapidly with well width. For instance, in the 40 Å well, the LH2 resonance is 173 meV above the bottom of the well. This resonance energy is based on the position of the maximum of the light-hole component of the wave function amplitude in the well, Fig. 31. Similarly, for a 35 Å well, the position of the LH2 resonance is found at 225 meV, and the resonance is even broader and weaker. By 30 Å, the light-hole resonance is so broad it is difficult to precisely determine the maximum amplitude position. For a 45 Å well, the maximum light-hole amplitude moves to the top of the well, is much sharper, and is three times stronger than for the 40 Å well. Since the response peak does not track the movement of the LH2 resonance to higher energy, the importance of including all of the continuum states in responsivity and absorption modeling becomes apparent.

Using our bound-to-continuum model, the calculated absorption spectra all peak very close to the top of the well. These results closely match the measured response peaks. For instance, for the 40 Å well, the calculated absorption spectrum peaks at 140 meV while the

TABLE 3
Structural Parameters for Samples.

Sample	L_W (Å)	L_b (Å)	x	Dopant	N_A (10^{18}cm^{-3})	Periods
A	30	500	0.30	Be	1	50
B	35	500	0.30	Be	1	50
C	40	500	0.30	Be	1	50
D	45	500	0.30	Be	1	50
E	50	500	0.30	Be	1	50

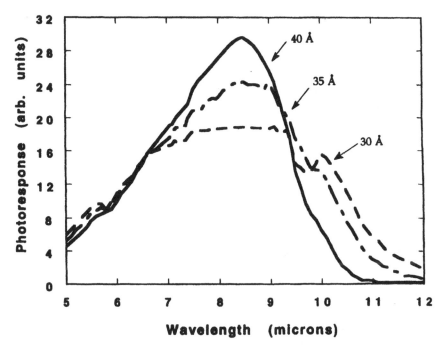

Fig. 30 The measured photoresponse spectra for three p-doped GaAs/Al$_{0.30}$Ga$_{0.7}$As MQWs at normal incidence as a function of photon wavelength.

light-hole incident state

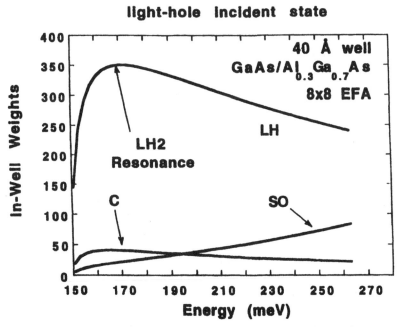

Fig. 31 In-well weights for the case of a light-hole incident state and a 40 Å p-doped GaAs/Al$_{0.30}$Ga$_{0.7}$As MQW, showing the presence of LH2 resonance.

photoresponse spectrum peaks at 146 meV. Not only does the photoresponse peak position match closely the calculated absorption results, but also the shape of the photoresponse spectra. Similar to the calculated absorption spectra (Fig. 29), the photoresponse spectra show that the 40 Å response is narrower and has a higher peak intensity than the other well widths studied (excluding the missing 45 Å well). The shift of the photoresponse cut-off wavelength to longer wavelengths also matches the shift in the calculated absorption onsets. These shifts reflect the movement of the HH1 state to higher energies, away from the well bottom, as the well width decreases.

A number of supporting measurements were performed in order to verify the structural and material parameters of the measured samples and to provide correct input into the theoretical models. Double crystal x-ray diffraction spectra were used to determine the aluminum alloy composition, superlattice period, and well width. The barrier composition was determined to be 30% aluminum in all three samples (A, B & C). The well widths were within 2 Å of the nominal widths. Similarly, photoluminescence measurements of the emission related to the HH1 to C1 transition provided a second check on the well widths, using the barrier composition determined by x-ray analysis. The photoluminescence peak energies closely matched the calculated peak positions expected for 30 Å, 35 Å, and 40 Å well widths to within a couple of angstroms. In addition, these spectra were used to assess the quality of the grown samples. In particular, the 40 Å sample had the sharpest luminescence and x-ray rocking curves, attesting to its quality. To determine the dopant concentration in the wells, Capacitance-Voltage (C-V) measurements were used to profile the upper wells. According to our C-V measurements, the doping density in the wells was $8 \times 10^{17} \pm 2 \times 10^{17}$ cm^{-3}, which is within 80% or better of the design value of 1×10^{18} cm^{-3}.

Since these photoresponse spectra were measured using a Fourier transform spectrometer, the intensity of the spectra were expressed in arbitrary units and do not accurately reflect the magnitude of the responsivity for a p-type QWIP. For quantitative responsivity measurements, a globar source, a monochromator, and a standard reference detector are used to determine the normalized responsivity wavelength spectrum. The absolute magnitude of the responsivity is then determined by measuring the photocurrent with a calibrated blackbody source [41]. Responsivity measurements have been made by Levine et al. [3, 13–15] on several different p-type MQWs with different well widths and Al alloy compositions. For well widths and compositions matching the above samples, the responsivity spectra show very similar peak and cut-off wavelengths.

As expected from the theoretical absorption spectra, the normal incidence photoresponse is larger than that for illumination coupled in through a polished 45° facet (Fig. 32) [13]. At normal incidence, both *x* and *y* polarizations contribute to the photoresponse. However, for a 45° facet, the polarization component that is not in the growth plane is broken into *x*- and *z* components. Since the oscillator strength for *z*-polarization is much weaker, the total absorption and photoresponse is decreased. In addition, using the 45° facet geometry, it was found that *s*-polarized light (*y*-polarized in our notation) has twice the photoresponse of *p*-polarized in *p*-type QWIPs with a 40 Å well width and 30% aluminum in the barrier alloy [13]. Again, this agrees with our model of the polarization behavior of the oscillator strengths.

The measured peak responsivity for a *p*-type QWIP, with a 40 Å well width and 30% Al in the barrier, is 28 mA/W at normal incidence [3]. This responsivity is an order of magnitude lower than the 0.42 A/W responsivity measured in an *n*-type MQW with a similar cut-off wavelength of 9.3 μm, using a 45° facet [37, 38]. The lower

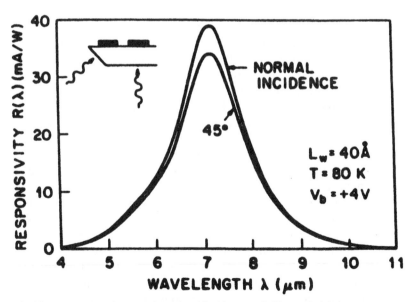

Fig. 32 Comparison between the normal incidence and 45° responsivity spectra measured at $T = 80$ K, $V_b = +4$V for a 40Å well GaAs/Al$_{0.30}$Ga$_{0.7}$As sample. The illumination geometry is shown in the inset. (Ref. 13).

responsivity of the p-type materials is a consequence of the heavier effective mass of the holes, which lowers the mobility (thus transport) of these charge carriers through the MQW stack. For instance, in n-type QWIPs, the low-field mobility is approximately 2,000 cm^2/V s, [42,43] while the mobility of p-type is probably closer to 200 cm^2/V s. The effects of the lower mobility and lifetime of the photo-excited holes are reflected in a parameter called the photoconductive gain (g).

The photoconductive gain, also referred to as optical gain, is the number of charge carriers that pass between the electrical contacts for each photo-excited carrier that is generated. The photoconductive gain is expressed as

$$g = \tau_L/\tau_T = v_d\tau_L/L, \tag{54}$$

where τ_L is the lifetime of the photo-excited carrier τ_T is the transit time across the detector active region, L is the total thickness of the MQW, and v_d is the drift velocity of the charge carrier under an applied bias, V_b. The drift velocity of the mobile charge carrier is given by

$$v_d \cong \left(\frac{\mu V_b}{L}\right)\left[1 + \left(\frac{\mu V_b}{v_s L}\right)^2\right]^{-1/2}, \tag{55}$$

where v_s is the saturation velocity of the carrier. It should be noted that this relationship for the gain predicts that responsivity and dark current will be independent of the number of quantum wells in the detector for $\kappa \alpha L \ll 1$. This result has been verified by Liu for n-type QWIPs [44]. From these expressions, we can see that an order of magnitude lower mobility will lead to an order of magnitude lower gain. In addition, a shorter free carrier lifetime will also reduce the gain of the detector. The optical gains, determined from noise and dark current measurements, for several different p-type QWIPs are in the range of 0.02 to 0.05 [3]. These gains are an order of magnitude lower than those in n-type QWIPs.

Another way of expressing the optical gain is in terms of the capture probability, p_c. This is the probability that a photo-excited carrier will be captured, i.e., will relax into the ground state of the well when crossing a single quantum well. Liu [45] has pointed out that for QWIPs, $g \approx (N_W p_c)^{-1}$, where N_W is the number of quantum wells in the detector. For p-type QWIPs, the capture probability for holes remains large even at high bias. For instance, p_c can be 50% or greater at $V_b = -6$ volts. This large capture rate is the reason for the small values of the optical gain and, thus, the responsivity.

6. DARK CURRENT

Dark current is defined as the current in the device that arises from sources other than photoexcitation. In quantum well infrared detectors, the dark current components are usually classified as either thermal or tunneling in origin. Three known components of the dark current are: thermionic emission, thermally-assisted tunneling, and temperature independent tunneling. Thermionic emission refers to the process in which carriers are thermally excited to current-carrying continuum states above the top of the well. The resulting current varies exponentially with temperature and is expected to dominate at temperatures above 50 K. Thermally-assisted tunneling is the term used for the mechanism in which carriers are thermally excited into higher states below the top of the well, and then tunnel through the triangular tip of the barrier (under an applied electric field) into the continuum states on the other side of the barrier. Thermally-assisted tunneling becomes significant, compared to thermionic emission, at relatively high electric fields [46]. It is a thermally activated process except at very high electric fields. Temperature-independent tunneling refers to the process in which carriers tunnel from the densely occupied ground subband of one well into a neighboring well. The temperature-independent tunneling component of the dark current has many of the features of sequential resonant tunneling [47]. It is observed at temperatures that are low enough to suppress thermionic emission.

The thermally-activated currents are modeled in terms of the density (n_t) of mobile carriers and their average drift velocity v_d, such that the dark current (I_D) is given by

$$I_D = e A_D v_d n_t, \tag{56}$$

where A_D is the detector area and e is the electronic charge [42]. The density of carriers which are thermally excited out of the well into the continuum transport states is given by

$$n_t = \left(\frac{m^* k T}{\pi \hbar^2 L_p}\right) e^{-(\Delta E - E_F)/kT}, \tag{57}$$

where m^* is the in-plane effective mass of the electron (hole), L_p is the MQW period, ΔE is the conduction (valence) band offset, E_F is the Fermi energy measured from the bottom of the band, and T is the temperature [42, 48]. This equation is valid for *n*-type MQWs because it is based on a density of states which assumes the bands are isotropic and parabolic. For p-type MQWs, this expression is an approximation which works near the bottom of the heavy hold band but probably overestimates the dark current at higher energies. Combining these two equations, the expression for the thermionic component of the dark current becomes

$$I_D = m^* v_d \left(\frac{e A_D}{L_p}\right)\left(\frac{kT}{\pi \hbar^2}\right) e^{-(\Delta E - E_F)/kT}. \tag{58}$$

To demonstrate the usefulness of this equation, consider the case of a p-type QWIP with a 200 μm diameter mesa with a 50 period MQW composed of a 40 Å well doped at $2 \times 10^{18} \text{cm}^{-3}$ and a 500 Å $Al_{.3}Ga_{.7}$ As barrier. Using a p-type mobility of 200 cm^2/Vs, the calculated dark current of this QWIP at 77 K with 2 volts applied bias is 2×10^{-7} amps. This value is in good agreement with the measured value of 5×10^{-7} amps for this detector configuration [3].

The magnitude of the thermionic emission is most sensitive to the energy difference between the band offset and the Fermi level. So, to compare the thermally activated dark current of a n-type and p-type QWIP, we will initially keep this energy difference constant. If the MQW period (L_p) and device area are also kept constant, then the only difference remaining between a n-type and p-type detector's thermionic dark current is the $m^* v_d$ product. For the same applied bias in the low field regime, this product will be about a factor of two smaller for the p-type QWIP. However, a more realistic comparison of these two types of devices would involve holding the cut-off wavelength constant. For a n- and p-type QWIP with a 40 Å well width, a doping density of $1 \times 10^{18} \text{cm}^{-3}$ and $\lambda_c = 9.3$ μm, the Fermi energy will be closer to the ground state in the p-type, $E_F = \text{HH1} + 3.47$ meV, than in the equivalent n-type, $E_F = \text{C1} + 10.8$ meV at $T = 77$ K. Thus, the p-type device will have about a factor of six lower dark current. However, there is a much larger advantage to be realized if we increase the doping density above $1 \times 10^{18} \text{cm}^{-3}$. For example, using the same λ_c as above but increasing the dopant concentration to $5 \times 10^{18} \text{cm}^{-3}$, the Fermi energy in the p-type QWIP will move to 11.4 meV above HH1, while in the n-type, E_F will move to 54 meV above C1. No exchange energy corrections have been applied to these energy estimates; however, since the exchange energy lowering is a function of sheet doping only, it will be the same for both n- and p-type QWIPs. The rapid increase in the Fermi energy in the n-type QWIP results in a dark current increase of nearly three orders of magnitude whereas, in the p-type QWIP, the dark current increases by only a factor of six. So, the doping density in a *p*-type QWIP can be increased beyond what is considered the optimum doping density for an *n*-type while still maintaining a reasonable thermally-activated dark current. Higher doping density in the wells can be used to increase the quantum efficiency and responsivity of the *p*-type QWIPs. In fact, a doping density as high as 2×10^{19} cm^{-3} has been proposed to increase the detectivity of *p*-type QWIPs up to the magnitude found in *n*-type QWIPs [49]. But, there is a trade off between higher quantum efficiency and increasing dark current that must be considered when increasing the dopant concentration. To determine what the optimum dopant concentration would be in the *p*-type quantum wells, the effects of dopant concentration on the quantum efficiency and on the dark current were calculated and used to compare the detectivity at 1×10^{18} cm^{-3} to the detectivities at higher doping densities. This calculation assumed an absorption coefficient of 2,500 cm^{-1} for $N_A = 1 \times 10^{18}$ cm^{-3}, no broadening of the absorption with increasing dopant concentration, and that the dark current at 77 K was due to

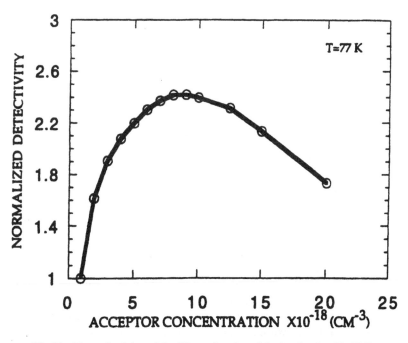

Fig. 33   Normalized detectivity D^* as a function of doping density; T = 77 K.

thermionic emission. The normalized peak detectivity as a function of dopant concentration is shown in Fig. 33. The normalized detectivity curve has a broad maximum from $5 \times 10^{18} \, \text{cm}^{-3}$ to $1.5 \times 10^{19} \text{cm}^{-3}$. In this range, the detectivity is increased by about a factor of 2.5.

While thermionic emission dominates the dark current at high temperatures, temperature-independent tunneling becomes significant at low temperatures and low bias. This component of the dark current could limit the performance of QWIPs in low background applications, which would necessitate a low infrared detector operating temperature. The proposed mechanisms for the temperature independent tunneling current are sequential resonant tunneling or defect-assisted tunneling [46, 50]. In p-type QWIPs, tunneling currents are low due to the large effective mass of holes in the AlGaAs barriers. In addition, with a p-type dopant, there are fewer defect levels in the AlGaAs barrier layer, i.e., no DX-centers formed due to dopant diffusion from the well. Deep Level Transient Spectroscopy measurements on several p-type MQWs revealed no detectable deep defect levels in these materials. With fewer deep levels, the defect-assisted tunneling is reduced.

Overall, the total dark current should be lower in p-type QWIPs than in the equivalent n-type QWIPs. A reduction in the dark current will in turn decrease the current noise in the detector when it is not background limited. The dominant noise associated with the dark current in quantum well infrared detectors is generation-recombination ($g-r$) noise, due to the escaping and trapping processes which control the dark current [51]. The noise current therefore should be given by the standard expression for $g-r$ or shot noise [52]

$$i_n^2 = 4eg I_d \Delta f; \tag{59}$$

here Δf is the measurement bandwidth, which is set to equal 1 Hz by convention. However, a correction to this classical form of shot noise to include the effects of capture probability (p_c) on current fluctuations, has been suggested. Beck[1] showed that, when taking the capture probability into account, the shot noise equation for quantum well detectors becomes

$$i_n^2 = 4eg I_d \Delta f (1 - p_c/2), \tag{60}$$

which reduces to the standard equation when $p_c \ll 1$. But the capture probability for p-type QWIPs is quite high, with p_c values of 50 to 80% [3]. Based on this expression, the lower dark current and gain, as well as the higher capture probability, found in p-type QWIPs serve to reduce the shot noise of the detector.

In practice, the noise current of a quantum well detector is measured using a spectrum analyzer [3]. The noise current is measured as a function of bias current and temperature. These noise measurements provide a means for optical gain (g) determination, which, in turn, determines the carrier capture probability. The measured current noise is also essential to calculations of the detectivity of an infrared detector.

7. DETECTIVITY

The standard figure of merit used for most single infrared detectors is the detectivity, which is a normalized signal-to-noise ratio. The peak detectivity D_λ^* is determined from

$$D_\lambda^* = R_P \sqrt{A_D \Delta f} / i_n, \tag{61}$$

where A_D is the infrared active area of the detector, $R_p (A/W)$ is the peak responsivity, i_n is the noise current ($A/\sqrt{\text{Hz}}$) and $\Delta f = 1\,\text{Hz}$. The noise current is composed of two components: dark current noise and noise due to fluctuations in the infrared photon background. When dark current noise dominates, the generation-recombination-limited detectivity can be expressed in terms of more fundamenetal parameters. Substituting the expressions for R_p, i_n and $I_d = n_t e v_d A_D$ into the above equation yields

$$D_\lambda^* = \left(\frac{\eta_a P_e}{2hv}\right) \sqrt{\frac{g}{v_d n_t}}. \tag{62}$$

This expression for D_λ^* is valid at low infrared illumination levels and high temperatures, such that the number of thermally excited carriers is greater than that of the background generated carriers. For high incident infrared power, the background photon noise dominates and the detectivity reaches background limited performance (BLIP).

Based on Eq. (62), for large D_λ^*, it is advantageous to have large values of η_a, P_e and g as well as a low density of thermally excited carriers n_t. For *p*-type QWIPs, the probability of escape is similar to *n*-type. The *p*-type quantum efficiency at normal incidence is higher than *n*-type with the incident radiation coupled in through a 45° facet or linear grating. In addition, the thermal carrier density should be lower for p-type QWIPs. The only parameter left that has a negative impact on the detectivity of p-type quantum well infrared detectors is the optical gain. So far, the optical gain in p-type QWIPs is an order of magnitude lower than in n-type QWIPs. Still, this difference in gain would cause only a factor of three difference in the detectivity. Typical values for the detectivity in p-type QWIPs range from 6.0×10^9 to 3.1×10^{10} cm $\sqrt{\text{Hz}}/\text{W}(\text{T} = 77\,\text{K}$ and $V_b = 2V)$ for cut-off wavelengths betwen 8 to $9\,\mu\text{m}$ [3,13]. These detectivities compare favorably to equivalent n-type QWIPs, for which reported detectivities are about $1 \times 10^{10}\text{cm}\sqrt{\text{Hz}}/\text{W}$ for $\lambda_c = 9.3\,\mu\text{m}$ at $T = 77\,\text{K}$ [40,42].

If we use the approximation for n_t given in Eq. (57), then we could again rewrite the expression for D_λ^* as

$$D_\lambda^* = \left(\frac{\eta_a P_e}{2hv}\right)\sqrt{\frac{g}{v_d}}\sqrt{\frac{\pi \hbar^2 L_p}{m^* kT}}\, e^{(\Delta E_v - E_F)/2kT}.\tag{63}$$

With this in mind, Levine [3] showed that the measured detectivities at $T = 77\,\text{K}$ of both n- and p-doped multi-quantum well detectors can be fitted as a function of the energy of the cut-off wavelength, $E_c = \Delta E_V - \text{HH}1,$ (Fig. 34). The best fit for the detectivity of an n-type QWIP, with a 45° polished input facet, is

$$D_e^* = 1.1 \times 10^6 \, e^{E_c/2kT} \, \text{cm}\sqrt{\text{Hz}}/\text{W},$$

while that for a p-type QWIP is

$$D_h^* = 2.0 \times 10^5 \, e^{E_c/2kT} \, \text{cm}\sqrt{\text{Hz}}/\text{W}.\tag{64}$$

Thus, for the same cut-off wavelength, the detectivity of n-QWIPs is about 5 times larger than the of p-QWIPs. However, none of the p-type structures were optimized for quantum efficiency or hole transport properties. There may still be ways to improve the detectivity of the p-type QWIPs. Also, we should keep in mind that, at low bias and low temperatures, this detectivity relationship does not apply. There are some applications for which the p-type QWIP may remain an attractive alternative [39].

8. SUMMARY

We have discussed in detail the physics behind normal incidence absorption in p-type GaAs/Al$_x$Ga$_{1-x}$As multiple quantum wells designed for infrared detection. An 8×8 envelope-function approximation

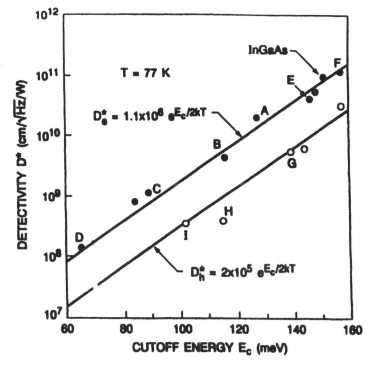

Fig. 34 Detectivity D^* (at $T = 77$ K) vs cutoff energy E_c for *n*-type QWIPs (solid circles) and *p*-type QWIPs (open circles). The straight lines are best fits to the measured data (Ref. 3).

was used to calculate the electronic structure, wave functions, and optical matrix elements for intervalence transitions in the case of coupled bands. The anisotropic nature of the light and heavy hole bands as well as the presence of *s*-like components in these subbands modify the selection rules for intersubband optical transitions to allow normal incidence absorption in valence band quantum wells. The bound-to-continuum absorption in *p*-type $GaAs/Al_xGa_{1-x}$ As quantum wells was also calculated with a model that includes true continuum states, including resonant extended states above the barrier. It was shown that the strongest bound-to-continuum optical transition, for the composition and well widths studied, is from the first heavy-hole subband to the second light-hole subband. Understanding the exact nature of the bound-to-continuum normal incidence absorption in these *p*-type quantum wells is crucial to optimizing the MQW design for infrared detector aplications.

P-type GaAs/Al$_x$Ga$_{1-x}$As multi-quantum well materials have been shown to have a strong normal incidence photoresponse. This photoresponse can be fairly broad and flat when the dominant transition is to regular continuum states instead of a resonant state. As the well width is increased, the photoresponse becomes stronger and narrower due to stronger coupling to the LH2 resonant subband as it approaches the top of the well. As our modeling has shown, the photoresponse spectra follow the same trends as the normal incidence absorption spectra, which are similarly enhanced by the LH2 resonant state. The presence of this light-hole resonant subband near the top of the well may also improve the character of the transport state.

We also addressed some of the trade offs between a *p*-type GaAs/Al$_x$Ga$_{1-x}$As QWIP and the conventional *n*-type QWIP. Without the complications of gratings or substrate thinning, an optimized *p*-type QWIP could attain quantum efficiencies of 60% for unpolarized, normally incident infrared radiation. However, the responsivity will still be lower than for *n*-type QWIPs unless the optical gain can be improved. For applications where the dark current must be kept low, the *p*-type structures should be advantageous. The thermionic emission should be lower for *p*-type, due to the lower Fermi energy, and tunneling currents should be lower, due to fewer defect levels in the barrier layer and the heavier effective mass of the holes in the ground state subband.

It is clear that there have been too few studies of *p*-type QWIPs to properly assess the full potential of this material. For example, the initial results implicated low hole mobility in the low gain figures for the structure. Therefore, it is important to perform measurements of above-the-barrier transport of holes. As we have shown, light holes are most important in this regard, and their mobility should be superior to that of the heavy holes. One can design structures in which well resonances are coupled to barrier resonances at the same energy, which would improve carrier mobilities and reduce the capture into the wells, thereby raising the gain. Such quantum engineering, coupled with improvements in the material quality, should go a long way in making *p*-type QWIPs competitive with *n*-type QWIPs.

APPENDIX A

The following input parameters were used in the calculation [53].

	GaAs	AlGaAs
m_c/m_0	0.0665	$0.0665 + 0.0835x$
γ_1	6.790	$6.790 - 3.0x$
γ_2	1.924	$1.924 - 0.694x$
γ_3	2.681	$2.681 - 1.286x$
E_g	1.519 eV	$1.519 + 1.247x$
$(2m/\hbar^2)P^2$	28.8 eV	28.8 eV
Δ	0.343 eV	$0.343 - 0.062x$

The modified Kohn-Luttinger parameters are related to the regular Kohn-Luttinger parameters as follows:

$$\frac{\hbar^2}{2m_0}\gamma_1' = \frac{\hbar^2}{2m_0}\gamma_1 - \frac{1}{3}\frac{P^2}{E_g}$$

$$\frac{\hbar^2}{2m_0}\gamma_2' = \frac{\hbar^2}{2m_0}\gamma_2 - \frac{1}{6}\frac{P^2}{E_g}$$

$$\frac{\hbar^2}{2m_0}\gamma_3' = \frac{\hbar^2}{2m_0}\gamma_3 - \frac{1}{6}\frac{P^2}{E_g}$$

$$\frac{1}{m_c'} = \frac{1}{m_c} - \frac{E_P}{3m_0}\left[\frac{2}{E_g} + \frac{1}{E_g + \Delta}\right].$$

We use a 60/40 conduction-to-valance band offset ratio.

APPENDIX B

Employing the phase conventions of Bir and Pikus [28], the wave functions for the six valence $(l = 1)$ bands are

$$v = 3 \quad u_{3/2,3/2} = \frac{1}{\sqrt{2}}[X + iY]\alpha,$$

$$v = 4 \quad u_{3/2,1/2} = \frac{i}{\sqrt{6}}[(X + iY)\beta - 2Z\alpha],$$

$$v = 5 \quad u_{3/2,-1/2} = \frac{1}{\sqrt{6}}[(X - iY)\alpha + 2Z\beta],$$

$$v = 6 \quad u_{3/2,-3/2} = \frac{i}{\sqrt{2}}[X - iY]\beta,$$

$$v = 7 \quad u_{1/2,1/2} = \frac{1}{\sqrt{3}}[(X + iY)\beta + Z\alpha],$$

$$v = 8 \quad u_{1/2,-1/2} = \frac{i}{\sqrt{3}}[-(X - iY\alpha + Z\beta],$$

where α and β are spin-up and- down spinors and X, Y, and Z are Bloch periodic functions which transform like x, y, z under the rotations of the tetrahedral point group T_d. For the wave functions at the bottom of the conduction band $l = 0$, and we choose

$$v = 1 \quad u_{1/2,1/2} = S\alpha,$$

$$v = 2 \quad u_{1/2,-1/2} = iS\beta,$$

where S indicates the s-like character under the rotations of the tetrahedral point group T_d. The first index on the u's gives the total angular momentum j and the second gives the projection of the total angular momentum along the z-axis. Index v gives the order in which these functions are used in the calculation of matrix elements.

References

1. B.F. Levine, *J. Appl. Phys.*, **74**, R1 (1993).
2. E. Rosencher, B. Vinter, and B.F. Levine (eds.), *Intersubband Transitions in Quantum Wells* (Plenum, New York, 1992).
3. B.F. Levine, A. Zussman, S.D. Gunapala, M.T. Asom, J.M. Kuo, and W.S. Hobson, *J. Appl. Phys.*, **72**, 4429 (1992).
4. D.D. Coon and R.P.G. Karunasiri, *Appl. Phys. Lett.*, **45**, 649 (1984).
5. L.C. West and S.J. Eglash, *Appl. Phys. Lett.*, **46**, 1156 (1985).
6. K.W. Goosen, S.A. Lyon, and K. Alavi, *Appl. Phys. Lett.*, **53**, 1027 (1988).
7. G. Hasnain, B.F. Levine, S. Gunapala, and N. Chand, *Appl. Phys. Lett.*, **57**, 608 (1990).
8. J.Y. Andersson and L. Lundqvist, *Appl. Phys. Lett.*, **59**, 857 (1991).
9. Y.C. Wang and S.S. Li, *J. Appl. Phys.*, **75**, 582 (1994).
10. Y.C. Chang and J.N. Schulman, *Phys. Rev.*, *B*, **31**, 2069 (1985).
11. G.D. Sanders and Y.C. Chang, *Phys. Rev.*, *B*, **36**, 4849 (1987).
12. Y.C. Chang and R.B. James, *Phys. Rev.*, *B*, **39**, 12672 (1989).
13. B.F. Levine, S.D. Gunapala, J.M. Kuo, S.S. Pei, and S. Hui, *Appl, Phys. Lett.*, **59**, 1864 (1991).
14. W.S. Hobson, A. Zussman, B.F. Levine, J. deJong, M. Geva, and L.C. Luther, *J. Appl. Phys.*, **71**, 3642 (1992).
15. J.M. Kuo, S.S. Pei, S. Hui, S.D. Gunapala, and B.F. Levine, *J. Vac. Sci. Technol.*, *B*, **10**, 995 (1992).
16. P. Man and D.S. Pan, *Appl. Phys. Lett.*, **61**, 2799 (1992).

17. D. Teng, C. Lee, and L.F. Eastman, *J. Appl. Phys.*, **72**, 1539 (1992).
18. G. Lucovsky, *Sol. St. Comm.*, **3**, 299 (1965).
19. H.C. Liu, *J. Appl. Phys.*, **73**, 3062 (1993).
20. K.K. Choi, *J. Appl. Phys.*, **73**, 5230 (1993).
21. J.M. Luttinger and W. Kohn, *Phys. Rev.*, **97**, 869 (1955).
22. M. Altarelli, in *Heterojunctions and Semiconductor Superlattices*, edited by G. Allan, G. Bastard, N. Boccora, and M. Voos, (Springer-Verlag, Berlin, 1986).
23. R. Eppenega, M.F.H. Schurrmans, and S. Colak, *Phys. Rev.*, **36**, 1554 (1987).
24. E.O. Kane, *J. Phys. Chem. Solids*, **1**, 82 (1956).
25. see Gerald Bastard, *Wave Mechanics Applied to Semiconductor Heterostructures*, (Halstead, New York, 1988).
26. F. Szmulowicz and Gail J. Brown, *Phys. Rev.*, *B*, **51**, 13203 (1995).
27. D.L. Smith and C. Mailhiot, *Phys. Rev.*, *B*, **33**, 8345 (1986).
28. G.L. Bir and G.E. Pikus, *Symmetry and Strain-Induced Effects in Semiconductors* (Wiley, New York, 1974).
29. L.C. Andreani, A. Pasquarello, and F. Bassani, *Phys. Rev. B*, **36**, 5887 (1987).
30. G.J. Brown, F. Szmulowicz, and S.M. Hegde, *J. Elec. Matls*, **24**, 559 (1995).
31. K.M.S.V. Bandara, D.D. Coon, and B.O, *Appl. Phys. Lett.*, **53**, 1931 (1988).
32. S.R. Smith, F. Szmulowicz, and G.J. Brown, *J. Appl. Phys.*, **75**, 1010 (1994).
33. D.V. Lang, in *Measurement of Band Offsets by Space Charge Spectroscopy*, edited by F. Capasso and G. Margaritondo (Elsevier, Amsterdam, 1987).
34. F. Szmulowicz, *Phys. Rev.*, *B*, **51**, 1613 (1995).
35. R.B. Feynman, *Phys. Rev.*, **56**, 340 (1939).
36. H. Hellman, *Einfuhrung in die Quantenchemie* (Deuticke, Leipzig, 1937).
37. B.F. Levine, C.G. Bethea, G. Hasnain, J. Walker, and R.J. Malik, *Appl. Phys. Lett.*, **53**, 296 (1988).
38. B.K. Janousek, M.J. Daugherty, W.L. Bloss, M.L. Rosenbluth, M.J. O'Loughlin, H. Kanter, F.J. DeLuccia, and L.E. Perry, *J. Appl. Phys.*, **67**, 7608 (1990).
39. R.L. Whitney, K.F. Cuff, and F.W. Adams, in *Semiconductor Quantum Wells and Superlattices for Long-Wavelength Infrared Detectors*, edited by M.O. Manasreh (Artech, Boston, 1993).
40. L. Lundqvist, J.Y. Andersson, Z.F. Paska, J. Borglind, and D. Haga, *Appl. Phys. Lett.*, **63**, 3361 (1993).
41. J.D. Vincent, *Fundamentals of infrared Detector Operation and Testing* (Wiley Interscience, New York, 1990) p. 173.
42. B.F. Levine, C.G. Bethea, G. Hasnain, V.O. Shen, E. Pelve, R.R. Abbott, and S.J. Hsieh, *Appl. Phys. Lett.*, **56**, 851 (1990).
43. A. Fraenkel, E. Finkman, S. Maimon, and G. Bahir, *J. Appl. Phys.*, **75**, 3536 (1994).
44. A.G. Steele, H.C. Liu, M. Buchanan, and Z.R. Wasilewski, *J. Appl. Phys.*, **72**, 1062 (1992).
45. H.C. Liu, *Appl. Phys. Lett.*, **60**, 1507 (1992).
46. E. Pelve, F. Beltram, C.G. Bethea, B.F. Levine, V.O. Shen, S.J. Hsieh, and R.R. Abbott, *J. Appl. Phys.*, **66**, 5656 (1989).
47. G.M. Williams, R.E. DeWames, C.W. Farley, and R.J. Anderson, *Appl. Phys. Lett.*, **60**, 1324 (1992).
48. M.A. Kinch and A. Yariv, *Appl. Phys. Lett.*, **55**, 2093 (1989).
49. B.W. Kim and A. Majerfeld, *Proceedings of the 20th Int'l Symp. on GaAs and Related Compounds*, Freiburg, Germany, Aug. 29-Sept. 2, 1993.
50. S.R. Andrews and B.A. Miller, *J. Appl. Phys.*, **70**, 993 (1991).
51. H.C. Liu, *Appl. Phys. Lett.*, **61**, 2703 (1992).
52. A. Rose, *Concepts in Photoconductivity and Allied Problems* (Wiley Interscience, New York, 1963).
53. L.W. Molenkamp, R. Eppenga, G.W. t'Hooft, P. Dawson, C.T. Foxon, and K.J. Moore, *Phys. Rev.*, *B*, **38**, 4314 (1988).

CHAPTER 6

N-Type III–V Multiple-Quantum-Well Detectors Exhibiting Normal-Incidence Response

E. R. BROWN, S. J. EGLASH, and K. A. McINTOSH

*MIT-Lincoln Laboratory, 244 Wood St., Lexington,
Massachusetts 02173-9108 USA*

1. INTRODUCTION TO QUANTUM-WELL INTERSUBBAND DETECTORS

1.1. Overview

Since the initial observation of strong intersubband absorption [1] and sensitive direct detection [2] in GaAs multiple quantum well (MQW structures), detectors made from these structures have been the prototype for a new class of infrared (IR) detectors. The advantages of the GaAs MQW detector over alternative IR detector technologies are well known and discussed at length in the chapter by H. C. Liu in this text. The primary benefits are (1) the peak wavelength λ_p of the spectral response curve can be "engineered" through control of the quantum-well width and the barrier height, and (2) GaAs is much more robust than the standard IR detector materials (e.g., HgCdTe) and is, there-fore, more amenable to fabrication in large focal-plane arrays. These advantages have been utilized to demonstrate a multiple-color detector [3] and a long-wavelength detector [4] ($\lambda_p \approx 18\,\mu$m) operating where intrinsic (cross-gap) semiconductors are very difficult to obtain.

The disadvantages of the GaAs MQW direct detector are also well known because they have been a major impediment to the implementation of these devices in IR staring arrays. One disadvantage is the polarization selection rule that results in a strong absorption of light only for the component of the incident IR electric field that is oriented perpendicular to the plane of the quantum wells. This requires that

two-dimensional grating structures be fabricated at the top interface between free space and the GaAs wafer so that randomly polarized radiation at normal incidence can be transformed into radiation that is
efficiently absorbed within the MQW structure. A second disadvantage is
that the dark-current density at 77 K is roughly 4 orders of magnitude
higher than in good HgCdTe detectors. This causes the specific detectivity D^* of the MQW detectors to be roughly 100 times lower under low
background illumination. The best detectivity at 77 K reported to date at
10-μm or longer wavelength [5] is $D^* = 1.6 \times 10^{10}\,\mathrm{cm\,Hz^{1/2}W^{-1}}$. As
will be discussed in Sec. 2, both of these disadvantages can, in principle,
be alleviated by MQW detectors designed with electrons confined to
ellipsoidal valleys in the quantum wells.

1.2. Operational Characteristics

The most successful quantum-well detectors studied to date have been
based on the bound-to-extended-state mode shown schematically in
Fig. 1. In this mode an incoming photon polarized perpendicular to
the plane of the quantum well excites an electron from the ground
state directly to the first extended state above the barriers. The electron then drifts away from the quantum well in response to the applied electric field and is collected at a contact. In GaAs detectors, the
responsivity spectrum is usually Gaussian in form with a peak energy
close to the bound-to-extended-state separation and a spectral energy
width roughly equal to one third of the peak energy.

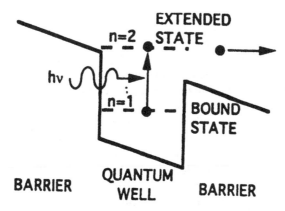

Fig. 1 Schematic diagram of bound-to-extended state intersubband transition of an
electron in a quantum well.

The polarization rules associated with intersubband absorption in ellipsoidal-valley and spherical-valley quantum wells can be understood from the following classical argument. In the same sense that atomic transitions were analyzed early in the 20th century, the intersubband transition in quantum wells can be modeled roughly as a classical charged oscillator. The restoring force on the oscillator is associated with the impulsive reflection from the walls of the quantum well. Hence, the acceleration of the oscillator, $\vec{a}$, that is required for electromagnetic absorption always occurs along the axis perpendicular to the plane of the quantum well. The relationship between the acceleration and the driving electric field $\vec{f}$ is given by Newton's second law, $\vec{a} = m^{-1} \cdot \vec{f}$, where m is the effective mass. In GaAs or any other quantum well in which the electron occupies a spherical constant-energy surface, m^{-1} is a scalar quantity so that $\vec{a}$ and $\vec{f}$ are necessarily colinear. In other words, absorption occurs only for the component of the electric field perpendicular to the plane of the quantum well. In materials having a nonspherical constant-energy surface, m^{-1} becomes a tensor quantity denoted as $\vec{w}$ (the reciprocal effective-mass tensor). In this case, $\vec{a}$ can couple to components of $\vec{f}$ lying in the plane of the quantum well provided that $\vec{w}$ has nonzero off-diagonal elements.

Most of the MQW detectors studied to date have been found to behave like classic photoconductors in the sense that the photocurrent I_0 can be expressed as

$$I_0 = \frac{\eta_0 e g P}{h\nu} = R_0 P, \tag{1}$$

where η_0 is the external quantum efficiency, e is the electron charge, g is the photoconductive gain, P is the optical power, h is Planck's constant, ν is the optical frequency, and R_0 is the external responsivity defined here as the change in photocurrent per unit change in optical power at a fixed bias voltage. As in other photoconductors, R_0 is expected to vanish at zero bias and to increase linearly with bias voltage. This increase follows from a decrease in transit time t_T, according to Eq. (1), and the definition $g \equiv \tau/t_T$, where τ is the photoelectron lifetime. Beyond some bias voltage, either t_T saturates because of a saturated drift velocity or τ decreases rapidly because of a increasing relaxation rate, leading to either a saturation or a decrease in the responsivity with further bias voltage.

In addition to the photocurrent, every MQW detector has a dark current I_D (i.e., current that flows in the absence of input radiation).

The dark current arises from several mechanisms, the most important of which is the thermionic emission of electrons out of the quantum wells. This mechanism tends to dominate the dark current at all temperatures above a value that depends on the barrier height relative to the ground state, the electron concentration in the quantum wells, and other factors to be discussed in Sec. 2.4. The dark current is important because it is always associated with electrical noise that limits the sensitivity of the MQW detector. The dark-current noise arises from the following two mechanisms: (1) fluctuations in the excitation of electrons into the extended states above the barriers in Fig. 1 because of randomness in the thermionic-emission rate, and (2) fluctuations in the relaxation rate of electrons back into the quantum wells because of the stochastic nature of electron scattering events. Note that the photocurrent exhibits an identical electrical noise phenomenology except that fluctuations in the excitation of photoelectrons arise from randomness in the photon flux.

To compute the power spectrum of the dark-current and photocurrent fluctuations, S_I, one utilizes the fact that the excitation and relaxation mechanisms in the two currents are statistically uncorrelated and are physically equivalent to the injection and collection mechanisms that govern shot noise in a junction (e.g., Schottky or *p-n*) diode. Hence, each mechanism should contribute a factor $2egI_O$ and $2egI_D$ to the photocurrent and dark-current noise, respectively. Each of these factors must be weighted to reflect the fact that noise current, like the continuous current of Eq. (1), is transferred to the external circuit only during the time between the excitation and relaxation events. This weighting factor is just the photoconductive gain, so that the power spectral density is given by

$$S_I = 4\,eg(I_D + I_O).\tag{2}$$

In addition to the dark-current and photocurrent noise, an MQW photoconductor must also generate thermal noise that becomes the only noise mechanism in the limit of zero bias (i.e., thermal equilibrium). Provided that the electrical and optical power are kept sufficiently low, the power spectral density of current fluctuations is given by the generalized Nyquist relation

$$S_I = 4kTG,\tag{3}$$

where k is Boltzmann's constant, T is the detector physical temperature, and G is the differential conductance measured at the bias point.

As in other infrared detectors, the most important measure of sensitivity is the specific detectivity D^*, which is proportional to the voltage or current signal-to-noise ratio measured at the detector output node. Assuming that detector noise is described by Eqs. (2) and (3) and that the detector output is connected to an ideal (i.e., noise-free) current amplifer, one can write

$$D^* = \frac{R_0}{\sqrt{4eg(J_D + J_0) + 4kTG'}} = \frac{\eta_0 eg}{2h\nu\sqrt{eg(J_D + J_0) + kTG'}}, \qquad (4)$$

where J is the current density and G' is the specific conductance (i.e., conductance per unit area). As an example, consider a GaAs spherical-valley detector containing 100 quantum wells and designed for a peak response at $\lambda = 10\,\mu m$ ($h\nu = 124$ meV). When operated at 77 K in the presence of low background radiation, one finds that the dark-current noise dominates the photocurrent and thermal noise. Therefore, in the highest-quality detectors having $\eta \approx 0.4$, $g \approx 0.1$, and $J_D \approx 0.01$ A cm^{-2}, one finds $D^* \approx 1.3 \times 10^{10}$ cm Hz$^{1/2}$W^{-1}. This is roughly 2 orders of magnitude less than the dark-current-limited D^* of a good HgCdTe photoconductive detector operating under the same conditions. This has been the greatest impediment to the application of GaAs MQW detectors in focal-plane staring arrays, particularly for low-background applications at 77 K operating temperature.

1.3. Ellipsoidal-Valley Materials

One class of materials in which $\bar{w}$ can have very large off-diagonal components is semiconductors and semimetals in which the conduction band edge lies at a point away from the center of the Brillouin zone. In most if not all of these materials, the electrons have ellipsoidal constant-energy surfaces (i.e., ellipsoidal valleys) rather than spherical valleys such as GaAs and other materials with the band edge at the center of the Brillouin zone. The two canonical ellipsoidal-valley semiconductors are the column-IV materials Si and Ge whose ellipsoids are centered near or at the X- and L-points, respectively, of the first Brillouin zone. Two promising III–V materials are $Al_xGa_{1-x}As$ and $Al_xGa_{1-x}Sb$. The former has Si-like ellipsoids for $x > 0.45$, and the latter has Ge-like ellipsoids for $0.18 < x < 0.55$ and Si-like ellipsoids for $x > 0.55$. They both have cubic crystal symmetry so that the position of their conduction valleys is designated in the same way: Γ-, X- and L-point. A promising, although unexplored, IV–VI material is $Pb_xSn_{1-x}Te$. This crystallizes in the wurtzite rocksalt structure. Although the crystal does not have cubic symmetry, it still yields highly eccentric ellipsoidal valleys for electrons in the conduction band.

Of the candidate ellipsoidal-valley materials, the III–Vs have an important advantage since they can be exactly lattice matched to the barrier material and to the substrate. In the case of $Al_xGa_{1-x}As$, lattice matching is automatic since GaAs and AlAs have nearly equal lattice constants. For $Al_xGa_{1-x}Sb$ lattice matching to GaSb can be achieved by incorporating a small fraction of As to made the quaternary $Al_xGa_{1-x}As_ySb_{1-y}$. Such lattice matching cannot be achieved within the column-IV materials. For example, a Ge quantum well can be confined by Si barriers but the lattice mismatch is 4%. Over the multitude of quantum wells required to make a useful detector, such mismatch could likely lead to lattice relaxation and the concomitant formation of crystal defects, such as misfit dislocations. Such defects are not desirable within a structure requiring very low levels of dark current. The disadvantage of IV–VI and similar ellipsoidal-valley materials is the difficulty in growing narrow quantum wells. Such wells require abrupt heterojunctions in a compositional sense. The heavier column-VI materials such as Ge and Pb have a strong tendency to diffuse into III–V or column-VI materials during growth, leading to a significant compositional smearing of the heterojunction and, in some cases, autodoping of adjacent layers. This is the reason that the very interesting structure of Ge quantum wells bounded by GaAs barriers is so difficult to realize.

Ellipsoidal-valley MQW detectors, like GaAs detectors, should behave like photoconductors in accordance with Sec. 1.2. However, the ellipsoidal-valley detectors have two characteristics that may lead to a significantly better detector sensitivity. First, they can provide a strong absorption of light at normal incidence. Because of this, ellipsoidal-valley MQW detectors are compatible with multiple-pass interference techniques, which can enhance the quantum efficiency per electron beyond the level attainable with spherical quantum wells. One promising enhancement technique, the resonant-optical cavity, is discussed in Sec. 6. Second, the ellipsoidal-valley MQW detectors should ultimately exhibit less dark current than spherical-valley detectors at a given operating wavelength, temperature, and doping density. This follows from the greater density-of-states effective mass of ellipsoidal-valley materials and is discussed in more detail in Sec. 2.4.

2. PRINCIPLES OF ELLIPSOIDAL-VALLEY QUANTUM-WELL DETECTORS

2.1. Intersubband Transition

Although the classical argument for absorption given in Sec. 1.2 explains the polarization rule, it cannot predict with satisfactory accuracy the absorption strength. For this, one must resort to quantum mechanics. In this section, a quantum-mechanical formalism is outlined that can be applied to any quantum well containing isolated conduction- or valence-band valleys describable by an effective mass tensor. This is particularly convenient because it then reduces to spherical-valley quantum wells as a special case, which allows the formalism to be tested against the plethora of GaAs quantum-well data. The formalism cannot be readily applied to holes because of the nonellipsoidal form of the constant-energy surfaces and because of the additional complication caused by light-hole and heavy-hole mixing.

To obtain the magnitude of the absorption, one starts with a powerful approximation in semiconductor physics known as the effective-mass theorem. In a generalized form first derived by Luttinger and Kohn [6], the quantum-mechanical (Schrödinger) equation of motion of an electron is given by

$$E(\mathbf{k}_0 + \mathbf{P}/\hbar - e\mathbf{A}/\hbar c) \cdot F(\mathbf{r}) = \varepsilon F(\mathbf{r}), \tag{5}$$

where $E(\mathbf{k})$ is the energy function for the electrons, $\mathbf{P}$ is the momentum operator, $\mathbf{A}$ is the vector potential, ε is the total-energy eigenvalue, $F(\mathbf{r})$ is the slowly varying (over a unit cell) envelope wave function, and all quantities shown in bold are three-dimensional vectors. To apply this to ellipsoidal-valley materials, one uses the following functional form for $E(\mathbf{k})$ valid around conduction-band minima:

$$E(\mathbf{k}) = E(\mathbf{k}_0) + \frac{1}{2} \sum_{m,n=1}^{3} (\mathbf{k}_m - \mathbf{k}_{0m}) w_{mn} (\mathbf{k}_n - \mathbf{k}_{0n}) \tag{6}$$

where w_{mn} is the reciprocal effective-mass tensor component equal to $\hbar^{-2} \partial^2 E(\mathbf{k}_0)/\partial k_m \partial k_n$. Substituting this into Eq. (5) and expanding the noncommuting quantum-mechanical operators, one obtains

$$\left\{ \frac{1}{2} \sum_{m,n} w_{mn} P_m P_n - \frac{e}{c} \sum_{m,n=1}^{3} \frac{w_{mn}}{2} \left([A_m, P_n]_+ - \frac{e}{c} A_m A_n \right) \right\} F(\mathbf{r}) \equiv$$
$$\{ H_0 + H_p \} F(\mathbf{r}) = \varepsilon F(\mathbf{r}) \tag{7}$$

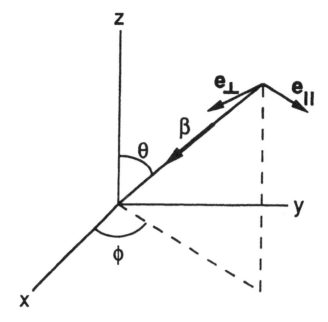

Fig. 2 Three-dimensional coordinate system used to analyze intersubband transitions in ellipsoidal-valley quantum wells.

where $[,]_+$ denotes anticommutation, H_0 and H_p represent the two summations in Eq. (7), and the symmetry of the reciprocal effective-mass tensor has been utilized. Provided that the magnitude of $\mathbf{A}$ is small, H_p can be treated like a perturbation with respect to H_0.

The next step in calculating the photon absorption is to obtain the envelope function as a solution to the eigenvalue equation, $H_0F(\mathbf{r}) = \varepsilon F(\mathbf{r})$. The solution in a quantum well is greatly simplified by the assumption that the envelope function vanish everywhere in the barrier region. This is formally equivalent to assuming that the quantum-well is infinitely deep, so that it is henceforth refered to as the infinite-well approximation. In the coordinate system of Fig. 2 with the quantum well lying in the x-y plane, it leads to the following solution in the jth subband,

$$F_j(\mathbf{r}) = \sqrt{\frac{2}{L_z}}\sin(\kappa_j z)\frac{\exp[i(k_x x + k_y y)]}{\sqrt{L_x L_y}}\exp[-i(k_{xy}z)], \qquad (8)$$

where L_z is the width of the quantum well, $\kappa_j = j\pi/L_z$, $k_{xy} = (w_{xz}k_x + w_{yz}k_y)/w_{zz}$, and the z axis is perpendicular to the quantum well. The eigenvalues are

$$\varepsilon = \varepsilon_j + \frac{\hbar^2}{2}\left[\left(w_{xx} - \frac{w_{xz}^2}{w_{zz}}\right)k_x^2 + 2\left(w_{xy} - \frac{w_{xz}w_{yz}}{w_{zz}}\right)k_xk_y + \left(w_{yy} - \frac{w_{yz}^2}{w_{zz}}\right)k_y^2\right], \quad (9)$$

where $\varepsilon_j = w_{zz}\hbar^2\kappa_j^2/2$.

Intersubband absorption is analyzed by adding to the vector potential in Eq. (5) a time dependence $e^{-i\omega t}$ and applying time-dependent perturbation theory. If the harmonic potential is present along with weak scattering events occurring at a rate τ_s^{-1}, perturbation theory yields the following expression for the probability transition rate from the jth to the kth subband:

$$R_{jk} = \frac{2\pi}{\hbar}|\langle F_k|H_p|F_j\rangle|^2\frac{\Gamma_S/2\pi}{(\varepsilon_k - \varepsilon_j - \hbar\omega)^2 + \Gamma_S^2/4}, \quad (10)$$

where $\Gamma_S = \hbar/\tau_s$ is the energy-broadening parameter. The validity of this expression requires that $\Gamma_S \ll \varepsilon_k - \varepsilon_j$.

Of greatest interest is the transition between the first and second subbands since this transition has the greatest rate per magnitude of vector potential. Substituting in the sinusoidal forms for $|F_1\rangle$ and $|F_2\rangle$ and integrating over all three coordinates, one obtains [7]

$$R_{12} = \frac{64\hbar e^2\kappa_1^2}{9\pi^2c^2}\frac{(A_xw_{xz} + A_yw_{yz} + A_zw_{zz})^2\Gamma_S}{(\varepsilon_2 - \varepsilon_1 - \hbar\omega)^2 + \Gamma_S^2/4}\delta_{k_{x2},k_{x1}}\delta_{k_{y2},k_{y1}}, \quad (11)$$

where the last two Kronecker delta functions enforce momentum conservation in the transverse plane.

The quantity R_{12} represents the transition rate for a single electron. To obtain the rate for a large population of electrons, S_{12}, one has to sum R_{12} over both subbands, weighting by the probability of occupancy (i.e., the Fermi factor) in the first subband and the probability of vacancy in the second subband. By assuming that the operating temperature is low enough that the second subband is unoccupied, one obtains the result $S_{12} = \sigma_1 R_{12}$, where

$$\sigma_1 = \frac{k_BT}{\pi\hbar^2}m_d^*\ln\{1 + \exp[(\varepsilon_F - \varepsilon_1)/k_BT]\}, \quad (12)$$

k_B is Boltzmann's constant, ε_F is the Fermi energy, and m_d^* is the density-of-states mass in the transverse plane of the quantum well for the conduction-band valley which the electrons occupy.

2.2. Absorption Strength

The experimentally observable quantity for intersubband transitions is the plane-wave fractional absorption, $\zeta(\omega, \theta, \phi)$, which is defined simply by $(I_i - I_t)/I_i$, where I_i and I_t are the incident and transmitted intensities, respectively, at frequency ω and incident orientation (θ, ϕ) with respect to the coordinate system in Fig. 2. The fractional absorption for one quantum well is evaluated by summing S_{12} over all ellipsoids and normalizing to the incident photon flux,

$$\zeta \equiv \frac{\sum_{\eta=1}^{M_c} S_{12}^{(\eta)}}{I/\hbar\omega} = \frac{128\hbar^2 e^2}{9\pi n\omega c} \sum_{\eta=1}^{M_c} \frac{(A_x w_{xz}^{(\eta)} + A_y w_{yz}^{(\eta)} + A_z w_{zz}^{(\eta)})^2}{|A|^2}$$

$$\times \frac{\kappa_{1\eta}^2 \sigma_\eta \Gamma_{s\eta}}{(\varepsilon_{2\eta} - \varepsilon_{1\eta} - \hbar\omega)^2 + \Gamma_{s\eta}^2/4}. \tag{13}$$

In this expression, $\kappa_1, \varepsilon_2, \varepsilon_1, \sigma$, and Γ_s are all written explicitly as functions of η because they generally depend on the ellipsoid orientation through the dependence of the energy in Eq. (9) on w_{zz}. A special but useful case is where the electrons occupy only the lowest-energy ellipsoid. This occurs at low temperatures or in quantum wells oriented along high-symmetry directions so that all of the ellipsoids share a common value of w_{zz}. In either case, M_c is the number of ellipsoids sharing the ground state (i.e., the ellipsoid degeneracy). Under this condition, one can apply the infinite-well approximation and symmetry considerations to transform Eq. (13) to

$$\zeta = \frac{256\hbar e^2 \omega_0 w_{zz} \sigma_T}{27\pi n\omega c} \cdot \frac{\Gamma_s}{(\hbar\omega_0 - \hbar\omega)^2 + \Gamma_s^2/4} \cdot G(\theta, \phi), \tag{14}$$

where σ_T is the total sheet charge density in the ground state, $\omega_0 \equiv (\varepsilon_2 - \varepsilon_1)/\hbar$ is the intersubband Bohr frequency, and

$$G(\theta, \phi) = \frac{w_{zz}^{-2}}{M_c} \cdot \sum_{\eta=1}^{M_c} \frac{(A_x w_{xz}^{(\eta)} + A_y w_{yz}^{(\eta)} + A_z w_{zz}^{(\eta)})^2}{|A|^2}. \tag{15}$$

The quantity defined by. Eq. (15) is useful in comparing different quantum-well materials and ellipsoid orientations. Before doing this, it is important to realize that $G(\theta, \phi)$ depends on both the propagation direction and the polarization of the incident radiation. To make this dependency clear, the incident radiation is characterized in the conventional way shown in Fig. 2. A linearly polarized **A** normal to the propagation vector **β** is resolved into two components, $A_\perp$ and A , perpendicular and parallel, respectively, to the plane of incidence, which is defined by **β** and the *z* axis.

Table 1 lists the form of G for some practical quantum-well orientations having high symmetry. The X-point material has six ellipsoids that contribute to the ground state in a (111) quantum well. Any one of these ellipsoids acting alone would exhibit a ϕ dependence, but acting together the ϕ dependence vanishes. Similarly, the L-point material has four ellipsoids that contribute to the ground state of a (100) well. Similarly, X-valley ellipsoids in a (100) quantum well and L-valley ellipsoids in a (111) quantum well have an equal effect.

2.3. Absorption Coefficient and Quantum Efficiency

Although the fractional absorption is the most sensible quantity to characterize the intersubband transition, it is sometimes convenient to use the absorption coefficient α since this is the more conventional measure of absorption strength in three-dimensional absorbing materials. To define α for intersubband transitions, one must introduce a length scale over which the absorption occurs. This length must be dependent on the orientation of β since the distance a plane wave traverses through the quantum well depends on this orientation. This

TABLE 1

The factor $G(\theta, \phi)$ as defined by Eq. (15) for some ellipsoidal-valleys in quantum wells of high symmetry. In this table, $m_{\parallel} = (m_l \cdot m_t)/(m_l - m_t)$, with $m_l > m_t$.

Ellipsoidal Valley	(100)			(111)		
	M_c	w_{zz}	G (Polarization)	M_c	w_{zz}	G (Polarization)
Γ	1	$\dfrac{1}{m^*}$	$0(A_\perp)$ $\sin^2\theta(A_\parallel)$	1	$\dfrac{1}{m^*}$	$0(A_\perp)$ $\sin^2\theta(A_\parallel)$
X	2	$\dfrac{1}{m_l}$	$0(A_\perp)$ $\sin^2\theta(A_\parallel)$	6	$\dfrac{2m_l + m_t}{3m_l m_t}$	$\dfrac{w_{zz}^{-2}}{9m_{\parallel}^2}(A_\perp)$ $\dfrac{w_{zz}^{-2}\cos^2\theta}{9m_{\parallel}^2} + \sin^2\theta(A_\parallel)$
L	4	$\dfrac{2m_l + m_t}{3m_l m_t}$	$\dfrac{w_{zz}^{-2}}{9m_{\parallel}^2}(A_\perp)$ $\dfrac{w_{zz}^{-2}\cos^2\theta}{9m_{\parallel}^2} + \sin^2\theta(A_\parallel)$	1	$\dfrac{1}{m_l}$	$0(A_\perp)$ $\sin^2\theta(A_\parallel)$

length scale has two well-defined limits. At normal incidence, it must equal L_z since the quantum well then looks like a thin absorbing sheet with equal effect on all parts of the plane wave. At parallel incidence (i.e., $\theta = 90°$), it must go to zero since the quantum well then has an infinitesimal absorption cross section (i.e., the filling factor is zero). A length scale that satisfies these conditions is $L_z\sec\theta$, and the absorption coefficient is given by

$$\alpha(\omega, \Omega) = \frac{\zeta(\omega, \Omega)}{L_z\sec\theta}. \tag{16}$$

In the present context the internal quantum efficiency η is the fraction of incident intensity I_i that is absorbed by the intersubband transition in an MQW detector rather than a single quantum well. Since η reduces to ζ in the limit of a single well, one has $I_{t1} = I_i(1 - \zeta)$ for a single well, $I_{t2} = I_i(1 - \zeta)^2$ for two wells, and by mathematical induction, $I_{tN} = I_i(1 - \zeta)^N$ for N wells. Then, using the definition $\eta = (I_i - I_{tN})/I_i$, one can write

$$\eta = [1 - (1 - \zeta)^N]. \tag{17}$$

As in bulk material, η increases sublinearly with increasing N. This is exemplified by the realistic case of an AlGaSb quantum well having ζ of 0.063% (see Sec. 5.1). For $N = 100, 200, 300, 400$, and 500, the internal quantum efficiency is 0.06, 0.12, 0.17, 0.22, and 0.27, respectively.

2.4. Dark Current and Detectivity

In addition to providing a more convenient absorption than GaAs quantum wells, ellipsoidal-valley quantum wells can, in principle, exhibit much less dark current. In a definitive paper by Kinch and Yariv [8], it was shown that a universal source of dark current in MQW structures is thermionic emission out of the wells. This mechanism has been shown to be the dominant component of the dark current in practically all GaAs/AlGaAs quantum wells at temperatures of 77 K and higher [9]. Using an energy model similar to that shown in Fig. 3, Kinch and Yariv derived the following expression:

$$J_D = \frac{m_d^* \, e v_d \, kT}{\pi\hbar^2 L_P}\exp\left[-(e\Phi_W - E_F - E_1)/kT\right], \tag{18}$$

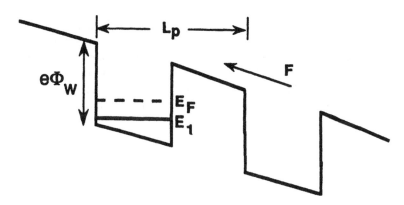

Fig. 3 Potential-energy diagram for an electron in an MQW structure in the presence of an applied electric field. The symbols are defined in the text.

where m_d^* is the density-of-states effective mass in the quantum well, v_d is the drift velocity through the MQW structure, T is the operating temperature, L_P is the period of the MQW structure, Φ_W is the depth of the well, E_F is the Fermi energy (relative to the ground state), and E_1 is the ground-state energy (referenced to the bottom right-side of the well in Fig. 3). The argument of the exponential in Eq. (18) is the separation in energy between the conduction-band edge of the barrier and the Fermi level, as one would expect intuitively.

In most MQW detectors operating at 77 K or below, the electron population in the wells is degenerate (i.e., every transverse state in the well has very close to unity probability of occupancy). In this case one can write from Eq. (12)

$$E_F - E_1 = \frac{\pi \hbar^2 \sigma_1}{m_d^*}. \tag{19}$$

Combining Eq. (19) with Eq. (18), one sees that the dark current increases exponentially with sheet density, which is much faster than the increase in quantum efficiency with sheet density. This is partly why the quantum wells cannot be doped arbitrarily high. However, the proportionality constant between E_F and σ in Eq. (19) is much smaller in ellipsoidal-valley quantum wells than in GaAs or other spherical-valley wells. In the latter type of wells, m_d^* is just equal to the Γ-valley effective mass, which tends to be small in all semiconductors (e.g., $m_\Gamma^* = 0.067 \, m_0$ in GaAs; $m_\Gamma^* = 0.042 \, m_0$ in GaSb). In ellipsoidal-valley quantum wells, $m_d^* = M_c\sqrt{m_1 m_2}$, where m_1 and m_2 are two character-istic masses of the transverse plane and M_c is the degeneracy factor defined in Sec. 2.2 [10]. For example, in the practical case of L-valley ellipsoids in a (100)-oriented quantum well, $m_1 = m_t, m_2 = (2m_l + m_t)/3$, and $M_c = 4$. Substitution of the following values for GaSb, $m_t = 0.10 m_0$ and $m_l = 1.25 m_0$, yields $m_d^* = 1.18 m_0$. This is 18 times greater than the Γ-valley m_d^* of GaAs and 28 times that of GaSb.

Using Eq. (18), one can make a direct comparison of the thermionic dark current in ellipsoidal-valley and spherical-valley MQW struc-tures. Each structure is assumed to have a common bound-to-exten-ded-state energy, MQW period, operating temperature, and sheet density (assumed degenerate at the operating temperature). Denoting the quantum-well type by a superscript s or e for spherical and ellip-soidal, respectively, one can write,

$$\frac{J_D^e}{J_D^s} = \frac{(m_d^* v_d)^e}{(m_d^* v_d)^s} \exp\left[(E_F^e - E_F^s)/kT\right] \approx \frac{(m_d^* \mu)^e}{(m_d^* \mu)^s} \exp\left[(E_F^e - E_F^s)/kT\right]. \quad (20)$$

The last step follows from the fact that in most MQW detectors studied to date the bias has been low enough that the internal electric field F is well below that required for a saturated velocity. Conse-quently, $v_d \approx \mu F$, where μ is the low-field mobility, and one gets the last expression in Eq. (20) by assuming that the internal field of the two detector types is the same.

A relevant example to test this expression is a GaSb ellipsoidal-valley well and a GaAs spherical-valley well, each containing a sheet density of 1×10^{12} cm^{-3} and having $e\Phi_W - E_1 \geqslant E_F$. For the GaSb well, $(m_d^*)^e = 1.18 m_0$ (L valley), $\mu^e \approx 7000$ (at 77 K), and $E_F^e = 2$ meV. For the GaAs well, $(m_d^*)^s = 0.067 m_0$ (Γ valley), $\mu^s \approx 20,000$ (at 77 K), and $E_F^s = 30$ meV. Hence, one finds from Eq. (20) $J_D^e/J_D^s \approx 0.1$. Physi-cally, J_D in the ellipsoidal-valley structure is lower because the elec-trons occupy a much narrower energy range in the transverse plane of the well. The narrower energy range corresponds to a reduction in E_F by the factor of 18 difference between the density-of-states masses. The exponential dependence on E_F in Eq. (20) offsets the linear dependence on m_d^* in the prefactor to yield the substantially lower dark current.

According to Eq. (4), a tenfold reduction in dark current and no change in η_0 and g would increase the dark-current-limited D^* more than three times that of the best GaAs quantum-well detector, assum-ing that there is negligible thermal noise. Because of this prediction and the possibility of enhancing η_0 substantially through multiple-pass interference techniques, several efforts have been made to develop the AlGaSb and other normal-incidence ellipsoidal-valley detectors. These efforts will be summarized in Sec. 5.

3. INTERSUBBAND DESIGN AND MEASUREMENT TECHNIQUES

3.1. Heterojunction Band Alignments

The foremost challenge in designing ellipsoidal-valley MQW detectors, or any MQW detector, is to position the intersubband energy in the desired spectral range. The first task in doing this is to define all of the band alignments at each heterojunction in the quantum well. While this exercise is straightforward and often taken for granted in spherical-valley quantum wells, it is nontrivial and must be carried out carefully for ellipsoidal-valley structures. Errors in this step can easily lead to a poor choice of materials parameters in the growth of the structure or false interpretations of experimental results.

A fundamental requirement in all band-alignment estimations are the band-gap energies E_G for each valley in the quantum-well and barrier materials. If the quantum well and barrier are lattice matched, the band-gap energies can be obtained in a straightforward fashion from measurements or theoretical calculations on bulk alloys. This is true for the III–V materials addressed in this chapter, AlGaAs and AlGaSb. Table 2 lists analytic expressions for their band gaps that provide very good fits to experimental data at room temerature [11].

TABLE 2
Properties of ellipsoidal-valley III–V ternary material systems.

Material	ΔE_V	Band-Gap Energy (eV)
$Al_yGa_{1-y}As$	0.50 eV (GaAs relative to AlAs)	$E_{G,\Gamma} = 1.424 + 1.247y,$ $(0.0 < y < 0.45)$ $E_{G,\Gamma} = 1.424 + 1.247y + 1.147(y-0.45)^2,$ $(0.45 < y < 1.0)$ $E_{G,L} = 1.708 + 0.642y$ $E_{G,X} = 1.900 + 0.125y + 0.143y^2$
$Al_yGa_{1-y}Sb$	0.40 eV (GaSb relative to AlSb)	$E_{G,\Gamma} = 0.726 + 1.129y + 0.368y^2$ $E_{G,L} = 0.799 + 0.746y + 0.334y^2$ $E_{G,X} = 1.020 + 0.492y + 0.077y^2$

For the column-IV material of relevance here, SiGe, the band gaps are not so straightforward because the lattice mismatch between the quantum well and the barrier usually causes a strain-induced shift in at least the energy gaps of the quantum well.

In estimating the band alignments in semiconductors, the most convenient energy reference is the valence band edge in the quantum well. This is partly because each of the ellipsoidal valleys within a given quantum-well or barrier material have a common valence band edge. Furthermore, the offset in the valence band edge, ΔE_V, of two such materials at their heterojunction is a quantity that can be derived with a satisfactory level of confidence. For a heterojunction between two ternary compounds having a single common cation (i.e., the heterojunctions considered in this chapter), one starts with theoretical or experimental values of ΔE_V for binary-compound pairs, denoted here by AC and BC. One then calculates ΔE_V for the heterojunction by linear interpolation. Explicitly, if one denotes the quantum-well material as $A_x B_{1-x} C$ and the barrier material as $A_y B_{1-y} C$, then the valence-band offset between these two compounds is given by

$$\Delta E_V \{ A_x B_{1-x} C / A_y B_{1-y} C \} = (y - x) \cdot \Delta E_V \{ BC/AC \}, \qquad (21)$$

where it is assumed that $y > x$ and the band gap of material AC is greater than that of BC. The interpolation procedure of Eq. (21) is known as Vegard's law [12].

Usually Eq. (20) is expressed in terms of a band offset rule. That is, given arbitrary barrier and quantum-well materials and a common ellipsoidal-valley type, some fraction δ_C of the band-gap difference ΔE_G occurs in the conduction band and some fraction $\delta_V = (1 - \delta_C)$ occurs in the valence band. From the parameters defined diagrammatically in Fig. 4, these fractions can be written

$$\delta_C = \frac{E_{G,B} - E_{G,W} - \Delta E_V}{E_{G,B} - E_{G,W}}, \qquad (22)$$

$$\delta_V = \frac{\Delta E_V}{E_{G,B} - E_{G,W}}. \qquad (23)$$

For example, GaAs and AlAs are known to exhibit $\Delta E_V \approx 0.50$ eV, [13–15] so that one obtains $\delta_C \approx 0.70$ using the room-temperature Γ band-gap energies listed in Table 2. This is the basis for the "70/30 rule" commonly used in analyzing GaAs/AlGaAs heterostructures. Similarly, GaSb and AlSb are known to exhibit $\Delta E_V \approx 0.40$ eV, so that one obtains $\delta_C \approx 0.73$ for the Γ band gaps.

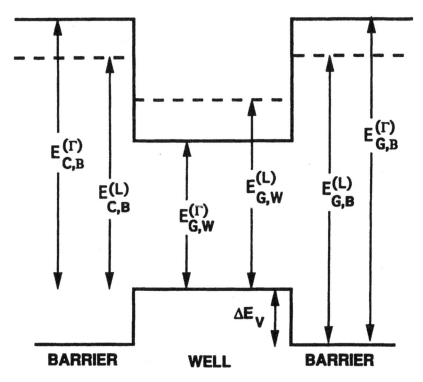

Fig. 4 Energy diagram for the valence and conduction bands in a quantum well. The solid lines indicate the energy at the Γ point of the Brillouin zone. The dashed lines indicate energy of an ellipsoidal valley, in this case the L valley.

Combining ΔE_V with the known band-gap energies of each material, one can determine the ellipsoidal-valley offsets. These offsets are hereby denoted by the quantity $E_{C,\,\mathrm{Position}}^{(\mathrm{Valley})}$, where the subscript Valley represents $\Gamma, X,$ or L, and Position represents W for quantum well or B for barrier. By definition, $E_{C,W}^{(\Gamma)}$, $E_{C,W}^{(L)}$, and $E_{C,W}^{(X)}$ are equal to $E_{G,W}^{(\Gamma)}$, $E_{G,W}^{(L)}$, and $E_{G,W}^{(X)}$, respectively, and the offsets of the valleys in the barriers are

$$E_{C,B}^{(\Gamma)} = E_{G,B}^{(\Gamma)} - \Delta E_V\{A_xB_{1-x}C/A_yB_{1-y}C\}, \tag{24}$$

$$E_{C,B}^{(L)} = E_{G,B}^{(L)} - \Delta E_V\{A_xB_{1-x}C/A_yB_{1-y}C\}, \tag{25}$$

and

$$E_{C,B}^{(X)} = E_{G,B}^{(X)} - \Delta E_V\{A_xB_{1-x}C/A_yB_{1-y}C\}. \tag{26}$$

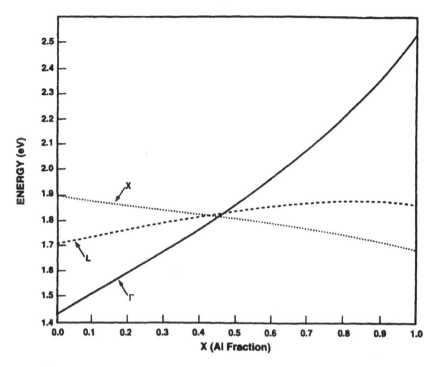

Fig. 5(a) Energy of the three primary conduction-band valleys of $Al_xGa_{1-x}As$ as a function of Al fraction. The energy is measured relative to the valence band edge of GaAs.

Figure 5 (a) displays the curves of Eqs. (24)–(26) for a GaAs quantum well with $Al_yGa_{1-y}As$ barriers ($0 < y < 1$), and Fig. 5 (b) shows the same for a GaSb quantum well with $Al_yGa_{1-y}Sb$ barriers ($0 < y < 1$). In Fig. 5 (a) the Γ valley has the lowest energy up to $y \approx 0.45$, beyond which the X-valley has the lowest. In fact, all three intersection points between the valley curves occur around $y = 0.45$, leading to the designation of this as the crossover point. Because the slope of the X-valley curve is negative, it is possible for an electron in an AlAs quantum well to be spatially quantized by the X valley in $Al_yGa_{1-y}As$ barriers for y as low as 0.45. This particular quantum well is the basis for a normal-incidence detector discussed in Sec. 6.1.

For the GaSb/$Al_yGa_{1-y}Sb$ curves in Fig. 5 (b), the Γ valley has the lowest energy only up to $y \approx 0.19$. Above this the L valley has the lowest energy up to $y \approx 0.56$ and the X valley is the lowest beyond this

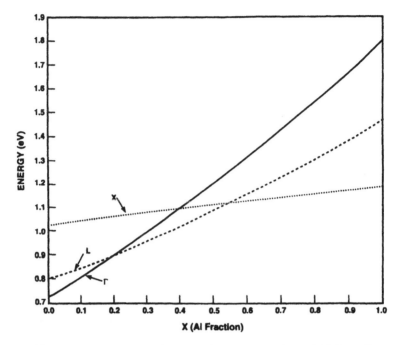

Fig. 5(b) Energy of the three primary conduction-band valleys of $Al_xGa_{1-x}Sb$ as a function of Al fraction. The energy is measured relative to the valence band edge of GaSb.

point. The wide range over which the L valley has the lowest energy is attractive from an experimental standpoint because it is evident from Sec. 2.3 that L-valley ellipsoids absorb normal-incidence radiation very strongly in (100)-oriented quantum wells. Such quantum wells are feasible to grow because of the prevalence of (100)-oriented III – V substrates. It is also important to note that the difference in energy offsets in Fig. 5 (b) between the L and Γ valleys at $x = 0$ is only about 70 meV. Because of the much larger w_{zz} of the L valley, ε_1 for the Γ valley can then lie above ε_1 for the L valley provided that the quantum well is sufficiently narrow. This is the basis for a normal-incidence detector discussed in Sec. 5.3.

3.2. Quantum-well Energy States

Knowing the band alignments, one can proceed with the calculation of the energy levels in the quantum well. In principle, this is a difficult task in ellipsoidal-valley quantum wells for several reasons. First, in real ellipsoidal-valley quantum wells with finite barriers, the form of the envelope function cannot be as simple as Eq. (8) since it must extend some distance into the barrier. Second, the energetically-favored envelope function in the barrier may not be associated with the same valley as that in the quantum well. In other words, the lowest-energy ellipsoidal valley in the barrier may be different than that in the well. A third complication is that the degeneracy for ellipsoids of common orientation discussed in Sec. 2 is apt to be broken in a real quantum well, again, because of the penetration of the wave functions in the barriers.

In practice, there are a few simplifying assumptions that make the task of calculating energy levels straightforward and surprisingly accurate. The first is to assume that the electron in the ellipsoidal-valley quantum well has a longitudinal effective mass that is a scalar quantity given by the reciprocal-effective-mass tensor component, W_{zz}^{-1}. This assumption is made plausible by its physical validity in the limit of an infinite well (see Sec. 2). The second assumption is that the wave function in the barrier occurs in the same type of ellipsoidal valley as in the quantum well, independent of its relative energy. This is made plausible by the fact that wave functions of electrons tend to maintain their Bloch component (i.e., the cell-periodic component) across ideal heterojunctions [16]. It is an important concept in the problem of heterobarrier tunneling, as occurs in resonant-tunneling and superlattice-transport devices.

A useful example is the energy-level calculation for an $Al_xGa_{1-x}Sb$ quantum well clad by AlSb barriers. Shown in Fig. 6 is the energy-level diagram for the case $x = 0.1$, a quantum-well composition that has been studied at Lincoln Laboratory. Shown in the far left and far right of the diagram are horizontal lines representing the conduction band edges of the Γ, L, and X ellipsoidal valleys in the AlSb barrier. Shown in the center are horizontal lines for the band edges in the quantum well, and four curves for the first two bound states in the Γ and L valleys. The latter curves are plotted against an inset axis of quantum-well width. Two interesting features are apparent. First, in spite of the fact that the Γ valley is the lowest-energy band edge in the wide quantum wells, the lowest energy state is associated with the L valley for well width less than approximately 80 Å. This arises from the greater quantum-size effect for the Γ valley state because of its lower effective mass. The other interesting feature is that the second energy state associated with the L valley, $E_2^{(L)}$, increases above the X-valley conduction-band-edge in the barrier for well widths less than

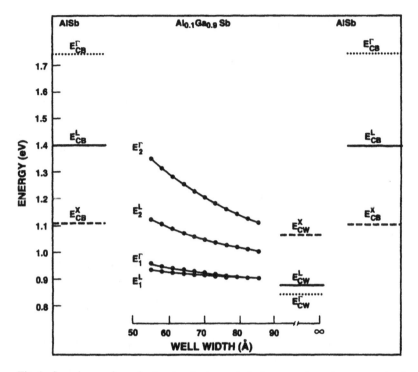

Fig. 6 Superimposed conduction-band-edge and electron-quantum-state energy level diagrams for an $Al_{0.1}Ga_{0.9}Sb$ quantum well confined by A1Sb barriers. The quantum-state energy is plotted as a function of well width.

approximately 55 Å. This feature will come up again in the experimental results of Sec. 5.

3.3. Intersubband Measurement Techniques

3.3.1. Characterization of Intersubband Absorption

The most common technique that has been used to characterize the intersubband absorption is infrared transmission. Since intersubband transitions can be engineered to occur throughout the middle infrared, the transmission is usually measured with a broadband Fourier-transform spectrometer (FTS). The FTS used in the experiments at Lincoln Laboratory covers the spectral range $400–4800 \, cm^{-1}$ ($2.1–25 \, \mu m$) and has two sample compartments and detectors. The primary sample compartment has fixed $f/3.5$ optics and a room-temperature pyroelectric detector, while the secondary compartment has adjustable $f/1$ optics and a liquid-nitrogen-cooled HgCdTe detector. Instrumental resolution is programmable over the range 2–32 cm^{-1}. Polarization of the infrared beam is controlled by polarizers located in the sample compartments.

To improve the signal-to-noise ratio in the data, multiple scans are coherently added. In addition, various smoothing, normalization, logarithmic-subtraction, and integration procedures are performed on the data to clarify the location and strength of absorption features such as those due to intersubband transitions. Interfering features in the spectral region near the intersubband transition are removed to help interpret the transmission spectra. These features include absorption from free-carriers in the substrate, higher-order phonons, surface oxide layers, contaminants on the samples, and Fabry-Perot interference fringes. Transmission measurements were made by normalizing the infrared signal measured with the MQW sample in the optical beam to that measured with a suitable bare substrate in the beam. As discussed in the following sections, spectra were typically recorded in either a single-pass or multiple-pass geometry, but other less-common geometries were also used.

3.3.2. Single-Pass Techniques and Sample Preparation

After growth of the MQW wafers, 1×2 cm samples were thinned by mechanically lapping and polishing the substrate side. This preparation is necessary to help eliminate scattering and free-carrier absorption artifacts in the transmission spectra of the samples. Single-pass data were collected over the temperature range of 20–290 K with the samples mounted in a closed-cycle helium refrigerator. Measurements were made with the samples oriented both at normal incidence and at Brewster's angle to the linearly polarized infrared radiation. This is shown schematically in Fig. 7(a). The sample area probed by the infrared beam was approximately 0.5 cm^2. For Brewster's angle measurements the infrared radiation was polarized in the plane of incidence (*p*-polarized) to eliminate interference fringes from reflections at the air-epitaxial and substrate-epitaxial interfaces. This is one of the standard geometries for studying intersubband transitions in spherical-valley quantum wells, since the radiation has a component of the electric field perpendicular to the plane of the quantum wells.

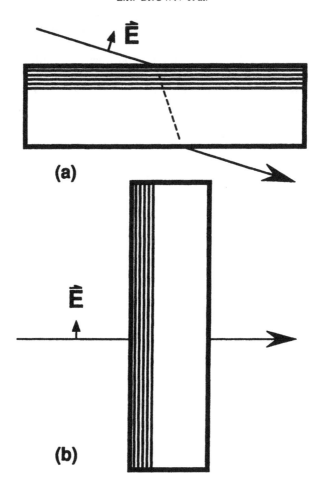

Fig. 7 Cross-sectional view of two common sample orientations for single-pass transmission measurement on MQW structures: (a) Brewster's angle of incidence, and
(b) normal incidence.

Brewster's angle measurements provide the best measure of the intersubband absorption strength and width, since there is no uncertainty
about the number of passes or interference state of the infrared
radiation.

To explore the polarization dependence of the absorption in ellipsoidal-valley quantum wells, single-pass measurements are also made
at normal incidence, as shown in Fig. 7(b). Interference fringes which

arise with the beam at normal incidence are difficult to completely eliminate, however, and can alias the true nature of the intersubband absorption. Two methods to overcome this problem allow accurate cancellation of the fringes. Low temperature measurements enable the study of persistent photoabsorption in quantum wells, a technique which can successfully eliminate all features in the spectrum unrelated to the electron intersubband transition [17]. In these measurements transmission spectra are obtained by normalizing spectra measured after white-light illumination of the sample to spectra taken before illumination. Photoionization of electron traps in the MQW increases the electron density in the wells and leads to a corresponding increase in the intersubband absorption. Although this technique helps to pinpoint the intersubband transitions and can be useful for relative polarization dependence studies, it is unable to yield absolute measurements of the intersubband absorption strength. Another technique for overcoming the fringing problem is to use a control sample which has an identical epitaxial layer thickness of a material with a dielectric constant equivalent to the average of the MQW layers.

3.3.3. Multiple-Pass Techniques and Sample Preparation

To more accurately study the polarization dependence of the intersubband absorption, multiple-pass internal reflection (MIR) measurements were made. Small ($3 \times 10 \, \text{mm}^2$) pieces were cleaved from each wafer and parallel 45° facets were lapped and polished on two opposing ends to allow multiple-pass transmission experiments. Multiple-pass spectra using the small 45° parallelepipeds more clearly show the intersubband absorption features due to the increased optical path length and the complete elimination of interference fringes in the spectra. The polarization of the optical beam, which was incident along the normal to one of the end facets as shown in Fig. 8(a), can be varied between parallel to the plane of the quantum well and at 45° with respect to the well. Again, transmission data are obtained by dividing the optical transmission spectrum by a corresponding spectrum taken with an identical MIR sample made from a bare substrate. Multiple-pass measurements are limited to MQWs on high resistivity substrates to avoid the possibility that strong free-carrier absorption in the substrate dominates the intersubband absorption. Although MIR measurements increase the strength of the intersubband absorption and can help to identify weak absorption features, uncertainties in the number of passes and in the intensity distribution of the

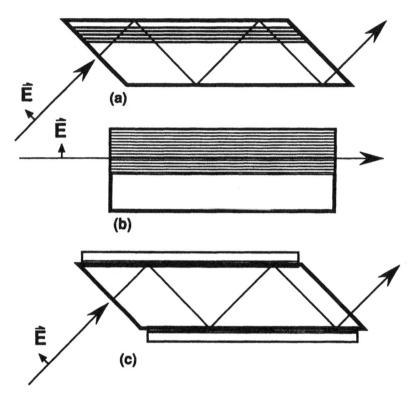

Fig. 8 Cross-sectional view of three common sample orientations for multiple-pass transmission measurement on MQW structures: (a) multiple-pass internal reflection (MIR), (b) cleaved-end-face lateral transmission and (c) attenuated total-internal reflection (ATR).

infrared beam at the quantum wells can lead to errors in measured absolute absorption strength.

Two additional geometries for measuring intersubband absorption are also shown in Fig. 8. The first drawn in Fig. 8(b), involves transmission of infrared radiation incident normally on a cleaved end-face through a slab of sample and out the opposite cleaved end. This technique suffers from the same pitfalls found in the MIR geometry, but has the advantage of needing less preparation work and the ability to vary the polarization from completely in the plane of the well to completely across it. However, coupling efficiency into these slabs is typically lower than with 45° MIR structures due to the smaller cross

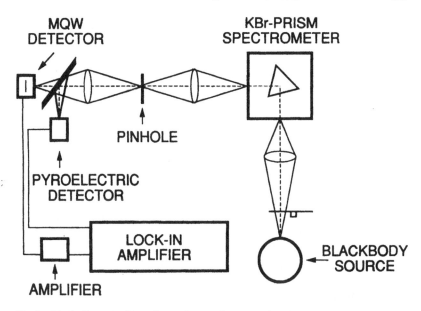

Fig. 9 Block diagram of experimental set up for measuring the external responsivity of MQW detectors in the middle-infrared spectral region.

section presented to the beam. The other unusual geometry used to study intersubband absorption, drawn in Fig. 8(c), involves the technique of frustrated or attenuated total-internal reflection (ATR) [18]. In this procedure the intersubband transitions are induced by the evanescent mode set up at the top surface of the epilayer when the infrared beam is incident at an angle greater than the critical angle from a higher dielectric material placed in contact with the sample. In practice, pieces of the MQW sample would sandwich a 60° parallelepiped of germanium, which has a dielectric constant near 16. By slightly varying the angle of incidence of the radiation, the pressure of the samples against the germanium ATR crystal, and the polarization of the radiation, it is possible to adjust the penetration depth of the evanescent mode into the MQW. In this way variations from quantum well to quantum well within a stack can be probed and one can avoid the problem of highly absorbing substrate layers.

3.4. Responsivity Measurement Technique

Although there are many ways to measure the external responsivity, the technique shown schematically in Fig. 9 was applied at Lincoln

Laboratory because of its accuracy and amenability to absolute calibration. The optical source for the measurements is a 1000 °C blackbody filtered by a prism spectrometer. The blackbody provides much greater luminosity than a popular alternative, the incandescent globar, particularly at longer infrared wavelengths. The resolution of the prism spectrometer varies from approximately 16 cm^{-1} at 1000 cm^{-1} (10 μm) to approximately 60 cm^{-1} at 2500 cm^{-1} (4 μm). The detectors were mounted on the cold finger of a closed-cycle helium refrigerator and radiation emitted from the spectrometer was focused onto the detectors at normal incidence.

For calibration purposes, a beamsplitter located between the spectrometer exit slit and the input window on the refrigerator directs a portion of the infrared beam onto a pyroelectric detector. The pyroelectric detector has a specially treated surface of known area for constant and calibrated R_0 vs wavelength throughout the near and middle infrared regions. Hence, the true spectral variation of R_0 for the MQW detector is obtained simply by normalization of the output of the MQW detector to that of the pyroelectric detector at all wavelengths. Absolute calibration of the MQW-detector R_0 is achieved by apodizing its input beam with a pinhole having a smaller area than the detector itself. The responsivity is then calculated using the incident optical intensity from the known responsivity and measured output of the pyroelectric detector.

4. FABRICATION OF NORMAL-INCIDENCE AlGaSb DETECTORS

4.1. Epitaxial Growth Techniques

One of the advantages of III–V MQW detectors over the conventional HgCdTe detectors is the robustness of the material. Be it AlGaSb or AlGaAs, the base material in normal-incidence MQW detectors is grown by conventional solid-source MBE in a manner very similar to the growth of the large variety of electronic and optical devices in the GaAs/AlGaAs material system. The sources are the group III and group V elements, which yield beams of Al and Ga atoms, and As$_4$ (or As$_2$, if a pyrolytic cracker is used) and Sb$_4$ molecules. At the substrate temperatures employed for the growth of MQW structures, the incorporation efficiency is unity for the group III atoms and less than unity for the group V molecules. The group

III fluxes were chosen to yield the desired Al/Ga ratios, with alloy growth rates of approximately 1 μm/h whenever possible. Because the sticking coefficient is much greater for Sb than As, high concentrations of Sb can be incorporated even though the As flux is much greater than both the Sb flux and the total group III flux [19]. For AlGaSb, the *n*-type dopant is Te provided by the sublimation of GaTe or PbTe. For AlGaAs, the *n*-type dopant is Si.

The condition for lattice matching of $Al_xGa_{1-x}As_ySb_{1-y}$ to GaSb substrates is $y = 0.08\ x$. By careful calibration of the As and Sb fluxes it is possible to obtain closely lattice-matched layers. GaTe was selected as the *n*-type dopant source for AlGaAsSb alloys, which yielded *n*-GaSb layers with carrier concentrations up to 1×10^{18} cm^{-3} at 300 K with a mobility of 1900 cm^{2}V^{-1}s^{-1} [19]. For a given GaTe source temperature, however, the electron concentration for *n*-AlGaAsSb was always lower than the value observed for GaSb.

4.2. Material Diagnostic Techniques

Layers were grown on commercial GaSb (100) substrates. The substrates were cleaned in solvent and etched, mounted with In on 75-mm Mo holders, loaded into a Varian Gen II modular MBE system, and heated in the presence of an Sb flux to desorb surface oxides and other contaminants. At this point a sharp, group-V-stabilized diffraction pattern could generally be observed by reflection high-energy electron diffraction (RHEED). The individual group III fluxes were calibrated by measuring the frequency of the RHEED intensity oscillations observed during the growth of test layers of GaSb and AlSb [20].

As an example of the use of RHEED intensity oscillations, a plot of RHEED intensity vs time for a GaSb layer grown on a GaSb substrate is shown in Fig. 10 (a). The corresponding power spectrum, which was obtained by using fast Fourier transform techniques [21], is shown in Fig. 10 (b). The fundamental oscillation frequency, which is given by the peak in the power spectrum, is the reciprocal of the time required for the deposition of one monolayer. Another example of RHEED oscillations is shown in Fig. 11 for the growth of $Al_{0.50}Ga_{0.50}As_{0.04}Sb_{0.96}$ on GaSb. The persistence of the oscillations is indicative of continuing layer-by-layer growth proceeding by the nucleation and growth of two-dimensional islands. Such layer-by-layer growth leads to high-quality quantum wells and abrupt interfaces.

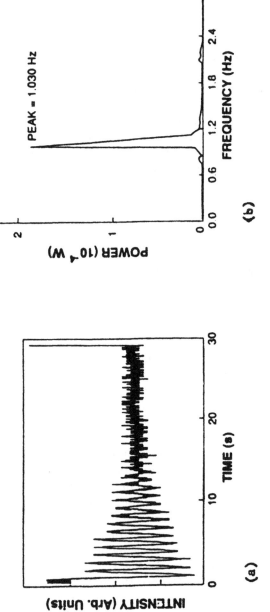

Fig. 10 (a) Plot of RHEED intensity vs time for a GaSb epitaxial layer during growth on a GaSb substrate. (b) Fourier transform of RHEED intensity in (a).

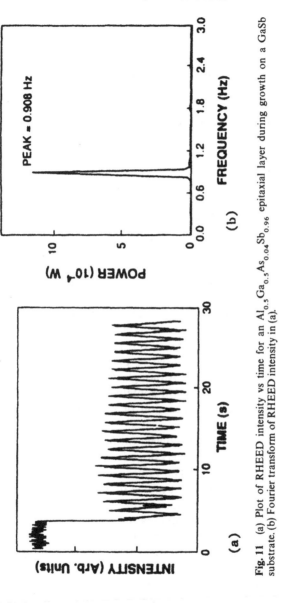

Fig. 11 (a) Plot of RHEED intensity vs time for an $Al_{0.5}Ga_{0.5}As_{0.04}Sb_{0.96}$ epitaxial layer during growth on a GaSb substrate. (b) Fourier transform of RHEED intensity in (a).

To determine the lattice constants of AlGaAsSb layers, the lattice mismatch between these layers and the GaSb substrate was found by double-crystal x-ray diffraction measurements, using a GaSb first crystal, of the (400) reflection. The alloy composition was measured by Auger electron spectroscopy, as calibrated by Auger analysis of AlSb, GaSb, and GaAs layers. For $Al_xGa_{1-x}As_ySb_{1-y}$ layers, the value of $1-y$ was determined from the measured lattice mismatch, on the assumption that the ratio of x to $1-x$ is equal to the ratio of the Al and Ga fluxes. This assumption is supported by the observation that the growth rate calculated from the total group III flux was equal to the value determined by cross-sectional thickness measurements on thick layers.

Nominally-undoped layers of GaSb and AlGaAsSb are p type. Silicon is commonly used for n-type doping of III–V semiconductors grown by MBE, but neither Si nor Sn is a donor in GaSb and AlGaAsSb, so the group VI elements S, Se, and Te are used instead. As described in Sec. 4.1, in this work the impurity used for n-type doping was Te provided by the sublimation of GaTe. The van der Pauw technique was used to measure the carrier type, concentration, and mobility in doped test layers, typically 1 to 2 μm thick, grown on semi-insulating GaAs substrates. For n-type AlGaAsSb, undoped Al-GaAsSb buffer layers were used to prevent the formation of a two-dimensional electron gas. Although GaSb and AlGaAsSb layers grown on GaAs substrates will contain a high density of misfit dislocations, only a slight degradation of the majority carrier transport properties is expected.

In Fig. 12, the electron concentration n at 300 K in GaSb and $Al_{0.50}Ga_{0.50}As_{0.04}Sb_{0.96}$ layers doped with GaTe is plotted vs inverse GaTe source temperature. The values of n for GaSb layers range from 8×10^{15} to 1×10^{18} cm^{-3}. At moderate GaTe source temperatures, n exhibits an Arrhenius dependence on temperature. The line drawn in Fig. 12 corresponds to a sublimation heat of 58 kcal/mol. At the highest GaTe source temperatures, the values of n fall below this line. The electron mobility of n-GaSb layers at 300 K is shown in Fig. 13. A mobility of 4000 cm^2 V^{-1}s^{-1} was obtained for $n = 4 \times 10^{16}$ cm^{-3}. In even more lightly doped layers, record mobilities as high as 7600 and 12,600 cm^2 V^{-1}s^{-1} were obtained at 300 and 77 K, respectively. As expected, lower mobilities are observed at higher values of n.

For comparison, a GaTe-doped GaAs sample was grown with a GaTe source temperature of 435 °C. The electron mobility is consistent with minimal compensation. The value of n is 5×10^{17} cm^{-3},

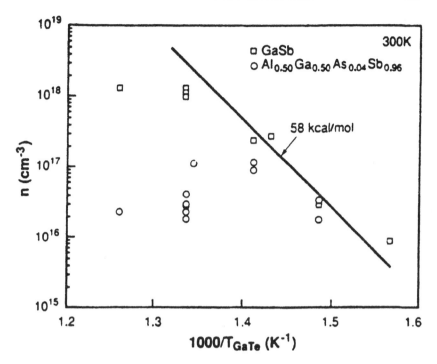

Fig. 12 Room-temperature free-electron concentration in Te-doped GaSb and $Al_{0.5}Ga_{0.5}As_{0.04}Sb_{0.96}$ epitaxial layers as a function of the reciprocal of the temperature of the GaTe source during growth. The straight line denotes the latent heat of sublimation of GaTe.

nearly the same as that obtained in layers of GaSb grown with the same GaTe source temperature. This result, together with the Arrhenius behavior described above, suggests that unity incorporation and doping efficiency of Te occur in GaSb layers grown at moderate GaTe source temperatures. For GaTe-doped $Al_{0.50}Ga_{0.50}As_{0.04}Sb_{0.96}$, values of n as high as 1×10^{17} cm^{-3} were obtained, as shown in Fig. 12. (The plotted values of n have not been corrected for simultaneous electron conduction in the Γ, L, and X conduction bands.) For GaTe source temperatures of 435 °C ($1000/T = 1.412$ K^{-1}) or lower, the results are relatively well behaved. For higher source temperatures, however, n decreases with increasing GaTe temperature, and the measured mobilities are anomalously high. Such anomalously high Hall mobilities may result from Te precipitates [22].

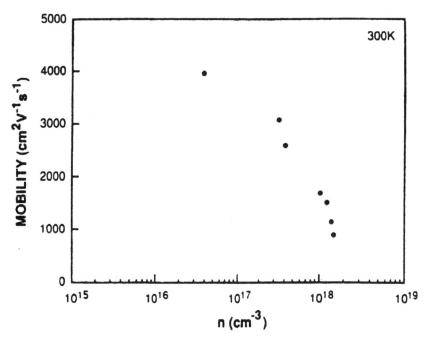

Fig. 13 Room-temperature electron mobility in GaSb epitaxial layer as a function of the free-electron concentration.

4.3. AlGaSb Fabrication Methods

Once the base material is grown, normal-incidence MQW detectors are obtained by straightforward steps in microelectronic fabrication. The fabrication sequence developed for the AlGaAsSb MQW detectors is outlined in Fig. 14. The first step, shown in Fig. 14 (b), is to spin on positive photoresist and pattern holes in the resist using optical lithography. Approximately 0.2 μm of ohmic metal is then deposited by electron-beam evaporation. The ohmic metal used in the Lincoln Laboratory process contains layers of Au, Sn, Ti, Pt, and Au of thickness 150, 150, 300, 500, and 2000 Å, respectively. After deposition, the underlying photoresist is dissolved in acetone, thereby lifting off the metal in all regions except the patterned holes. The remaining metal dots will become the top ohmic contacts.

The next step in the fabrication is to define the MQW detector mesa. Starting in Fig. 14 (c), disks of photoresist are patterned in such

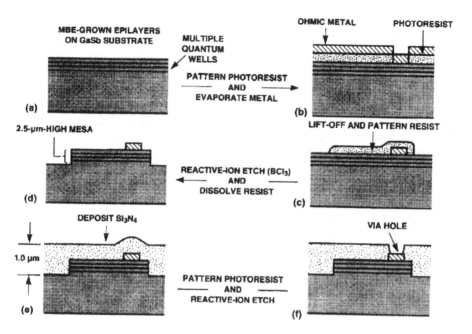

Fig. 14 Fabrication sequence for AlGaSb MQW detectors. The fabrication steps are described in the text.

a way that the ohmic dot is encompassed near one edge. The AlGaAsSb wafer is then put in a plasma etching chamber where it is exposed to a mixture of argon gas and BCl₃. The dc and rf bias conditions in the chamber were 450 V and 50 W, respectively. The total gas pressure was approximately 5 mTorr. The chlorine species in the plasma form volatile compounds (e.g., GaCl) on the semiconductor surfce, leading to a fast etching rate of approximately 800 Å/min or more [23]. Since the photoresist does not react with the BCl₃, it is relatively unaffected and a circular mesa forms underneath. Etching is continued to a depth just below all of the quantum wells, as shown schematically in Fig. 14 (d). For the AlGaAsSb MQW structures developed at Lincoln Laboratory, this depth was approximately 2.5 μm.

The final step in the process was surface passivation and electrical contact. The AlGaAsSb structures have two features that warrant passivation. First, the GaSb material used in the top and bottom ohmic-contact layers is similar to InSb in that it can have an electrically conducting surface because of the variety of oxides that form there. Second, the barrier layers in the MQW structure have such a high Al fraction that they can degrade upon exposure to air, leading to physical degradation of the detector mesa. To prevent these effects, the mesas were cleaned, lightly etched, and covered with approximately 1000 Å of Si_3N_4, as shown schematically in Fig. 14 (e). The Si_3N_4 was deposited by a plasma-enhanced chemcal-vapor-deposition system at a temperature of approximately 200 °C. It was chosen over other possible inorganic dielectrics (e.g., SiO_2) because of its ability to form smooth films at low-deposition temperatures, and it was chosen over possible organic dielectrics (e.g., polyimide) because of its chemical purity and inertness. After the Si_3N_4 was deposited, photoresist was spun on top and a round hole was patterned directly above the top ohmic contact. Then the wafer was exposed to the reactive-ion gas (DE-100) to etch a via hole through the exposed Si_3N_4 using the photoresist as a mask. Finally, the photoresist was stripped off, rendering the structure shown in Fig. 14 (f). Electrical bias was provided to the detector by a wire bond through the via hole to the top ohmic contact.

5. EXPERIMENTAL RESULTS FOR NORMAL-INCIDENCE QUANTUM-WELL STRUCTURES

5.1. Absorption Measurements

Because of the resident expertise in the molecular-beam epitaxy of AlGaAsSb described in Sec. 4.1, this was the material system that was investigated at Lincoln Laboratory as a proof of normal-incidence intersubband absorption in ellipsoidal-valley quantum wells. Initially, three MQW samples were grown, each containing 100 $Al_{0.09}Ga_{0.91}Sb$ quantum wells separated by $AlAs_{0.08}Sb_{0.92}$ barriers. The designation and some material properties of the three samples are listed in Table 3. Samples 1 and 2, which had nominal well widths of 70 and 64 Å, respectively, were grown on (100)-oriented n-GaSb substrates. Sample 3 which had a 70 Å well width, was grown on a (100) semi-insulating (SI) GaAs substrate. The SI GaAs substrate was used, in spite of its much greater lattice mismatch, because it has a much lower background electron concentration than any GaSb substrate and, therefore, exhibits much less free-carrier absorption in the 10 μm region. In sample 1, the barriers and quantum wells were doped uniformly with a nominal Te concentration of $2.5 \times 10^{17} cm^{-3}$. In samples 2 and 3,

TABLE 3
Material properties of three AlGaSb MQW structures measured in absorption.

Sample	Substrate	Well Width (Å)	Well Doping (cm^{-3})	Barrier Width (Å)
1	(100) n-GaSb	70	$\approx 1 \times 10^{18}$	200
2	(100) n-GaSb	64	$\approx 1 \times 10^{18}$	200
3	(100) SI-GaAs	70	$\approx 1 \times 10^{18}$	200

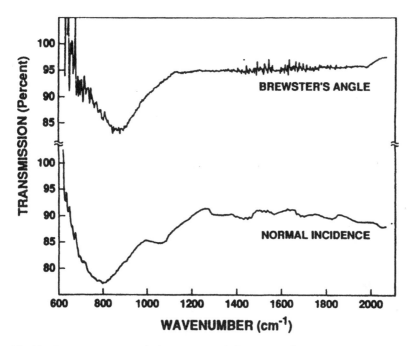

Fig. 15 Room-temperature single-pass transmission spectra for sample 1 with radiation at Brewster's angle and normal incidence.

the barriers were undoped, while the quantum wells were doped to $1 \times 10^{18}\,\mathrm{cm}^{-3}$. The outermost AlAsSb barriers in each structure were made 1000 Å thick to help confine most of the extrinsic electrons to the quantum wells.

The absorption strength was determined by Fourier-transform spectroscopy over the range 4 to 16 μm. Figure 15 shows the Brewster's-angle and normal-incidence single-pass transmission spectra measured for sample 1 at room temperature. A transmission minimum is observed at about 870 cm^{-1} (11.5 μm), which is approximately 10% deep and has a full width at half-maximum (FWHM) of roughly 200 cm^{-1}. The Brewster's-angle data shows more random noise than the normal-incidence curve because of the weak transmission, but does not display the systematic undulations at higher wavenumbers that arise from multiple-pass interference at normal incidence. In both cases, the data are unreliable at the lowest wavenumbers because of strong free-electron absorption in the n-GaSb substrates.

Sample 3 provided an opportunity to observe the intersubband absorption with little obscuration from free-electron absorption. Fig. 16 shows the Brewster's-angle single-pass transmission plotted in units of absorbance, $A \equiv -\log_{10}(I_t/I_i)$, where I_t and I_i are the transmitted and incident intensities, respectively. The maximum absorbance A_{max} of 0.036 occurs at approximately 860 cm^{-1}, and corresponds to a minimum transmission of 92%. Evidently the intersubband-absorption spectrum of sample 3 is nearly identical in all

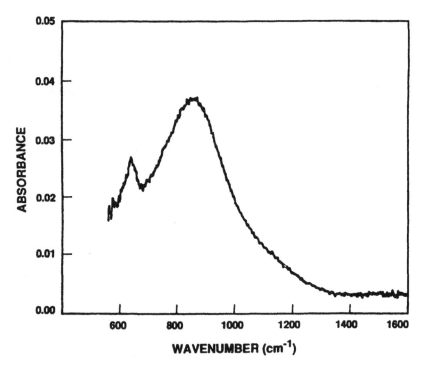

Fig. 16 Room-temperature single-pass transmission (in absorbance units) through sample 3 at Brewster's angle of incidence.

respects to that of sample 1, indicating that the crystal defects introduced by lattice-mismatched growth have little effect. This is consistent with the results for GaAs/AlGaAs MQW structures grown on
silicon substrates [24]. A_{max} is related to the peak fractional absorption per well, ζ_p, by

$$A_{max} = -\log_{10}[(1 - \zeta_P)^{N \cdot P}], \tag{27}$$

where N is the number of wells and P is the number of passes of the
radiation. With $N = 100, P = 1$, and $A_{max} = 0.036$, one finds $\zeta_p =$
8.1×10^{-4} for sample 3.

The peak fractional absorption can be compared directly with the
theoretical formalism of Sec. 2. From Eq. (14), one can write (in CGS
units)

$$\zeta_p \equiv \zeta(\omega = \omega_0) = \frac{1024\,\hbar e^2 w_{zz}\,\sigma_T}{27\pi\,nc\Gamma_S} \cdot G(\theta,\phi). \tag{28}$$

For a (100)-oriented *L*-valley quantum well, Table 1 shows

$$G = \frac{w_{zz}^{-2}\cos^2\theta}{9m_{lt}^2} + \sin^2\theta. \tag{29}$$

Assuming that the $Al_{0.09}Ga_{0.91}Sb$ quantum wells have the same
L-valley properties and refractive index as GaSb, one has $w_{zz}^{-1} =$
$0.144\,m_0$, $m_{lt} = 0.109\,m_0$, and $n = 3.8$ near a wavelength of $10\,\mu m$. It is
assumed that Γ_S is the experimental FWHM of $240\ cm^{-1}$, and σ_T is
the product of the Te doping concentration $(1 \times 10^{18}\ cm^{-3})$ and the
well width (70Å), or $\sigma_T = 7 \times 10^{11}\ cm^{-2}$. The polar angle θ is determined by Snell's law and Brewster's formula as $\theta = \sin^{-1}[n^{-1}$
$\sin(\tan^{-1}n)] = 14.7°$. With all of these parameters, one finds
$G = 0.246$ and $\zeta_p = 8.4 \times 10^{-4}$, which is just 4% greater than the experimental value.

A more convincing feature of an ellipsoidal-valley intersubband
transition is absorption of radiation at normal incidence, since this is
not observed in spherical-valley quantum wells such as GaAs. Unfortunately, the intersubband absorption at normal incidence in a single-
pass structure is greatly complicated by standing-wave effects between
the front and back surfaces of the MQW structure. To avoid this
problem, transmission measurements were carried out on sample 3 in
an MIR configuration, as described in Sec. 3.3. The incident beam
from the thermal source of the spectrometer was focused on one of the

45° end facets of the MIR parallelepiped. The linear polarization of the incident beam was oriented either in the plane of the quantum wells (i.e., equivalent to normal-incidence absorption) or 45° with respect to the plane. The beam transmitted through the opposite facet was processed by an FTS and normalized for each polarization with respect to the corresponding transmitted beams through an MIR parallelepiped made from SI GaAs only (i.e., no epitaxial layers).

Shown in Fig. 17 is the normalized transmission for the two polarizations plotted in absorbance units. Both spectra display absorbance maxima at 890 cm^{-1} having a FWHM of approximately 240 cm^{-1}, in good agreement with the single-pass results of Fig. 16. Although the

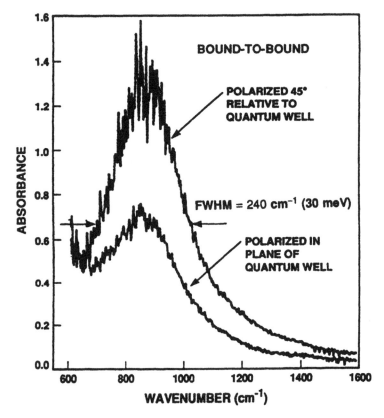

Fig. 17 Room-temperature multiple-pass transmission spectra for sample 1 with radiation incident normal to a 45° facet and polarized either in the plane of the quantum wells or at 45° with respect to the quantum wells.

in-plane-polarized absorbance feature is the weaker of the two, it is well out of the instrumental noise (rapidly varying fluctuations in Fig. 17) and, therefore, is regarded as proof of an ellipsoidal-valley intersubband transition. Both spectra are superimposed on a background that slowly decreases with increasing wavenumber, and is attributed to free-carrier absorption of in-plane-polarized radiation by electrons in the quantum wells. By subtracting out this background numerically, one finds a maximum absorbance of 0.41 and 1.02 for the in-plane and 45° polarizations, respectively.

To convert the absorbance maxima in Fig. 17 to ζ_p, one can apply Eq. (27) directly, assuming that the intensity in the MQW region is uniform and that no interference is occurring near the top surface. For the MIR structure of sample 3, $P \approx 15$, and one finds $\zeta_p \approx 6.3 \times 10^{-4}$ and 1.56×10^{-3} for in-plane polarized radiation and 45°-polarized radiation, respectively. To compare this with theory, one can write from Eqs. (29) and (28) $G = 0.194$ and 0.597, and $\zeta_p = 5.6 \times 10^{-4}$ and 1.72×10^{-3} for the in-plane and 45°-polarized radiation, respectively. Remarkably, experiment and theory agree within 12%.

Combining the in-plane, Brewster's angle, and 45°-polarized radiation, one can plot the angular dependence of the peak fractional absorption as shown in Fig. 18. Superimposed on this plot is the theoretical curve of Eq. (28). Note that the theoretical ζ_p is approximately 4 times stronger at 90° than it is at normal incidence, which the experimental data appear to support. This exemplifies the point that the coupling of the intersubband transition to the in-plane field (as measured by the off-axis components of the reciprocal effective-mass tensor) is necessarily weaker than the coupling to the perpendicular field (as measured by the on-axis components) . Nevertheless, the in-plane absorption is very attractive in devices like detectors, modulators, and lasers because it can be enhanced greatly by multiple-pass interference techniques such as those discussed in Sec. 6.

It is also informative to compare the experimental data to the analytical model developed in Sec. 3. Shown in Fig. 19 is the peak of the first-to-second-state intersubband transition (in units of wavenumber) as a function of well width between 4.5 and 8.5 nm. The lower theoretical curve was computed for an L-valley barrier height, $E_{C,B}^{(L)} - E_{C,W}^{(L)}$, of 0.642 eV, $m_l = 1.25m_0$, $m_t = 0.10\,m_0$, and $w_{zz}^{-1} = 0.144\,m_0$. This curve yields much better agreement with the experiment. The top theoretical curve in Fig. 19 was computed for $E_{C,B}^{(L)} - E_{C,W}^{(L)} = 0.642$ eV, $m_l = 1.20$ m_0, $m_t = 0.08\,m_0$, and $w_{zz}^{-1} = 0.118\,m_0$, which are effective-mass values obtained from a theoretical $\mathbf{k \cdot p}$ calculation [25]. The discrepancy is

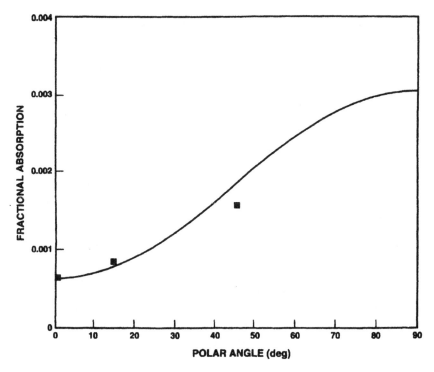

Fig. 18 Fractional absorption per quantum well deduced from room-temperature transmission measurements on sample 3 at three different angles of incidence. The solid curve applies to the theoretical expression (28) in the text.

caused primarily by the difference in m_t. The values $m_l = 1.25\, m_0$ and $m_t = 0.10\, m_0$ are averages between the corresponding $\mathbf{k} \cdot \mathbf{p}$ values and the experimental values $m_l = 1.30\, m_0$ and $m_t = 0.12\, m_0$ obtained by piezoelectric measurements [26].

5.2. Detector Responsivity and Detectivity Measurements

In order to maintain the same materials as successfully demonstrated in Sec. 4.4 but obtain a high responsivity, the first detector sample at Lincoln Laboratory was made with $Al_{0.09}Ga_{0.91}Sb$ quantum wells and $Al_{0.1}Ga_{0.9}As_{0.08}Sb_{0.92}$ barriers, but the quantum wells were reduced in width to 49 Å. In reference to Fig. 6, the strategy was to

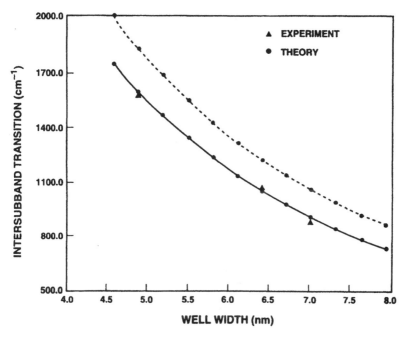

Fig. 19 Comparison of experimental data points (solid triangles) and theoretical predictions (solid circles) for $Al_{0.1}Ga_{0.9}Sb$ quantum wells as a function of well width. The dashed line connects the theoretical points for the following quantum-well parameters: $m_1 = 1.20\,m_0$, $m_t = 0.08\,m_0$, and $w_{zz}^{-1} = 0.118\,m_0$. The solid line connects the theoretical points for the following parameters: $m_1 = 1.25\,m_0$, $m_t = 0.10\,m_0$, and $w_{zz}^{-1} = 0.144\,m_0$. The assumed barrier height of 0.642 eV is the same in both cases.

push the second *L*-valley state above the *X*-valley conduction band edge in the barriers so that an electron excited to the second *L*-valley state could transfer into the *X* valley of the barrier and then drift to the collecting contact as an *X*-valley electron. The desired transport mechanism is similar to what was observed in GaAs/$Al_yGa_{1-y}As$ MQW detectors with the Al fraction in the barriers ($y = 0.55$) exceeding the crossover value [27]. The complete material structure is shown in Fig. 20. In addition to the quantum-well and barrier layers, a 0.5-μm-thick *n*-type GaSb layer and a 1.0-μm n^+ GaSb layer were deposited below and above the MQW structure, respectively, to provide for low-resistance ohmic contacts.

After growth, the sample was tested for absorption strength using the single-pass normal-incidence technique described in Sec. 3.3.2. Surprisingly, the absorption in the sample was imperceptibly low

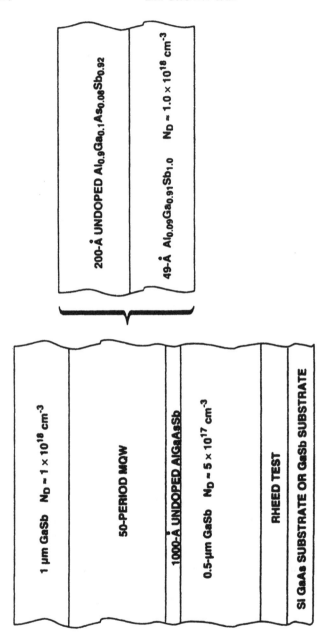

Fig. 20 Compositional profile of 50-period MQW detector made from 4.9-nm-wide $Al_{0.09}Ga_{0.91}Sb$ quantum wells and $Al_{0.9}Ga_{0.1}As_{0.08}Sb_{0.92}$ barriers.

between 4 and 16 μm. Being that the minimum detectable absorbance in the experiment was roughly 0.002, the minimum detectable fractional absorption (absorption coefficient) per well was about 0.0001 (188 cm⁻¹). After absorption measurements, detectors were fabricated according to the procedure outlined in Fig. 14 with a mesa diameter of 300 μm. After fabrication, the detectors were tested for external responsivity by the procedure described in Sec. 3.4. For these tests, they were mounted on the copper cold finger of the helium refrigerator and maintained at temperatures between 10 and 290 K. The spectral response curve of this AlGaSb MQW detector at 77 K and zero bias is shown in Fig. 21. The peak of the response curve is at a wavelength of 6.5 μm (192 meV, 1550 cm⁻¹), and the long- and short-wavelength 3-dB-down points are at 7.6 μm (163 meV, 1311 cm⁻¹) and 5.2 μm (237 meV, 1912 cm⁻¹). The normalized R_0 is plotted as a function of bias voltage in Fig. 22. In this plot the spectrometer was set to a fixed frequency at the peak of the spectral response curve. Clearly, the responsivity is maximum at zero bias and drops rapidly beyond $+/-30$ mV, completely disappearing by $+/-500$ mV. The absolute magnitude of responsivity at zero bias and the peak wavelengh was determined to be 0.9 mA/W by the calibration procedure described in Sec. 3.4.

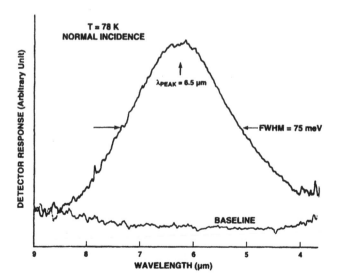

Fig. 21 Spectral responsivity of AlGaSb MQW detector of Fig. 20 at a temperature of 78 K and radition at normal incidence.

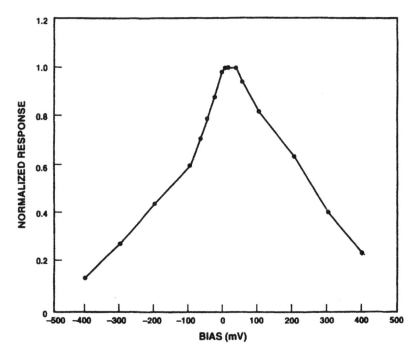

Fig. 22 Responsivity vs dc bias voltage for the MQW detector in Fig. 21 at the peak of the spectral responsivity curve ($\lambda = 6.5 \, \mu m$).

Since current noise is nonexistent at zero bias, the intrinsic detectivity was computed from Eq. (4) with only the thermal-noise term included. The specific differential conductance of $4.7 \, mS/cm^2$ was taken directly from the current-voltage and resistance-voltage curves shown in Fig. 23. The resulting detectivity was $2.1 \times 10^8 \, cm \, Hz^{1/2} \, W^{-1}$. This was obviously not a competitive result, and it led to some serious questioning of the observed detection mechanism. One clue was that the peak of the observed spectral response was consistent with a L-valley bound-to-bound transition, but the width was more consistent with a bound-to-extended transition. To elucidate this point, a data point for the peak of the spectral response at $6.5 \, \mu m \, (1550 \, cm^{-1})$ is shown in Fig. 19. Clearly, it nearly falls on the lower theoretical curve, which yielded the excellent fit to the experimental bound-to-bound transitions in the 6.4 and 7.0 nm wells. In contrast, a bound-to-extended transition would likely occur at a much longer wavelength

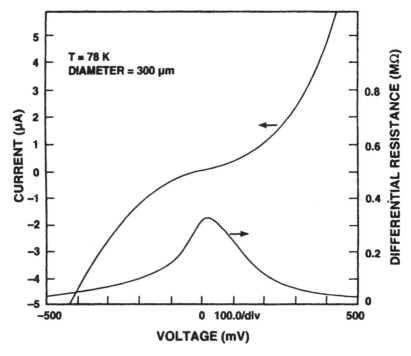

Fig. 23 Current-voltage and differential-resistance curves for the AlGaSb MQW detector of Figs. 19 to 22.

because of the tendency for the second state to hang up at the conduction band edge as it is being squeezed out of the quantum well. On the other hand, the 75 meV FWHM of the observed detector response is 2.5 times greater than the bound-to-bound absorption features. This is typical of the difference in absorption width between bound-to-bound and bound-to-extended transitions in GaAs/AlGaAs quantum wells.

The tentative explanation of the low detectivity is that mixing or scattering between the *L* valley in the quantum well and the *X* valley in the barrier greatly weakens the first-to-second state *L*-valley intersubband transition. This is consistent with the unmeasurable absorption in this sample. It is also consistent with the weak absorption coefficient of 185 cm^{-1} observed in the GaAs/Al$_{0.55}$Ga$_{0.45}$As MQW detectors mentioned above [27]. The rapid decay of the responsivity with bias voltage is still a mystery, but may be explained through the work of other groups discussed below on similar quantum-well samples.

5.3. AlGaSb Quantum-Well Results from other Research Groups

5.3.1. University of California, Santa Barbara

Long before the discovery of intersubband absorption at normal incidence, the research group of Prof. H. Kroemer at the University of
California, Santa Barbara (UCSB) was very active in the materials
science of AlGaSb quantum wells. Recently they pursued normal-incidence transitions in GaSb quantum wells because the absence of DX
centers allowed them to dope these wells much more heavily than
$Al_yGa_{1-y}Sb$ wells having $y \geqslant 10\%$. They relied on the fact that the
small electron effective mass of the Γ valley in GaSb tends to push the
first Γ-valley state above the first L-valley state. Furthermore, even if
the Γ-valley state was slightly lower in energy, its much smaller density of states would yield a relatively low sheet density in the Γ valley
of a heavily doped quantum well. In this case, most of the electrons
spill over into the L valley, even at cryogenic temperatures.

The UCSB reasoning proved to be accurate, and very strong normal-incidence absorption was observed in a variety of GaSb quantum
wells clad by AlSb barriers [28]. A typical set of results for 8.0-nm-
wide wells is shown in Fig. 24. This sample was grown on a (100)
SI-GaAs substrate and contained 50 GaSb quantum wells doped with
Te to an electron sheet density of approximately $1.6 \times 10^{12} \, cm^{-2}$ (as
determined by Hall measurements). The peak absorbance in Fig. 24
at 300 K corresponds to a peak fractional absorption and absorption
coefficient of 0.0056 and 7000 cm^{-1}, respectively. The FWHM for this
sample was approximately 20 meV [29]. In contrast, the theoretical
expressions of Eqs. (29) and (28) yield $G = 0.194$, $\zeta_p = 0.0023$, and
$\alpha = 2870 \, cm^{-1}$, assuming $\sigma_T = 1.6 \times 10^{12} \, cm^{-2}$, $w_{zz}^{-2} = 0.144 \, m_0$, $m_{lt} =$
$0.109 \, m_0$, and $\Gamma_s = 20$ meV. The over twofold discrepancy between ex-
periment and theory is not presently understood; however, it clearly
bodes well for the experiments.

Interestingly, the attempts to make detectors from these GaSb
structures met with similar results as the Lincoln Lab effort [30]. A
broad-spectral-response feature of roughly 60 meV FWHM was observed at 77 K in a GaSb/AlSb MQW detector containing a
3.5-nm-wide well. The peak absorption coefficient associated with this
feature was only about 1000 cm^{-1}. The best detectivity was observed
at low bias voltage, and the signal-to-noise ratio became impractically
low at bias voltages above $+/-0.4$ V. Somewhat better performance
was observed with 5.2-nm-wide GaSb wells clad by $Al_{0.75}Ga_{0.25}Sb$
barriers, as shown in Fig. 25. In this case, the absorption coefficient

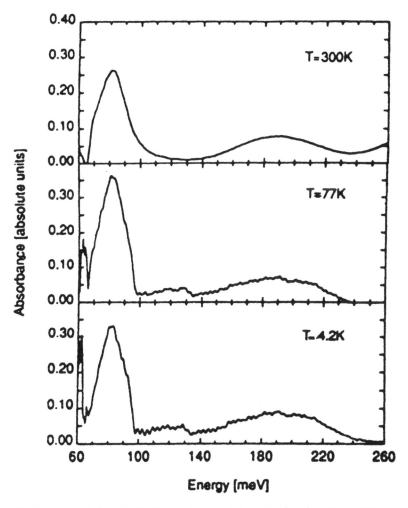

Fig. 24 Transmission spectra (in absorbance units) measured at three temperatures on an MQW structure containing 8.0-nm-wide GaSb quantum wells and AlSb barriers (after Ref. 28).

increased to approximately $3000\,\mathrm{cm}^{-1}$, and the spectral width of the absorption and the detector responsivity were both much less than in the sample with AlSb barriers. Note, however, that the detector displayed two spectral response features, one around 145 meV photon energy and the other around 300 meV. The former response behaved like the canonical bound-to-extended photoconductor, whereas the

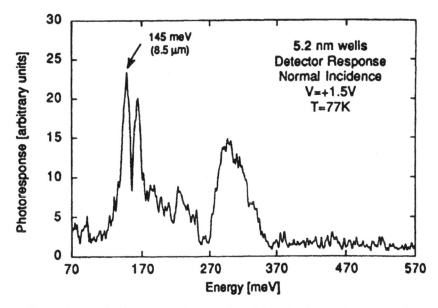

Fig. 25 Normal-incidence spectral responsivity at 77 K of an MQW detector contain-ing 50 5.2-nm-wide GaSb quantum wells and $Al_{0.75}Ga_{0.25}Sb$ barriers (after Ref. 30).

latter response was akin to the photovoltaic response of the 3.5-nm-well detector and the Lincoln Lab detector. At the time of this writing, absolute responsivities and detectivities on the UCSB de-tectors remain unreported.

5.3.2. Columbia University

A major breakthrough in the normal-incidence GaSb detectors was reported in 1993 by the research group of Prof. W. I. Wang [31]. Like the UCSB group, the Columbia group used as samples GaSb quan-tum wells doped heavily ($\sigma_T = 1.2 \times 10^{12}\,cm^{-2}$) with Te. However, they introduced the following two important changes: (1) growth along the (311) direction in addition to the (100) direction, and (2) reduction of the Al fraction in the barriers while maintaining lattice match by the use of $Al_{0.4}Ga_{0.6}Sb_{0.9}As_{0.1}$ barriers. The experimental normal-incidence absorption for (311)-oriented quantum wells grown on GaSb substrates is shown in Fig. 26. The peak occurs at 7.8 μm and has a magnitude of 9100 cm^{-1}, which is comparable to the UCSB result. The (100)-oriented quantum wells yielded a peak absorption of

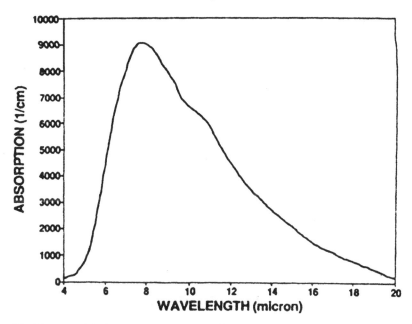

Fig. 26 Normal-incidence absorption derived from experimental measurements on an MQW sample containing GaSb quantum wells and $Al_{0.4}Ga_{0.6}Sb_{0.96}As_{0.1}$ barriers (after Ref. 31). Unlike the other samples discussed in the present article, this one was grown on a (311)-oriented GaAs substrate.

8100 cm^{-1} at 7.4 μm. This confirmed a previous theoretical calculation predicting that orientations other than (100), namely (211), (311), (110), and (511), would yield stronger normal-incidence absorption in L-valley quantum wells [32].

The most significant result of the Columbia work was the detector performance. Shown in Fig. 27 is the normal-incidence absolute responsivity measured for a (311)-oriented detector containing five quantum wells and operating at 20 K [31]. The peak responsivity of 310 mA/W is substantially higher than that reported for other normal-incidence AlGaSb detectors, or any other n-type normal-incidence quantum-well detector for that matter. Although a reason for the superior response is not given, it is quite plausible that the relatively low Al fraction in the barriers is significant. As evidence for this, notice in Fig. 5 (b) that for an Al fraction of 0.4 the L valley has the lowest energy of all three valleys. This fact would translate to the $Al_{0.4}Ga_{0.6}Sb_{0.9}As_{0.1}$ barriers of the Columbia samples provided that

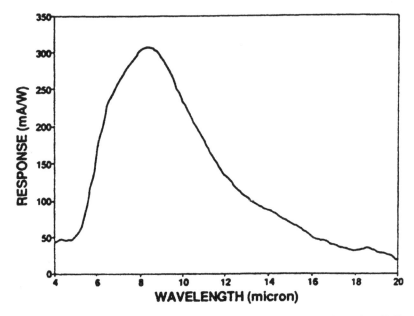

Fig. 27 Spectral responsivity of an MQW detector at 20 K containing five GaSb quantum wells and $Al_{0.4}Ga_{0.6}Sb_{0.9}As_{0.1}$ barriers on a (311)-oriented GaAs substrate (after Ref. 31).

the 10% As does not significantly alter Fig. 5 (b). In this case, an electron excited from the first L-valley state in the GaSb quantum wells to the second L-valley state would remain at the L point of the Brillouin zone after transferring to the barriers. This would tend to avoid the capture of photoelectrons by traps located at the X point, such as the well-known DX center. In addition, it avoids the process of intervalley transfer, which is likely to lower the mobility in the barriers and can also lead to charge-domain formation, negative differential resistance, and other spurious effects detrimental to low-noise-detector performance.

5.4. Normal-Incidence Results in AlGaAs

5.4.1. University of Florida

The other n-type III – V quantum well that has been explored is AlAs confined by AlGaAs barriers. As discussed in Secs. 1.3 and 2.2, in order to get normal-incidence absorption in this quantum well, it is necessary to grow the quantum wells along a crystallographic direction other than [100]. The first experimental report was made by the University of Florida group under Prof. S. S. Li [33]. They fabricated MQW detector samples containing AlAs quatum wells and $Al_{0.5}Ga_{0.5}$ As barriers on (110)-oriented GaAs substrates. The quantum wells were 3.0 nm wide and were doped uniformly n-type at a concentration of $2 \times 10^{18} cm^{-3}$, corresponding to a sheet density of $6 \times 10^{11} cm^{-2}$. The peak intersubband absorption of $450 cm^{-1}$ occured at a wavelength of approximately 13.8 μm. The FWHM of this absorption feature was approximately $200 cm^{-1}$.

The absorption characteristics can be compared directly with the theoretical formalism of Sec. 2 using the following expression for the geometrical factor of the X valleys in a (110)-oriented quantum well [7]:

$$G = \frac{w_{zz}^{-2}\cos^2\theta \sin^2\phi}{4m_{lt}^2} + \sin^2\theta. \tag{30}$$

This expression applies to the X-valley ground state which four of the six X-valley ellipsoids occupy, and it applies to radiation polarized parallel to the plane of incidence. For the material properties of AlAs, one has $w_{zz}^{-1} = 0.36\,m_0$, and $m_{lt} = 0.25\,m_0$, so that $G = 0.26$ for normal incidence and random polarization (i.e., the $\sin^2\phi$ term averages to 1/2). Substitution of the values $n = 3.3, \Gamma_S = 200\,cm^{-1}$, and $\sigma_T = 6 \times 10^{11}\,cm^{-2}$ yields $\zeta_P = 0.00035$ and $\alpha = 1182\,cm^{-1}$. This absorption is approximately 2.6 times greater than the experimental result.

Having confirmed the normal-incidence absorption, the Florida group commenced with the fabrication of 210×210 μm detectors and responsivity measurements were carried out at 25 K. The peak responsivity of 24 mA/W was observed near a wavelength of 12.9 μm. The different spectral peaks displayed by the absorption and responsivity was not explained in this work; however, other response features were observed including a peak at 0.8 μm and another at 2.22 μm. The short wavelength spectral response features were attributed to the type-II cross-interface transition and to the intersubband transition of electrons confined to the second X-valley state.

5.4.2. Columbia University

More recently, the AlAs quantum-well detector has been investigated by the Columbia group [34]. They studied samples containing 303.5-nm-wide wells on both (115)- and (113)-oriented GaAs substrates. The nominal electron sheet density in the wells was $5 \times 10^{11} \text{cm}^{-2}$. The barriers consisted of $\text{Al}_{0.45}\text{Ga}_{0.55}\text{As}$ and were 50 nm thick. At 68 K, the absorption was spectrally broad and displayed a peak of 1900 cm^{-1} at 12.2 μm. Normal incidence detectors were made having a mesa diameter of 500 μm and an operating temperature of 68 K. The peak responsivity of the detector on the (115) substrate ocurred at 11.7 μm and was down 50% at 9.5 and 14.3 μm, corresponding to a FWHM of 353 cm^{-1} (44 meV). The dark current of the detector at a bias voltage of 1 V was 1 μA, corresponding to a current density of 5×10^{-4} A cm^{-2}. This is comparable to the lowest current density of GaAs/AlGaAs MQW detectors at this operating temperature. Unfortunately, the researchers did not report the absolute responsivity, so that the sensitivity cannot be estimated.

6. FUTURE IMPROVEMENTS IN NORMAL-INCIDENCE MULTIPLE-QUANTUM-WELL DETECTORS

6.1. Resonant Optical Cavities

As emphasized in Sec. 1, a key advantage of normal-incidence absorption in quantum wells is the ability to enhance that absorption by multiple-pass interference techniques. In this section, a type of multiple-pass-interference structure will be addressed known as a resonant optical cavity (ROC). This ROC was first applied to homogeneous absorbing HgCdTe [35]. However, it is even better suited to quantum-well absorbers because it can be tailored to provide intensity maxima at the physical location of the quantum wells and much lower intensity away from the wells. Hence, the fractional absorption per quantum well can be raised substantially over that in a single-pass coupling structure.

The ROC consists of a semiconductor slab of thickness h having a metallic reflecting mirror on one face and the semiconductor-air interface at the other face. A cross-sectional view is shown in Fig. 28. A thin quantum well is assumed to be located at an arbitrary position relative to the metallic mirror. It will be assumed that an optical plane wave of wavelength λ and E_A is incident normally on the semiconductor-air interface. One wants to know the resulting field amplitude $E(z)$ at an arbitrary plane z in the slab. By adding successive passes of the incident wave through the slab and including a π phase shift ($e^{i\pi}$ amplitude factor) for reflection from the metal, one finds

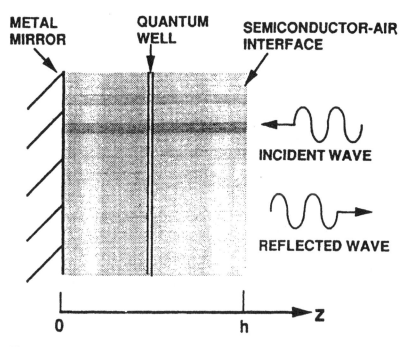

Fig. 28 Cross-sectional view of resonant optical cavity consisting of a quantum well embedded in a semiconductor slab with a metal mirror at one surface and air at the other surface.

$$E(z) = E_A[te^{ik_S(h-z)} + te^{ik_S(h-z)} e^{2ik_Sz}e^{i\pi}$$

$$+ te^{ik_S(h-z)} e^{2ik_Sz}e^{i\pi}re^{2ik_S(h-z)}$$

$$+ te^{ik_S(h-z)} e^{2ik_Sz}e^{i\pi}re^{2ik_S(h-z)}e^{2ik_Sz}e^{i\pi} + \dots, \qquad (31)$$

where t is the field transmission coefficient at the semiconductor-air interface for radiation incident from the air side, r is the field reflection coefficient at the same interface for radiation incident from the dielectric side, and $k_S = 2\pi n/\lambda$ is the wave vector in the semiconductor. Eq. (31) can be written as the sum of two terms, $E(z) = S_1 + S_2$, where

$$S_1 = tE_A e^{ik_S(h-z)} [1 - re^{2ik_Sh} + r^2 e^{2ik_Sh} - r^3 e^{2ik_Sh} + \dots] \qquad (32)$$

$$S_2 = tE_A e^{ik_S(h-z)} [- e^{2ik_Sz} + re^{2ik_Sz}e^{2ik_Sh}$$

$$- r^2 e^{2ik_Sz}e^{4ik_Sh} + r^3 e^{2ik_Sz}e^{6ik_Sh} - \dots]. \qquad (33)$$

By summing the infinite series, one can write these two equations as

$$S_1 = \frac{tE_A e^{ik_s(h-z)}}{1 + re^{2ik_sh}} \tag{34}$$

$$S_2 = \frac{-tE_A e^{ik_s(z-h)}}{1 + re^{2ik_sh}}. \tag{35}$$

Combining these two sums and taking the modulus, one obtains the intensity,

$$I \equiv E^*(z)\,E(z) = \frac{2TE_A^2(1 - \cos 2\,k_s z)}{1 + R + 2r\cos 2k_s h}, \tag{36}$$

where $T \equiv t^*t = t^2$ and $R \equiv r^*r = r^2$.

For all possible values of T, R, and h, the intensity displays a maximum at values of z satisfying $\cos 2k_s z = -1$, or $z = \lambda/4n$, $3\lambda/4n$ $5\lambda/4n...$, up to values $\leqslant h$. At these planes the absolute maximum intensity is obtained for a cavity length satisfying $\cos 2k_s h = -1$, which is the optimum ROC condition. The absolute maximum intensity is given by

$$I_{max} = \frac{4E_A^2 T}{(1 - r)^2}. \tag{37}$$

For a lossless dielectric, $r = (n-1)/(n+1)$ and $T = 1 - r^* r = 4n/(n+1)^2$, so that one obtains

$$I_{max} = 4nE_A^2. \tag{38}$$

Physically, Eq. (38) means that the maximum intensity in the dielectric is $4n$ times larger than the incident intensity. Since the fractional absorption of a quantum well depends linearly on the net intensity, a quantum well located in an ROC at a peak position will absorb $4n$ times more power than in a single-pass arrangement having an anti-reflection-coated semiconductor-air interface.

The optimum cavity condition, $\cos 2k_s h = -1$, is satisfied by a sequence of possible cavity widths given by $(2m-1)\lambda_0/4n = (2m-1)$ $\lambda/4 = h$, where $m = 1, 2, 3,...$ is called the order of the cavity. Fig. 29 shows the intensity maxima in the case of a $3\lambda/4\,(m=2)$ cavity. The sinusoidal curve is the intensity in the cavity relative to the incident intensity. One important feature of this plot is the significant spatial extent over which the internal intensity is enhanced. For example, the

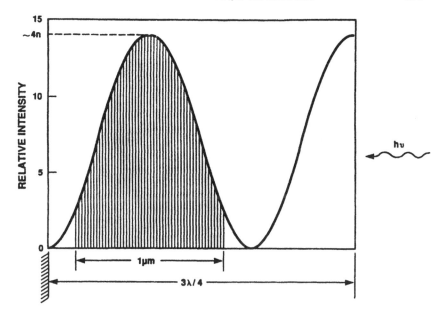

Fig. 29   Internal intensity profile relative to the incident intensity of a resonant optical cavity having a semiconductor slab thickness of $3\lambda/4$, where λ is the wavelength in the semiconductor. The vertical lines denote a judicious distribution of quantum wells.

enhancement is 10 times or greater over approximately 26% of the cavity width. This would correspond to a spatial extent of approximately $0.5\,\mu$m in a cavity operating at $\lambda_0 = 10\,\mu$m. For the type of detectors studied in Sec. 5, ten or more quantum wells could readily be located within such a spatial extent. A second feature is that the optimum cavity always has an intensity maximum at the semiconductor-air interface. This intensity maximum is not particularly useful for intersubband detectors because the quantum wells must be positioned substantially below the surface to avoid depletion of the free carriers by Fermi-level-pinning effects.

It is of great practical interest to know how the intensity at a fixed plane in an optimum ROC varies with the wavelength of the incident radiation. The internal planes of maximum intensity satisfy $\cos 2k_s z = -1$. Substitution of this and the optimum-cavity condition into Eq. (36) leads to

$$I = \frac{4TE_A^2}{1 + R + 2r\cos(2m - 1)\pi\lambda_0/\lambda}, \tag{39}$$

which reduces to $4nE_A^2$ for $\lambda = \lambda_0$. One can characterize the variation in I with λ by determining the change in λ required to decrease I by 50%. Setting the right side of Eq. (39) equal to $2nE_A^2$ and substituting for R and r, one finds the implicit equation,

$$\cos(2m - 1)\,\pi\lambda_0/\lambda = \frac{3 - n^2}{n^2 - 1}. \qquad (40)$$

Introducing the wavenumber variable $\sigma = \lambda^{-1}$, one can write

$$\frac{\sigma}{\sigma_0} = \frac{1}{(2m - 1)\pi}\cos^{-1}\frac{3 - n^2}{n^2 - 1}. \qquad (41)$$

As an example, consider a lossless semiconductor having $n = 3.8$ (e.g., GaSb near $\lambda = 10\,\mu m$). In this case, $(3 - n^2)/(n^2 - 1) = -0.85$, and one finds $\cos^{-1}(-0.85) = \pi(1 \pm 0.176)$. Therefore, the wavenumbers at the two 50% points are given by

$$\left(\frac{\sigma}{\sigma_0}\right)_\pm = \frac{1 \pm 0.176}{2m - 1}.$$

The FWHM, $\Delta\sigma/\sigma_0$, is given by $2(0.176)/(2m - 1)$, so that $\Delta\sigma/\sigma_0 = 0.35$ for $m = 1$, $\Delta\sigma/\sigma_0 = 0.12$ for $m = 2$, $\Delta\sigma/\sigma_0 = 0.07$ for $m = 3$, and so on.

6.2. Multicolor detectors

A particularly useful implementation of the normal-incidence MQW detectors and ROC structures is multicolor detection. By this, it is meant that the detector can respond selectively to radiation in different infrared spectral regions, or bands. The most important spectral regions correspond to the well-known infrared atmospheric windows: the 3 to 5 μm band, the 8 to 12 μm band, and to a lesser extent, the 16 to 20 μm band.

The manner in which two of these bands could be detected simultaneously is shown schematically in Fig. 30. An MQW stack that responds primarily to the shorter-wavelength band (centered about λ_1) is placed near the top of a mesa at a physical distance $\lambda_1/4n$ away from the top metallization. In this way a standing wave resonance for this radiation occurs at the center of the MQW stack and an enhancement of the quantum efficiency occurs consistent with the analysis given above. A second MQW stack that responds primarily to the longer-wavelength band (centered about λ_2) is placed below the first structure at a distance $\lambda_2/4n$ away from the top metallization. Under

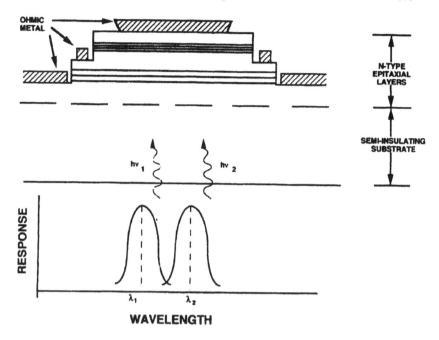

Fig. 30 Cross-sectional view (top) of a possible two-color detector made from two different MQW sections having different quantum-well thicknesses. Idealized normal-incidence spectral responsivity (bottom) of two-color MQW detector.

this condition, a standing wave resonance for the λ_2 band occurs at the center of the second MQW stack. As in the basic ROC, the bottom semiconductor-air interface acts as the second mirror of the cavity.

To discriminate radiation in one band from another, three ohmic contacts are provided at positions above the λ_1 stack, between the two MQW stacks, and below the λ_2 stack. The top ohmic contact consists of a heavily doped *n*-type layer with alloy metallization deposited in a small round area near the middle of the mesa and pure highly reflective metal deposited around it. Photocurrent is read out of the detector through the small round contact. The intermediate ohmic contact is formed by doping this region heavily *n*-type and then depositing a nonalloyed metallization using photolithography and lift-off techniques. Although the resulting contact resistance will be high by electronic device standards, the current flowing through the MQW detectors is so small that partially resistive contacts should have little detrimental effect on the detector performance. Finally, the bottom ohmic contact is formed by doping the bottom semiconductor layer heavily n-type and depositing alloy metal to this layer from one side of the mesa (the side opposite to the intermediate ohmic contact).

One particularly attractive aspect of the multicolor ROC is that the thickness can be designed to be optimum for any two or all three of the aforementioned atmospheric bands. For example, with the cavity of Fig. 30 and GaSb ($n = 3.8$) quantum wells, one could have an optimum resonance at $\lambda_1 = 5\,\mu m$ and $\lambda_2 = 8.33\,\mu m$, with the λ_1 stack separated from the top mirror by $\lambda_1/4n = 0.33\,\mu m$, the λ_2 stack separated from the top mirror by $\lambda_2/4n = 0.55\,\mu m$, and an overall thickness of the detector of $h = 5\lambda_1/4n = 3\lambda_2/4n = 1.64\,\mu m$. The fabrication of such a detector would be challenging but not at all beyond the capability of modern III–V microelectronic techniques.

References

1. L.C. West and S.J. Eglash, *Appl. Phys. Lett.*, **46**, 1156 (1985).
2. B.F. Levine, C.G. Bethea, G. Hasnain, V.O. Shen, E. Pelve, R.R. Abbott, and S.J. Hsieh, *Appl. Phys. Lett.*, **56**, 851 (1990).
3. H.C. Liu, J.R. Thompson, Z.R. Wasilewski, M. Buchanan, J. Li, and J.G. Simmons, 1993 Device Research Conference Digest, Paper VIB-2.
4. B.F. Levine, *Intersubband Transitions in Quantum Wells* (Plenum, New York, 1992), p. 43.
5. V.S. Swaminathan, DARPA IR Focal Plane Array Program Review, McLean, VA, Dec. 1992.
6. J.M. Luttinger and W. Kohn, *Phys. Rev.*, **97**, 869 (1955).
7. E.R. Brown and S.J. Eglash, *Phys. Rev.*, *B*, **41**, 7559 (1990).
8. M.A. Kinch and A. Yariv, *Appl. Phys. Lett.*, **55**, 2093 (1989).
9. K.K. Choi, M. Dutta, P.G. Newman, M.L. Sanders, and G.J. Iafrate, *Appl. Phys. Lett.*, **57**, 1348 (1990).
10. F. Stern and W.E. Howard, *Phys. Rev.*, **168**, 816 (1967).
11. H.C. Casey and M.B. Panish, *Heterostructure Lasers, Part B* (Academic, New York, 1978), Sec. 5.3.
12. L. Vegard, Z. Krist, **67**, 239 (1928).
13. J. Tersoff, *Phys. Rev.*, *B*, **30**, 4874 (1984).
14. F.L. Schuermeyer, P. Cook, E. Martinez, and J. Tantillo, *Appl. Phys. Lett.*, **55**, 1877 (1989).
15. Y. Tsou, A. Ichii, and E.M. Garmire, *IEEE J. Quantum Electron.*, **5**, 1261 (1992).
16. C. Mailhoit and T.C. McGill, *J. Vac. Sci. Technol.*, *B*, **1**, 439 (1983).
17. K.A. McIntosh, J.W. Bales, E.R. Brown, and T.C.L.G. Sollner, Solid State Research Report (Lincoln Laboratory, MIT, 1987:3), p. 56.
18. N.J. Harrick, *Internal Reflection Spectroscopy* (Interscience, New York, 1967).
19. S.J. Eglash, H.K. Choi, and G.W. Turner, *J. Cryst. Growth*, **111**, 669 (1991).
20. J.H. Neave, B.A. Joyce, P.J. Dodson and N. Norton, *Appl. Phys.*, **A31**, 1 (1983).
21. G.W. Turner, B.A. Nechay, and S.J. Eglash, *J. Vac. Sci. Technol.*, **B8**, 283 (1990).
22. C.M. Wolfe, G.E. Stillman, and J.A. Rossi, *J. Electrochem. Soc.*, **119**, 250 (1972).
23. S.J. Pearton, U.K. Chakrabarti, W.S. Hobson, and A.P. Kinsella, *J. Vac. Sci. Technol.*, *B*, **8**, 607 (1990).

24. E.R. Brown, F.W. Smith, G.W. Turner, K.A. McIntosh, and M.J. Manfra, *Proc. SPIE* **1735**, 228 (1992).
25. H. Hazama, Y. Itoh, and C. Hamaguchi, *J. Phys. Soc. Jpn.*, **54**, 269 (1985).
26. M. Averous et al., *Phys. Status Solid*, **37**, 807 (1970).
27. B.F. Levine, S.D. Gunapala, and R.F. Kopf, *Appl. Phys. Lett.*, **58**, 1551 (1991).
28. L.A. Samoska, B. Brar, and H. Kroemer, *Appl. Phys. Lett.*, **62**, 2539 (1993).
29. L.A. Samoska, private communication.
30. L.A. Samoska, B. Brar, and H. Kroemer, *Proc.* SPIE **2021**, 149 (1993).
31. Y. Zhang, N. Baruch, and W.I. Wang, *Appl. Phys. Lett.*, **63**, 1068 (1993).
32. J. Katz, Y. Zhang, and W.I. Wang, *Appl. Phys. Lett.*, **61**, 1697 (1992).
33. L. Yu, S.S. Li, and P. Ho, *Electron. Lett.*, **28**, 1468 (1992).
34. Y. Zhang, N. Baruch, and W.I. Wang, *Electron. Lett.*, **29**, 213 (1993).
35. D.L. Spears, MIT Lincoln Laboratory, private correspondence.

CHAPTER 7

Infrared Detectors Based on GaInSb/InAs Superlattices

R.H. MILES and D.H. CHOW

Hughes Research Laboratories, 3011 Malibu Canyon Road, Malibu, California 90265 USA

1. INTRODUCTION

Development of $Ga_{1-x}In_xSb/InAs$ strained layer superlattices has largely been motivated by the promise of overcoming fundamental and practical limitations of conventional, high performance infrared (IR) detectors such as those based on HgCdTe or extrinsic Si [1,2]. A mature superlattice technology could ease detector cooling requirements or give higher performance at a fixed operating temperature, and could demonstrate both uniformity and producibility superior to that of other intrinsic detectors. While these advantages are more pronounced at longer wavelengths, they appear to hold for a wide range of cutoff wavelengths; it is hoped that $Ga_{1-x}In_xSb/InAs$ superlattices would span the 3–20 µm spectral range.

In the longer term, this class of structures offers potential for direct integration with III–V read-out electronics, avoiding reliability problems inherent to detectors bump-bonded to readouts, particularly in the presence of large thermal strains (e.g., between HgCdTe and Si). Compatibility with a wide array of III–V materials and devices also suggests a number of schemes for achieving multi-spectral detection and/or on-chip signal processing. However, these benefits are unlikely to be important until improvements over conventional detectors have been demonstrated, hence work to-date has focused on validating and realizing predictions of improved operating temperature, performance, and uniformity.

In this chapter we review the current status of $Ga_{1-x}In_xSb/InAs$ superlattices. With long wave infrared detectivities currently of the order of $10^{10} \, cm\sqrt{Hz}/W$ at 77 K, apparently limited by extrinsic material properties rather than passivation, development of materials remains the most active area of research for this technology. Accordingly, much of this chapter is devoted to a discussion of intrinsic and extrinsic material properties, and device results are only summarized. However, we begin by presenting theoretical performance advantages of this and other classes of superlattices, and reviewing the properties underlying these advantages.

1.1. Distinction between Superlattice and Multi-Quantum Well Detectors

Both superlattice and multi-quantum well (MQW) structures have been proposed as alternatives to conventional infrared detector materials. As was first discussed by Kinch and Yariv [3], the two approaches employ considerably different IR transitions, resulting in significantly different levels of performance. As illustrated in Fig. 1, IR detection in a superlattice is accomplished by photoexciting electrons from a valence to a conduction band state, whereas MQW detectors

Superlattice
(intrinsic)

QWIP
(extrinsic)

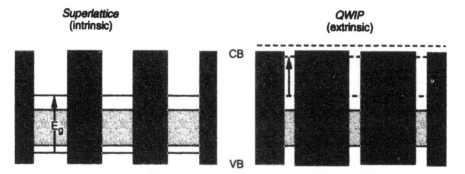

Fig. 1 Excitation processes employed in superlattice and MQW detectors (QWIPs). The valence-to-conduction band nature of the superlattice transition results in comparatively long intrinsic exicited state lifetimes (of the order ns-μs), while the intersubband transition of the MQW is characterized by fast relaxation processes (≈ ps).

(sometimes referred to as quantum-well intersubband photodetectors, or QWIPs) employ transitions between states within either the conduction or valence bands. Optical absorption selection rules often dictate use of diffraction gratings in the MQW detectors. No such measures are required in the superlattice. In either case, IR detection arises from a photocurrent associated with drift or diffusion of photoexcited carriers. The intrinsic nature of the relevant superlattice transition makes these detectors analogous to those based on HgCdTe, whereas MQW devices are extrinsic, corresponding more closely to Si:X detectors.

Critical to the performance of the two classes of detectors are their characteristic photogenerated carrier lifetimes. In the case of ideal, thermally limited performance, detectivity can be related to excited state lifetime [4] through

$$D^* = \frac{\eta}{2h\nu}\sqrt{\frac{\tau}{n_m t}},$$

where τ is the photoexcited state lifetime, η is the quantum efficiency, $\check{n}_m$ is the minority carrier concentration, and $h\nu$ reflects the cutoff wavelength of the detector. In the case of photoconductive devices, t is the thickness of the active region, yielding $D^* \propto \tau^{1/2}$, while for high performance photovoltaics, t corresponds to the minority carrier diffusion length, $L_p = \sqrt{D\tau}$, yielding $D^* \propto \tau^{1/4}$. As superlattices are limited to excited state lifetimes of the order of μs, while hot carriers relax to ground states on ps timescales in multi-quantum well structures, it is

readily apparent that detectivities (or, equivalently, operating temperatures, in the case of thermally limited operation) will be dramatically different for the two classes of detectors.

Figure 2 shows thermal generation currents for MQW and HgCdTe
detectors [3], illustrating the marked superiority in fundamental detectivity of the latter. This is primarily a consequence of the considerably longer excited state lifetime of HgCdTe. However, in contrast to
the MQW case, calculated detectivities for superlattices can appreciably *exceed* those for HgCdTe [5]. This subject will be examined in
Sec. 1.3. It should be emphasized that all of these quoted theoretical
numbers reflect nearly ideal performance; they have not and will not
be greatly exceeded in practice. As a consequence, applications for
which MQW detectors are credibly proposed are those which do not
place a premium on detectivity or operating temperature, benefiting
instead from features such as the high speed or unusual spectral

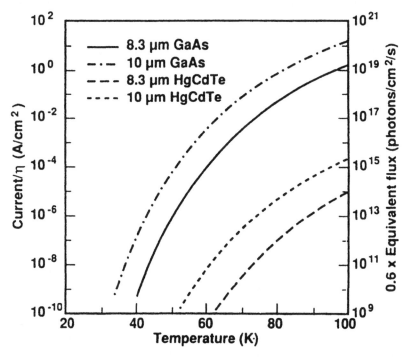

Fig. 2 Comparison of calculated thermal generation currents in AlGaAs MQW and HgCdTe IR detectors. Fast transitions in the MQW lead to insensitivity to low photon fluxes, relative to HgCdTe, as indicated in the figure. From Ref. [3].

response of these devices [6]. No such restrictions apply to the super-lattice detectors.

1.2. Superlattice Approaches to Infrared Detection

The capability to tailor the properties of superlattices through choices of layer thicknesses and compositions has prompted work on several III–V and II–VI systems. In the case of the III–V's, staggered band offsets allow intrinsic long wavelength superlattices to be fabricated where no long wavelength alloy exists (excluding comparatively unexplored alloys such as InTlSb and InSbBi, the narrowest gap III–V alloy, InSbAs, has a gap of 12 μm at room temperature but a gap of only 9 μm at 77 K). In addition, work on both the III–V and II–VI superlattice systems has been motivated by hopes of engineering structures with properties superior to those of conventional infrared materials. While the intrinsic properties of an alloy such as $Hg_{1-x}Cd_xTe$ are fixed by the choice of cutoff wavelength (corresponding to a particular x value), the additional degrees of freedom available to a superlattice allow several characteristics to be adjusted simultaneously. Of particular importance to IR detectors is the promise of tailoring intrinsic transport and/or recombination properties, as well as the hope of improving uniformity and producibility (i.e., sensitivity of these critical material properties to deviations from intended growth parameters), independent of cutoff wavelength. Performance and/or operating temperature improvements have been predicted based on the suppression of band-to-band Auger recombination and reduction of tunnel currents, achieved through increased effective masses and judicious choices of energy minigaps and band splittings. Promise of improved uniformity and/or producibility derives both from growth-technique specific benefits to controlling layer thicknesses rather than alloy compositions, and from the existence of robust regions of parameter space in which small deviations from intended superlattice layer thicknesses or compositions have little effect on key properties of the structure.

Several caveats regarding projected benefits of superlattices should be stated. While superlattices afford greater freedom of device design than alloys, it is clearly impossible to simultaneously optimize all properties of importance. Choice of the optimum layer thicknesses and compositions of a superlattice will be rigidly dictated by system needs; placing a premium on a figure such as operating temperature will inevitably reduce gains in others such as uniformity. Further, contrary to some blanket claims, superlattices are not necessarily panaceas for the shortcomings of alloys. Just as properties can be improved through appropriate choices of structures, they are readily degraded through poor choices. Additional concerns are of greater substance. Superlattices have been found to display a wealth of extrinsic characteristics distinct from those of alloys. Reduction of extrinsic effects such as efficient Shockley-Read-Hall recombination is essential to realizing calculated *intrinsic* benefits. While most calculated intrinsic properties have been substantially confirmed by experiment, perpendicular transport characteristics remain unknown and are critical to mesa devices. Growth of these structures relies on the techniques of molecular beam epitaxy (MBE) or chemical vapor deposition (CVD), which have only in the last year yielded conventional IR detector materials competitive with those grown by bulk methods [7]. Last, as has already been mentioned, no superlattice has yet demonstrated performance competitive with HgCdTe or extrinsic Si.

Despite the comparative immaturity of superlattice approaches to infrared detection, the potential benefits are considerable. Following is a brief review of the promise and status of approaches explored to date.

HgCdTe

Although predated by research into narrow gap GaSb/InAs heterostructures [8], long wavelength HgTe/HgCdTe superlattices were the first to be proposed for application in IR detectors. The original proposal, by Schulman and McGill in 1979 [9], was followed by predictions of advantages of uniformity arising from control of layer thicknesses rather than alloy composition, and of increased perpendicular effective masses, promising reductions in tunnel currents for mesa devices [10]. Both of these benefits are of increased importance at longer wavelengths, promising to extend the spectral range of these detectors beyond those practical with HgCdTe alloys. While these II–VI superlattices have been fabricated and their optical properties studied [11], structures with crisp cutoff wavelengths or long minority carrier lifetimes have not yet been realized [12]. Although this system continues to merit attention, it appears to face the same material and processing complexities that have slowed the development of HgCdTe detectors and other II–VI devices.

GaSb/InAs

Narrow gap GaSb/InAs structures were among the first superlattices studied [8]. Magnetic, structural, and optical properties of these structures were reported by Esaki and co-workers starting in 1978

[13]. It was established that long wavelength gaps could be obtained by virtue of the "broken gap" type-II band alignment in this system, which places the valence band edge of GaSb *above* the conduction band edge of InAs. It was demonstrated that gaps throughout the infrared could be obtained; increasing constituent superlattice layer thicknesses to approximately 100 Å was found to reduce quantum confinement energies sufficiently to yield a semimetallic (zero gap) structure at cryogenic temperatures [14].

In 1984, Arch et al., examined optical absorption coefficients of long wavelength InAs/GaSb superlattices, but found them too small to be practical for IR detectors [15]. Although not universal, this drawback is common in type-II structures and doping superlattices (nipi's); the band alignment reduces oscillator strengths by localizing electrons and holes in different layers of a superlattice, an effect which is amplified as constituent layers are made thicker.

InAsSb

In 1984, Osbourn suggested straining the narrowest gap III–V alloy, $InAs_{0.39}Sb_{0.61}$, to extend infrared response of that system to 12 μm at 77 K [16]. Ensuing work has emphasized attaining LWIR response through use of the type-II band offsets available to $InAs_{1-x}Sb_x/InSb$ structures. While these superlattices should benefit from reduced tunnel currents and improved uniformity relative to HgCdTe, experimental studies have revealed significant drawbacks of this system. As with InAs/GaSb superlattices, the thick layers necessary to achieve long wavelength response result in very low LWIR absorption coefficients (< 1000 cm^{-1}) [17]. Perhaps more significantly, this particular system suffers a variety of materials problems, including a miscibility gap, which leads to pronounced compositional fluctuations [18,19], and absence of a lattice matched binary substrate, which leads to microcracking incompatible with fabrication of two-dimensional detector arrays [20]. Further, consideration of band alignments in this system suggests that Auger mechanisms will *reduce* ultimate performance in this system relative to HgCdTe.

77 K zero bias resistances of 9Ω-cm^2 have been reported for 10 μm diodes fabricated from $InAs_{0.13}Sb_{0.87}/InSb$ superlattices, from which a detectivity slightly greater than 10^{10} cm$\sqrt{Hz}$/W has been calculated [21]. While research into the material properties of MBE-grown InAsSb is continuing, it is unclear that ideal superlattices would compete with conventional detectors on the basis of operating temperature or performance.

1.3. Promise of GaInSb/ InAs Superlattices

Projected Device Benefits

Calculations of the performance and operating temperatures of detectors based on GaInSb/InAs superlattices reveal significant improvements relative to HgCdTe, and even greater benefits relative to extrinsic Si. Figure 3 compares limiting photovoltaic detectivities of 11 μm HgCdTe with those of two superlattices of the same gap [5]. A 25 Å/41Å $Ga_{0.75}In_{0.25}Sb$/InAs superlattice is considered, as this particular structure has been extensively studied experimentally, as well as a 15Å/39.8Å $Ga_{0.6}In_{0.4}Sb$/ InAs superlattice chosen for optimal performance at this wavelength, within practical constraints dictated by crystal quality (only (100) structures were considered in these

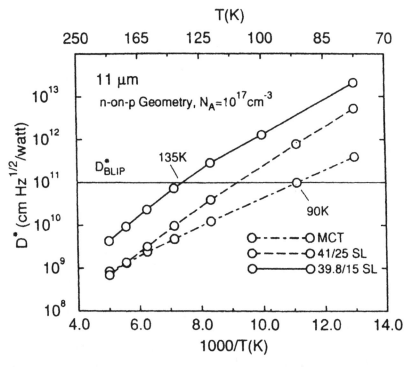

Fig. 3 Comparison of ideal detectivities of GaInSb/InAs superlattices and HgCdTe of equal band gap and doping. At $D^* = 10^{11}$ cm$\sqrt{Hz}$/W, corresponding approximately to the background limit for a 300 K photon flux with a 60° field of view, an optimized superlattice can operate at 135 K while HgCdTe must be cooled to 90 K. From Ref. [5].

calculations, and it was assumed that InSb fractions $x > 0.4$ were inconsistent with high structural quality, due to the $> 2.5\%$ strain of resulting $Ga_{1-x}In_xSb$ alloys relative to a (100) GaSb substrate). Doping of 10^{17} cm^{-3}, p-type, is assumed for each of the structures. This is the lightly doped side of the juction under consideration (leakage currents from the more heavily doped n-type side of the juction have been assumed to be negligible).

As is apparent from the figure, a detectivity of 10^{11} cm$\sqrt{Hz}$/W is achieved at roughly 90 K in HgCdTe, and at 110 K and 135 K for the two superlattices considered here. Improvements at higher detectivities are seen to be greater than 50% in detector operating temperature for 10^{17} cm^{-3} p-type doping. It should be noted that 11 μm detectors doped n-type or more lightly doped p-type generally show lesser improvements than illustrated in the figure, while the performance of HgCdTe devices are less sensitive to these changes [22].

In general, advantages relative to conventional detectors are amplified at very long wavelengths, and diminish in the mid-wave infrared. Calculations show that the performance of 4.5 μm HgCdTe devices currently operated at 180 K might be achieved at temperatures 20–40 K higher in GaInSb/InAs superlattices. While less dramatic than the gains quoted at 11 μm, such an improvement could reduce the number of stages required of a thermoelectric cooler, and hence would appreciably lower the size, weight, and cost of high performance MWIR systems. Much greater gains are to be achieved at longer wavelengths, where the superlattice also benefits from improved uniformity. Figure 4 illustrates detectivity versus temperature for thermally limited GaInSb/InAs photovoltaic detectors of varying cutoff wavelengths, assuming 10^{15} cm^{-3} p-type doping [22]. While the operating temperature of the superlattice is only slightly greater than that of HgCdTe at 11 μm for, e.g., $D^* \approx 10^{14}$ cm$\sqrt{Hz}$/W (a detectivity consistent with low photon background, longer wavelength applications), it drops very little as the cutoff wavelength is increased; comparisons to HgCdTe or extrinsic Si greatly favor the superlattice in this regime.

System cost, size, and weight often place a premium on operating temperature, but there are applications in which it may be desirable to operate a detector well below that nominally required to achieve background limited performance. It has been proposed that in missile seeker applications, for example, it may be beneficial to replace a "BLIP" LWIR HgCdTe array with one based on a higher performance technology. The suggestion is that the superior array could yield

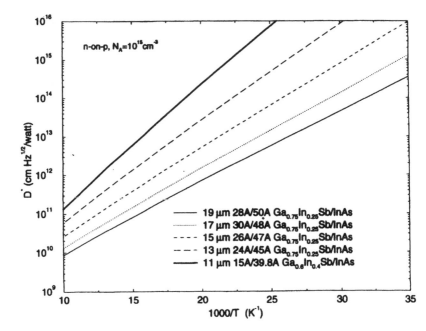

Fig. 4 Detectivities as a function of temperature for GaInSb/InAs superlattices of 11–19 μm cutoff wavelengths. Unlike HgCdTe, detectivity falls off slowly in the superlattice, due primarily to the suppression of Auger processes. (C.H. Grein, et al., unpublished [22].

a system more tolerant of operating temperature variations resulting from rapid cooldown, and that the greater headroom available to the detectors would reduce the incidence of "blinkers," pixels exhibiting marginal or unstable performance.

It is hoped that a mature GaInSb/InAs technology would benefit not only from improved performance, but also from advantages of III–V semiconductors over II–VI's. The mechanically robust nature of these materials should yield devices highly stable under thermal cycling. Ultimately, compatibility with other III–V materials and devices could lead to a "smart focal plane", integrating detectors, readouts, and signal processing electronics. Perhaps in the nearer term is the hope of employing nearly lattice-matched materials such as AlGaSb to achieve superior heterojuction devices. For example, use of this material as the middle layer of an *n-p-n* (LWIR-SWIR-MWIR) detector stack effectively blocks gain in this device, allowing sequential 2-color detection with excellent color separation without use of a

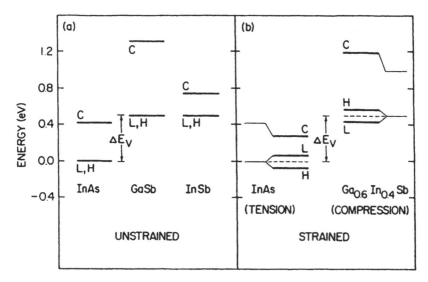

Fig. 5 Band alignments in the (InAs, GaSb, InSb) system. The type-II, broken gap band alignment is enhanced by the introduction of strain, placing the valence band edge of the antimonide well above the conduction band edge of the arsenide. From Ref. [1].

base contact. Last, a clear advantage of a successful GaInSb/InAs detector would be use of conventional III–V processing techniques, leveraging off a technology developed for a wide variety of electronic and optoelectronic devices.

Electronic Band Structure underlying Performance Benefits

Performance advantages of detectors based on GaInSb/InAs superlattices come from the electronic band structure of the active layer, which is qualitatively distinct from that of HgCdTe [1, 2, 23]. As with GaSb/InAs structures, long wavelength photoresponse is derived from a broken gap, type-II band alignment, in which the valence band edge of $Ga_{1-x}In_xSb$ lies below the conduction band edge of InAs. As illustrated in Fig. 5, the benefit of alloying the antimonide layer is to enhance the strain in the superlattice, increasing the energy separating the InAs conduction band and GaInSb valence band edges. Realizing a positive energy gap with such a band alignment requires the use of thin constituent superlattice layers to achieve large quantum confinement energies, as shown in Fig. 6; in practice, GaInSb and InAs layer thicknesses rarely exceed 35 Å and 60 Å, respectively. While the type-II band offset tends to localize electrons in the InAs layers and holes

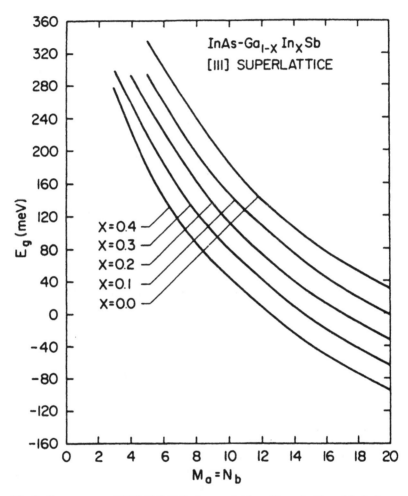

Fig. 6 Energy gaps of (111) GaInSb/InAs superlattices. $Ga_{1-x}In_xSb$ and InAs layers are assumed to be of equal thickness. (Layer widths are expressed in "monolayers"; in the (111) case, 1 ml ≈ 5.2 Å.) From Ref. [1].

in the GaInSb, the localization of electrons is weak for layers this thin. Thus, electron-hole overlap is significant in the antimonide layers despite the type-II band alignment, resulting in appreciable optical matrix elements for the superlattice valence-to-conduction band transitions.

As shown in Fig. 7, the combination of strain and quantum confinement effects in the superlattice results in a band structure unlike that

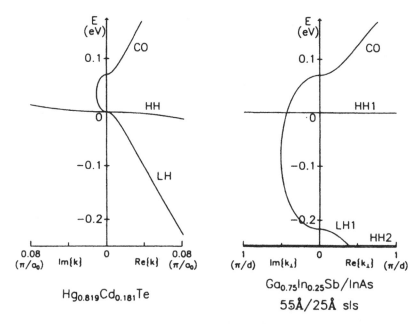

Fig. 7 Comparison of band structure in a VLWIR HgCdTe alloy and a GaInSb/InAs superlattice. The considerable magnitude of the imaginary perpendicular wavevector in the superlattice gap suppresses tunneling, while the splitting of the $k = 0$ light and heavy hole band energies suppresses Auger recombination. From Ref. [23].

of HgCdTe. Zone-center heavy- and light-hole valence band states are split in the superlattice, often by an energy appreciably greater than the band gap, and dispersion of the bands is reduced relative to HgCdTe, except in the case of the heavy hole band in the in-plane direction (not illustrated in the figure). The consequence is a greatly increased joint density-of-states for valence-to-conduction absorption processes, which offsets a reduction in oscillator strength due to spatial separation of electrons and holes in the type-II structure, yielding an absorption coefficient very similar to that of HgCdTe near the band gap. Calculated absorption coefficients are illustrated in Fig. 8.

As has been mentioned, performance and/or operating temperature benefits of the superlattices are expected based on suppressed band-to-band Auger recombination and tunneling processes. Reduction in tunneling [1, 2, 23] is readily understood by referencing the band structures of Fig. 7. While imaginary wavevectors at energies in the gap are small in the case of HgCdTe, the zone-center splitting of the light-hole

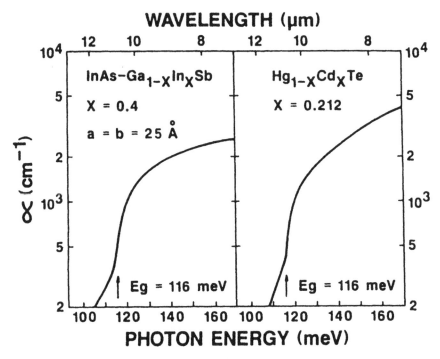

Fig. 8 Calculated absorption coefficients for a HgCdTe alloy and a GaInSb/InAs superlattice of equal band gap. From Ref. [1].

and electron states greatly increases the magnitude of these wavevectors in the superlattices, reducing the probability of tunneling between electron and hole states in a depletion region. (In the W. K. B. approximation, the tunnel current across a forbidden barrier can be related to the imaginary component of the wavevector by $I \propto e^{-2\{\text{Im}\{k(x)\}dx\}}$.)

Suppression of band-to-band Auger recombination is expected in both p-[24, 25] and n-type [26] GaInSb/InAs superlattices. Placement of a minigap between superlattice subbands an energy gap above the conduction band or below the valence band edge (for n-and p-type processes, respectively) suppresses Auger processes for which the in-plane wavevector is small ($k_{\parallel} \approx 0$). The dominant Auger channel in p-type material at low temperatures is depicted in Fig. 9. Although energy and momentum conservation require that an electron recombine with a hole having a sizable in-plane wavevector, dispersion in $k_{\parallel}$

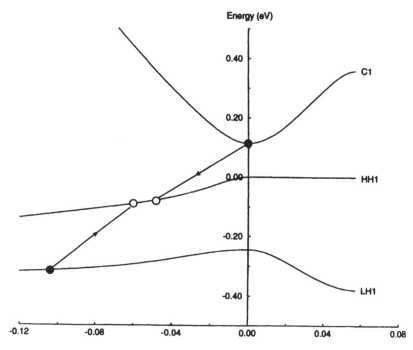

Fig. 9 The dominant p-type Auger recombination process in a tailored superlattice requires significant thermal excitation of holes. From Ref. [25] and C.H. Grein, et al., unpublished [22].

makes such a state unlikely to be occupied by a hole at low temperatures. This is in sharp contrast to the p-type case in HgCdTe, in which
comparatively little dispersion in the heavy hole band allows momentum conservation through transitions to regions of high occupation
probability, while excitations to a light hole band with high dispersion
facilitate energy conservation. The efficiency of this process means
that Auger processes often dominate radiative in HgCdTe, whereas
the opposite is often true in the GaInSb/InAs superlattice. The case of
n-type Auger recombination in the superlattice is qualitatively similar
to that of p-type structures; suppression derives both from the introduction of a minigap in $k_\perp$ and from dispersion of the heavy hole band
in $k_\parallel$ [26].

Although detectivities published to data have pertained only to
(100)-oriented superlattices, calculations have shown that (111) structures can have greater absorption coefficients than (100) superlattices

of identical composition and layer thicknesses. This fact has stimulated hopes of superior performance for (111) superlattices. However, a more legitimate comparison is between (100) and (111) structures of equivalent *strain*. Such a comparison suggests that the two cases are roughly equivalent in intrinsic properties, with the caveat that the constraint of lattice match is more readily achieved in the (111) orientation, and that extrinsic factors may be found to favor this case.

Practical Advantages

Benefits of working with III–V materials and employing a superlattice, rather than an alloy, are several. GaInSb/InAs superlattices are well lattice matched to GaSb substrates; growth on such a template balances compressive strain in the GaInSb layers with tensile in the InAs. Substrates of structural quality comparable to GaAs are readily available. However, bulk semi-insulating GaSb has not been achieved, hence free carrier absorption requires that such substrates be thinned for use in backside illuminated focal plane arrays, as is the case for conventional InSb detectors. Both p-and n-type doping are readily achieved, and wet and dry processing similar to that developed for GaAs devices has been demonstrated. GaInSb and InAs are mechanically robust materials, stable to temperatures considerably higher than those to which HgCdTe can be heated. While the thermal expansion coefficients appear no better matched to those of Si readouts than the II–VI's, it is likely that thinning the detectors will greatly alleviate difficulties of delamination, as it has for InSb arrays [27][1]. Further, stability of the superlattices at comparatively high temperatures may facilitate outgassing of dewars in applications employing modestly sized arrays.

Last, attaining a particular uniformity specification in a GaInSb/InAs superlattice optimized for performance requires a degree of growth control similar to that for HgCdTe, at a cutoff wavelength of approximately 10 μm [28]. However, while the uniformity of the II–VI degrades rapidly with increasing cutoff wavelength, that of the superlattice drops slowly, due to a reduction in band gap variations

---

[1]To our knowledge, II–VI arrays have not been successfully thinned, owing to the poor mechanical strength of these materials.

[2]Although the superlattice band gap changes less than that of HgCdTe for a fixed relative change in composition or thickness of a particular layer, summing contributions from each degree of freedom in the superlattice yields a net error equivalent to that of HgCdTe at 10 μm.

arising from InAs layer thickness fluctuations, suggesting significant potential for VLWIR applications. Further, the superlattice can be tailored for improved uniformity at shorter wavelengths, at the expense of performance and/or operating temperature.

1.4. Outline of Chapter

The remainder of this chapter addresses the present status of GaInSb/InAs superlattices, both from a material and a device standpoint. In Sec. 2 we review growth techniques and resulting structural properties. Issues of compositional abruptness of the group-V sublattice are addressed, relating both to As incorporation in antimonide layers and to the possibility of Sb segregation during growth. As in other mixed arsenide/antimonide heterostructures, interfacial stoichiometry is found to affect the properties of these superlattices. Control of this composition is discussed. The section closes with a review of p-and n-type modulation doping methods and results.

Measured characteristics of GaInSb/InAs superlattices relevant to IR detectors are presented in Sec. 3, largely validating calculated detectivities based on intrinsic properties but showing that extrinsic recombination processes have limited the performance of devices fabricated to date. In Sec. 4 we summarize device processing techniques and figures of merit obtained to date.

2. GROWTH AND STRUCTURAL PROPERTIES

The first superlattices and heterostructures based on combinations of arsenides and antimonides were grown by molecular beam epitaxy (MBE) approximately twenty years ago [29]. Since that time, advances in MBE technology have enabled the growth of device quality arsenide/antimonide heterostructures, including state-of-the-art 2–5 μm lasers [30, 31], resonant tunneling devices [32], and field effect transistors [33–35]. Unlike the situation for arsenide/phosphide heterostructures, which have been grown successfully by metal organic chemical vapor deposition, chemical beam epitaxy, and gas source MBE, it has proven difficult to grow high quality antimonide/arsenide heterostructures (particularly quantum well and AlSb-containing structures) by techniques other than MBE. The greatest outstanding issue for the growth of these structures by gas and metal organic source techniques is the development of precursors compatible with low substrate temperatures. Although it seems likely that these issues will be resolved in the future, MBE is the technique of choice at present for growth of infrared detectors based on $Ga_{1-x}In_xSb/InAs$ superlattices.

2.1. Growth by Molecular Beam Epitaxy

As described in the previous section, the existence of lattice-matched conditions for $Ga_{1-x}In_xSb/InAs$ superlattices grown on GaSb substrates over a range of layer thicknesses and x-values is advantageous for the fabrication of materials with good structural properties. As compressive strain in the $Ga_{1-x}In_xSb$ layers is balanced by tensile strain in the InAs layers, appropriate choices of layer thickness and composition enable the growth of thick, coherently strained superlattices with low threading dislocation densities. Figure 10 is a plot of the $Ga_{1-x}In_xSb/InAs$ layer thickness ratio needed to achieve net

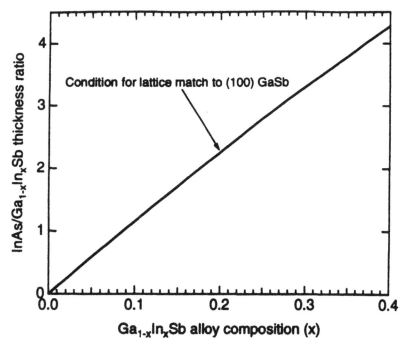

Fig. 10 $Ga_{1-x}In_xSb/InAs$ layer thickness ratio required for net-lattice match to a GaSb substrate. The effects of interfacial composition are neglected.

lattice match to a (100) GaSb substrate as a function of x. The effects of interfacial composition (to be discussed in detail in Sec. 2.3) were neglected in generating this plot.

Growth on Mismatched (100) Substrates

Initial exploration of growth conditions for $Ga_{1-x}In_xSb/InAs$ super-lattices was performed on thick GaSb buffer layers depositied on semi-insulating, (100)-oriented, GaAs and InP substrates. Although this approach results in high threading dislocation densities due to relaxation of the GaSb buffer layer, the availability of IR transparent, insulating substrates enables the straightforward application of certain material diagnostics, such as Hall measurements and optical transmission spectra. GaAs substrates have primarily been employed for these purposes, as they are less expensive than InP and are more straightforward to prepare for growth of the relaxed GaSb buffer layer. For GaAs (InP) substrates, a typical buffer layer growth sequence is as follows: a 1000–2000 Å GaAs ($In_{0.53}Ga_{0.47}As$) layer is deposited at 600°C (520°C), followed by a 10-period, 1-monolayer/1-monolayer GaSb/GaAs (GaSb/$In_{0.53}Ga_{0.47}As$) short-period super-lattice deposited at 480°C, and a 0.5–2 μm thick GaSb buffer layer. The GaSb buffer is grown at a substrate temperature at which the surface reconstruction (as determined by reflection high energy electron diffraction, RHEED) changes from 1×3 to 1×5 when the Ga-flux is interrupted and the surface is left in an Sb_2-flux with an approximate beam equivalent pressure of 1.0×10^{-6} Torr [36]. This substrate temperature is estimated to be 430°C based on optical pyrometer readings. More detailed analyses of the 1×3 to 1×5 transition in reconstruction of the GaSb surface have been reported recently [37,38], demonstrating an exponential dependence on $1/kT$ for the Sb-flux needed to produce the 1×3 to 1×5 transition, where T is the substrate temperature.

A range of substrate temperatures (350°C–450°C) has been explored for MBE growth of $Ga_{0.75}In_{0.25}Sb/InAs$ superlattice layers. The samples exhibit shiny, smooth surfaces for substrate temperatures below 400°C, while higher substrate temperatures result in hazy, rough surfaces. Consistently sharp x-ray diffraction peaks are obtained by setting the substrate temperature at the point where the 1×3 $Ga_{0.75}In_{0.25}Sb$ surface reconstruction changes to a 1×5 reconstruction under an Sb-flux of approximately 1.0×10^{-6} Torr; this temperature is approximately 380°C as measured by an optical pyrometer. Figure 11 displays $\theta/2\theta$ x-ray diffraction taken from a 25 Å/41 Å,

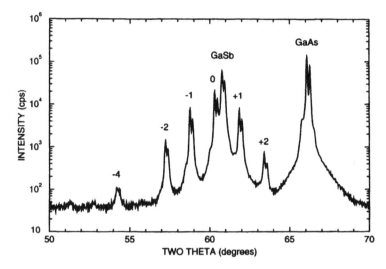

Fig. 11 (400)-like $\theta/2\theta$ x-ray diffraction from a $Ga_{1-x}In_xSb/InAs$ Superlattice grown on a stress-relaxed GaSb buffer layer on a GaAs substrate. High order superlattice satellites and well resolved Cu-Kα doublets are observed.

$Ga_{0.75}In_{0.25}Sb/InAs$ superlattice grown on a GaSb buffer and (100)-oriented GaAs substrate. High order superlattice satellites and well-resolved Cu-Kα doublets are observed, indicating a highly regular superlattice period with little interdiffusion [39]. The use of a cracked arsenic source appears to be necessary for obtaining this level of structural quality. Although a cracked antimony source is less essential for growth of $Ga_{0.75}In_{0.25}Sb/InAs$ superlattices, the lowest background doping levels reported in these structures were obtained in an MBE system equipped with an Sb cracker cell [40].

The period and average interatomic spacing of $Ga_{1-x}In_xSb/InAs$ superlattices can be determined from x-ray diffraction data, such as that shown in Fig. 11, by measuring the satellite spacings and zeroth order peak position, respectively. This information is sufficient to compute the In and Ga fluxes used during growth (two measured quantities, two determined variables), assuming that cross-incorporation of group-V species is limited (to be discussed in Sec. 2.2), and interfacial composition is known (to be discussed in Sec. 2.3). Cross-sectional transmission electron microscopy images taken of 0.5 μm thick $Ga_{1-x}In_xSb/InAs$ superlattices deposited on GaAs or InP substrates using the procedure described above reveal highly planar superlattice layers. However, a high density of threading dislocations (approximately 10^8 cm^{-2}) is observed. These dislocations commence at the GaSb/GaAs interface and propagate through the entire superlattice [36].

Growth on (100) GaSb Substrates

As it is generally accepted that the performance of minority-carrier devices is degraded by the presence of threading or misfit dislocations, elimination of these defects may be crucial to obtaining a competitive $Ga_{1-x}In_xSb$/InAs superlattice infrared detector [41, 42]. Techniques for deposition of $Ga_{1-x}In_xSb$/InAs superlattices on (100)-oriented GaSb substrates have been developed to achieve this goal. Pre-growth preparation of low etch pit density ($< 10^4$) GaSb wafers consists of *ex situ* chemical etching, followed by *in situ* outgassing at 250°C, and subsequent heating in an Sb-flux to the oxide desorption temperature (approximately 530°C) [43]. The substrates are radiatively heated to prevent warping and/or cracking under thermal stresses resulting from indium bonding. The previously outlined MBE growth conditions for the $Ga_{1-x}In_xSb$/InAs superlattice can be employed following deposition of a thin 1000Å GaSb buffer layer on the GaSb substrate.

Figure 12 is a cross-sectional transmission electron micrograph (TEM) from a $Ga_{1-x}In_xSb$/InAs superlattice grown on a (100)-oriented GaSb substrate. The micrograph is a two-beam bright-field image, with the beam direction tilted a few degrees from the [011] zone axis and a [200] diffraction vector operating. Also shown, as an inset in the figure, is a [011] zone axis electron-diffraction pattern from the specimen, exhibiting clear superlattice satellites around {200}, {111} and {022} diffraction spots. The micrograph reveals superlattice layers that are highly planar, regular in thickness, and free of threading and misfit dislocations. A thorough search for defects over several regions of the superlattice sample revealed no dislocations, yielding an upper bound of 10^7 cm^{-2} for its dislocation density. The actual dislocation density is likely significantly lower, but could not be determined through TEM.

An x-ray rocking curve for a $Ga_{1-x}In_xSb$/InAs superlattice grown on a GaSb substrate is shown in Fig. 13(a). The monochromatic Cu Kα_1 source used in this measurement provides an extremely stringent test of diffraction peak width relative to $\theta/2\theta$ scans such as that shown in Fig. 11. For comparison purposes, a rocking curve from a $Ga_{1-x}In_xSb$/InAs superlattice grown on an InP substrate is provided in Fig. 13(b); this superlattice yields data comparable to that shown in

Fig. 12   Cross-sectional transmission electron micrograph (TEM) from a $Ga_{1-x}In_xSb/$ InAs superlattice grown on a GaSb substrate. From Ref. [43].

Fig. 11 in a conventional $\theta/2\theta$ scan. The outstanding structural quality of the $Ga_{1-x}In_xSb/InAs$ superlattice grown on GaSb is evidenced by the considerable amplitude of higher-order superlattice satellite peaks, and by the absence of appreciable broadening of the satellites. The lower plot in Fig. 13(a) displays the x-ray diffraction pattern calculated by a kinematical model for a 28Å/33Å $Ga_{0.8}In_{0.2}Sb/InAs$ superlattice. Analysis of the diffraction is complicated substantially by the uncertain nature of interfaces in the superlattice (to be discussed in Sec. 2.3). The kinematical model used here explicitly sums scattering amplitudes from each superlattice layer, allowing the changing form factors and interplanar spacings of interfacial layers to be accounted for. As demonstrated in Fig. 13(a), the model provides an outstanding fit to experimental x-ray diffraction data.

Figure 14 displays further verification of the agreement between theory and experiment through a comparison of calculated and measured diffraction in the vicinity of the zeroth order superlattice peak. Measured and calculated Pendellösung fringes are displayed in the figure;

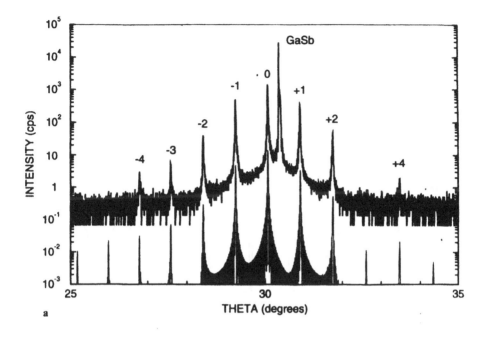

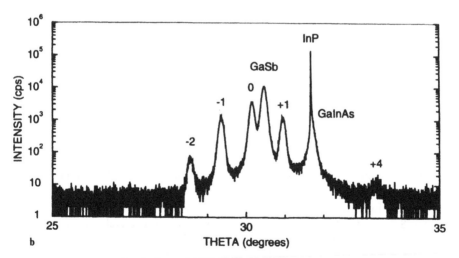

Fig. 13 (a) X-ray rocking curve from a $Ga_{1-x}In_xSb/InAs$ superlattice grown on a GaSb substrate. Also plotted is the x-ray diffraction pattern calculated by a kinematical model for a 28 Å/33 Å $Ga_{0.8}In_{0.2}Sb/InAs$ superlattice. (b) X-ray rocking curve from a $Ga_{1-x}In_xSb/InAs$ superlattice grown on an InP substrate. Adapted from Ref. [43].

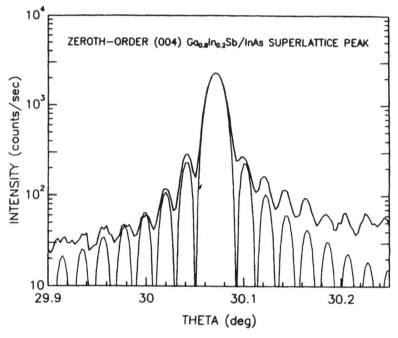

Fig. 14   X-ray rocking curve from a $Ga_{1-x}In_xSb/InAs$ superlattice taken in the vicinity of the zeroth order superlattice peak. Measured and calculated Pendellösung fringes are displayed.

the spacing of these fringes is consistent with the interference pattern expected from coherent x-ray scattering from the top and bottom interfaces of a 40-period, 28Å/33Å, $Ga_{0.8}In_{0.2}Sb/InAs$ superlattice.

Growth on (111) Substrates

Preliminary results have been presented by one group for growth of (111)-oriented GaInSb/InAs superlattices on both GaSb and GaAs substrates [44]. Growth on (111) A and (111) B surfaces was investigated, for both on-axis and 1° and 2° off-axis substrates. 3 × 3 and 8 × 8 Sb- and Ga-stabilized reconstructions, respectively, were observed at substrate temperatures close to 420°C, which was found to yield the best structural quality. Although growth on GaAs substrates was typically observed to yield films of poor morphology and structural quality, better results were obtained for growth on GaSb. X-ray diffraction from a superlattice grown on a 1° off-axis (111) B GaSb substrate is shown in Fig. 15. Although broadening of the peaks is a

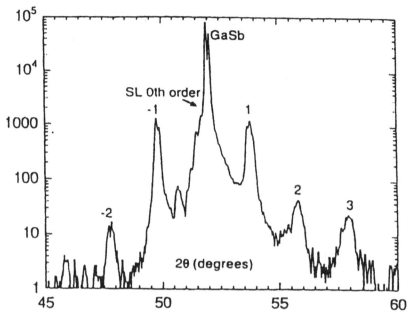

Fig. 15 $\theta/2\theta$ x-ray diffraction for a $Ga_{0.8}In_{0.2}Sb/InAs$ superlattice grown on a 1°-off (111) B GaSb substrate. From Ref. [44].

clear indication of structural imperfections, the amplitude of the diffraction and presence of several satellites indicates successful growth of a superlattice of fair quality. Initial TEM data reveal that interdiffusion may be a greater concern in (111) grown material, but the preliminary nature of work in this area suggests that significant improvements may be made.

2.2. Cross-incorporation of Group-V Species

A significant issue for the growth of arsenide/antimonide heterostructures is the degree to which As and Sb are unintentionally incorporated into antimonide and arsenide layers, respectively. In the case of $Ga_{1-x}In_xSb/InAs$ superlattices, significant cross-incorporation of group-V species can be expected to dramatically alter energy gaps and absorption coefficients due to large variations in constituent-material band-edge positions. Although a modest and reproducible degree of

arsenic/antimony mixing on the group-V sublattice does not funda-
mentally preclude the development of high performance Ga_{1-x}
$In_xSb/InAs$ superlattice detectors, it is highly desirable for both basic
(e.g., absorption coefficients) and practical (e.g., control of lattice mismatch) reasons to minimize cross-incorporation.

There exist two important mechanisms for cross-incorporation of
group-V species in arsenide/antimonide heterostructures. The first is
the incorporation of As or Sb molecules (monomers, dimers, or tetramers) that bounce off or reevaporate from a closed shutter. These
"background" molecules, which are abundant when the group-V sources are hot, may collide with chamber walls until they reach the
growth surface. The second mechanism is the incorporation of excess
group-V species remaining on the surface of a previous layer. The first
mechanism is particularly important for unintentional incorporation
of As in antimonide layers, while the second is more relevant to the
case of Sb in arsenide layers.

Background As-incorporation in Antimonide Layers

A set of GaSb layers deposited under varying conditions has been
investigated to determine the dependence or arsenic cross-incorporation in GaSb on Sb_2-flux, As background pressure, and substrate
temperature. In all of the samples, GaSb layers were grown in the
presence of an As background pressure created by a hot source shielded from direct line of sight to the substrate by a closed shutter. The
As_2 source used was a cracking effusion cell (i.e., As_2 emanates from
the source). However, it is not clear whether background As is predominantly in the form of tetramers or dimers (monomers are not
likely to be present). Appreciable As incorporation is observed for a
variety of growth conditions (i.e., $GaSb_{1-x}As_x$ alloys are grown). As
incorporation is observed to increase with As background pressure,
but is not affected by Sb_2 flux. Further, the As mole fraction increases
sharply at high substrate temperatures; As fractions as high as
$x = 0.25 \pm 0.05$ have been observed at a growth temperature of 520°C.
By contrast, As-mole fractions $x < 0.01$ are obtained at low substrate
temperatures (< 450°C), despite the presence of a significant As background pressure. These results indicate that As competes favorably
with Sb for lattice sites in $GaSb_{1-x}As_x$, consistent with the greater
bond strength of GaAs relative to GaSb. The increase in As incorporation with substrate temperature is likely due to the increased volatil-
ity of Sb atoms on the growth surface, which enhances the probability

that As atoms find available lattice sites. Last, we note that thick InAs layers grown in the presence of an Sb background pressure (created by a hot source shielded from line of sight to the substrate by a closed shutter) display no signs of group-V alloying as determined by x-ray diffraction.

The findings described above show that low substrate temperatures and As-background pressures are desirable for reducing As incorporation in $Ga_{1-x}In_xSb$ layers during the growth of $Ga_{1-x}In_xSb$/InAs superlattices. These conditions tend to limit the InAs growth rate (we generally use ≈ 0.2 monolayers per second) as modest arsenic fluxes are needed to keep As background pressures low. However, the availability of commercial valved cracking effusion cells for arsenic now offers the potential for rapidly and controllably varying As-fluxes, providing a possible means of raising the InAs growth rate above those used in the past. (Use of such cells may not enable higher growth temperatures to be employed, however, due to the problems with surface morphology described in Sec. 2.1).

Sb-surface Riding

Group-V-stabilized surface conditions are generally used for MBE growth of III–V semiconductors to prevent group-III droplet formation. In the case of $Ga_{1-x}In_xSb$ /InAs superlattices, these conditions can result in the presence of excess group-V species at the growth front of a constituent layer. These excess atoms may either be incorporated at an interface to a subsequent layer (to be discussed in more detail in Sec. 2.3), or ride the growth front to be gradually incorporated into the "bulk" of the subsequent layer. Detailed RHEED studies by Collins et al. [45] suggest that excess Sb on the surface of a $Ga_{1-x}In_xSb$ layer will be incorporated over several monolayers of a subsequent InAs layer (i.e., a graded InAsSb alloy will be formed). In contrast, excess As atoms on the InAs growth front appear to be incorporated at the interface to a subsequent $Ga_{1-x}In_xSb$ layer. Cross-sectional scanning tunneling micrographs (STM) have further shown evidence of asymmetry between the $Ga_{1-x}In_xSb$-on-InAs and InAs-on-$Ga_{1-x}In_{1-x}Sb$ interfaces, which can be attributed to pronounced surface riding[3] of excess Sb [46, 47]. These data are consistent with the larger size and weaker chemical bonding of Sb to group-III atoms, both of which tend to favor surfactant behavior.

---

[3]An alternate explanation is presented in the next section.

2.3. Interface Composition

As each interface in a $Ga_{1-x}In_xSb/InAs$ superlattice delineates a change in composition on both group-III and group-V sublattices, it is possible to obtain two distinctly different interfacial bonding configurations. These two bonding configurations are depicted schematically in Fig. 16. If an InAs layer is terminated with a final monolayer of In, the adjoining $Ga_{1-x}In_xSb$ layer will commence with a monolayer of Sb (assuming that group-III and group-V atoms are restricted to their respective sublattices), leading to InSb bonds across the interface (an "InSb-like" interface). Conversely, if the InAs layer is terminated with a final monolayer of As, the adjoing $Ga_{1-x}In_xSb$ layer will commence with a monolayer of $Ga_{1-x}In_x$, resulting in $Ga_{1-x}In_xAs$ bonds across the interface (a "$Ga_{1-x}In_xAs$-like" interface). Intermediate interfacial compositions are also possible. For example, a non-abrupt group-V sublattice interface may consist of a random As_zSb_{1-z} layer sandwiched between In and $Ga_{1-x}In_x$ layers, resulting in both $Ga_{1-x}In_x$ As and InSb bonds (along with $Ga_{1-x}In_xSb/InAs$ superlattices of interest for infrared detection, it is reasonable to expect that key material properties, such as optical absorption coefficients, will depend strongly upon interfacial composition (discussed in

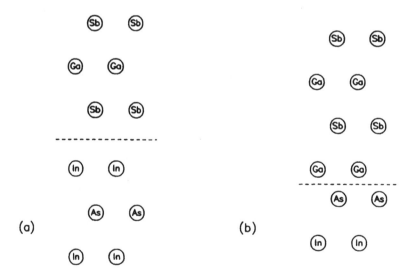

Fig. 16 Schematic diagram depicting two distinct bond configurations for a $Ga_{1-x}In_xSb/InAs$ interface. Adapted from Ref. [48].

more detail in Sec. 3). Furthermore, the high strain energy associated with both InSb-like and $Ga_{1-x}In_xAs$-like interfaces may have implications for the formation of interfacial defects.

A set of 8 ml/13 ml $Ga_{0.75}In_{0.25}Sb$/InAs superlattices has been grown with varying shuttering schemes to investigate the degree to which interfacial composition can be controlled [40]. Figure 17 depicts the interfacial shuttering sequences used to produce superlattices with $Ga_{0.75}In_{0.25}As$-like and InSb-like interfaces. Solid lines in the figure represent time periods during which shutters are open. X-ray diffraction data reveal that superlattices with $Ga_{0.75}In_{0.25}As$-like interfaces possess significantly smaller zeroth order d-spacings (average lattice spacing in the superlattice) than superlattices with InSb-like interfaces. This shift in d-spacing is consistent with the tremendous difference between the interfacial bond lengths of $Ga_{0.75}In_{0.25}As$ and InSb. Figure 18 is a plot of the average lattice spacing in five superlattices grown with different shuttering sequences. In the figure, the superlattice with InSb-like interfaces is assigned a nominal interfacial composition of 1, while the superlattice with $Ga_{0.75}In_{0.25}As$-like interfaces is

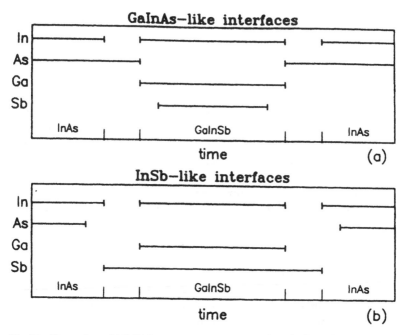

Fig. 17 Illustration of MBE shutter sequences used to deposit $Ga_{1-x}In_xSb$/InAs superlattices with (a) GaInAs-like interfaces, and (b) InSb-like interfaces. From Ref. [40].

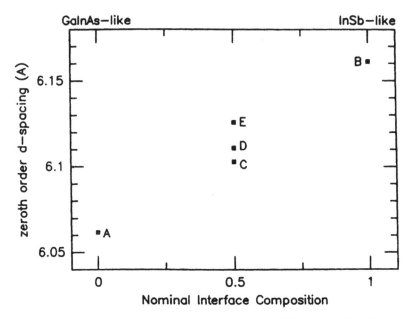

Fig. 18   Plot of zeroth order d-spacing vs. nominal interfacial composition for five $Ga_{1-x}In_xSb/InAs$ superlattices. Sample A: all GaInAs-like interfaces, Samble B: all InSb-like interfaces when growing GaInSb on InAs, Sample D: opposite of Sample C, Sample E: intermediate interfacial composition (intentionally nonabrupt group-V sublattice). From Ref. [40].

assigned a value of 0. A clear correlation between average interatomic spacing and nominal interfacial composition is observed, consistent with expected interfacial bond length differences. Although Raman studies performed on a similar set of superlattices by Sela et al. [49] did not yield well-resolved phonon modes corresponding to InSb and GaInAs interfacial bonds, subsequent Raman experiments by Waterman et al. [50] performed on GaSb/InAs superlattices confirmed distinct InSb and GaAs phonon modes in superlattices grown with InSb-like and GaAs-like interfaces, respectively.

Although the MBE shuttering schemes described in Fig. 17 are obvious choices for realizing GaInAs-like and InSb-like interfaces, they expose the growth surface to naked group-III flux at each interface, creating the possibility of In or Ga droplet formation. For this reason, the vast majority of $Ga_{1-x}In_xSb/InAs$ superlattice samples described in this chapter were grown with a group-V shutter open at all times. Control of interfacial stoichiometry in these samples was attempted only by employing a 5s Sb soak at the termination of each constituent layer. This scheme has the disadvantage that interfacial composition depends upon the competition between Sb and As for lattice sites during and immediately after the soak.

X-ray diffraction data from $Ga_{1-x}In_xSb/InAs$ superlattices grown using the 5s Sb-soak procedure are consistent with a net 70–80% InSb-like character to the interfaces. While this reflects an average of the $Ga_{1-x}In_xSb$-on-InAs and InAs-on-$Ga_{1-x}In_xSb$ interfacial bond lengths, it is unlikely that the compositions of the two sets of interfaces are the same. Specifically, Sb probably does not fully replace As on the InAs surface, and excess Sb on the GaInSb surface may ride into the InAs layer [46, 47]. That the two sets of interfaces are different in composition and/or that Sb segregation is appreciable is evident from the cross-sectional STM image of a GaInSb/InAs superlattice shown in Fig. 19 [47]. GaInSb layers are bright in the image and InAs dark. The GaInSb-on-InAs interfaces appear to be abrupt, likely reflecting the presence of GaInAs interfacial bonds and an abrupt compositional change. In contrast, the InAs-on-GaInSb interfaces are less distinct, probably reflecting a greater fraction of InSb interfacial bonds as well as an Sb surface riding effect.

Finally, x-ray photoelectron spectroscopy (XPS) measurements by Wang et al. [51] have demonstrated that exchange of the As-atoms on the InAs surface for Sb-atoms increases with Sb-cracking zone temperature and soak time. However, the effects of these parameters on interfacial composition in as-grown $Ga_{1-x}In_xSb/InAs$ superlattices has not been reported to date.

2.4 Intentional Doping

Techniques for controllably doping an infrared material both n- and p-type are essential for the realization of photovoltaic detectors. A significant benefit to using III–V semiconductors (as compared to II–VI compounds) is the greater availability of well-behaved substitutional impurity dopants. For the case of III–V MBE growth, silicon and beryllium are by far the most popular dopants. Beryllium is known to be an acceptor for all of the conventional III–V compounds. Hence, it should be possible to dope $Ga_{1-x}In_xSb/InAs$ supperlattices p-type by codeposition of Be during growth of either or both consitituent layers. To date, p-type doping levels up to $1 \times 10^{18}cm^{-3}$ have been demonstrated through Be-doping. It has been demonstrated that silicon is a well-behaved p-type dopant for

growth direction

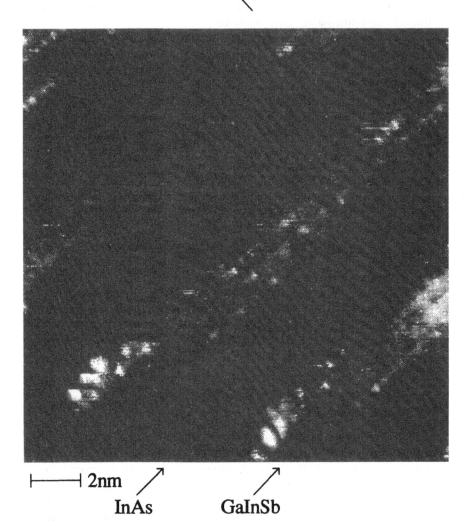

Fig. 19 Cross-sectional STM image of a 25 Å/33 Å $Ga_{0.75}In_{0.25}Sb/InAs$ superlattice. Asymmetry is apparent in the dark InAs layers, reflecting a greater InSb-bond character for the InAs-on-GaInSb interfaces and/or an Sb surface riding effect. From Ref. [47].

GaSb, [52], while it is an *n*-type dopant for InAs. Therefore, codeposition of Si during growth of InAs layers should result in an *n*-type superlattice (assuming little diffusion of the silicon atoms occurs). Hall measurements verify that this selective silicon doping technique yields substitutional *n*-type doping up to levels of $1 \times 10^{18} \text{cm}^{-3}$. Higher levels of *n*-type doping have not been attempted, but are unlikely to be needed for the growth of photovoltaic detector structures.

3. OPTICAL AND ELECTRONIC CHARACTERISTICS

Many predicted optical and electronic properties of GaInSb/InAs superlattices have now been established experimentally. Parameters measured to date which are of particular importance to IR detectors include the magnitude and cutoff of intrinsic optical absorption, carrier recombination rates, in-plane carrier mobilities and effective masses, and background doping levels. Most notable among those that have not been measured are mobilities and effective masses in the growth direction, which will greatly influence the performance of conventional mesa photojunctions.

Measured intrinsic parameters have in all cases been in good agreement with theory, supporting the contention that ideal detectors based on these structures would appreciably out-perform ideal HgCdTe-based devices. However, as extrinsic characteristics have typically been observed to dominate intrinsic, GaInSb/InAs superlattice detectors competitive with established technology have not yet been demonstrated. Perhaps foremost amongst the limitations of current materials are the excited-state carrier lifetimes; recombination times greater than 10 ns have not yet been demonstrated, while LWIR HgCdTe routinely exhibits lifetimes in excess of 1 μs. Nevertheless, considerable improvements have been made since the first superlattices were grown in 1989. Most significant among these is the very recent realization of GaInSb/InAs superlattices of high optical quality. While the carrier lifetimes in these structures have not yet been measured, dramatic improvements in band-to-band optical recombination efficiencies have led to the demonstration of stimulated emission in these superlattices [53]. Achieving radiatively dominated recombination had been widely regarded as the most significant hurdle facing these materials; the recent observation of lasing suggests that competitive infrared detector and laser material may already have been realized.

In this section we summarize work on optical and electrical properties of GaInSb/InAs superlattices to date. Emphasis is placed on basic properties of immediate importance to detectors, but complications such as effects of interfacial stoichiometry are addressed insofar as they relate to these parameters.

3.1 Recombination Properties

Photoluminescence and Photoconductivity

The dependence of superlatttice band gap on layer thicknesses and $Ga_{1-x}In_xSb$ alloy composition was first investigated by photoluminescence (PL) spectroscopy [54]. The case of GaSb/InAs superlattices had been studied earlier, revealing double-peaked PL spectra attributed to a band-to-band and an impurity or defect-related recombination process [55]. While such features have been reproduced in GaSb/InAs structures grown under the same MBE conditions as used for GaInSb/InAs superlattices, only a single peak has been seen from the latter class of superlattices. Representative spectra are illustrated in Fig. 20. As in Ref. [55], we associate the peaks seen in the GaInSb/InAs cases with the conduction-to-valence band transitions, varying in energy in reasonable accordance with theoretical

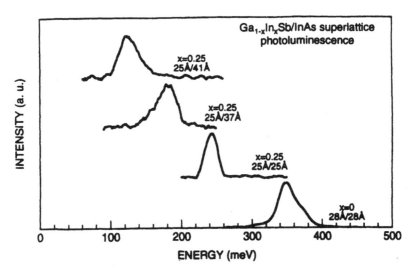

Fig. 20 5 K photoluminescence spectra from GaInSb/InAs superlattices, illustrating reduction in band gap with increasing InSb content or InAs layer thickness.

expectations of the superlattice band gap. Spectra presented in Fig. 20 clearly illustrate the utility of introducing InSb or increasing the InAs layer thickness to push the intrinsic response of the superlattice to longer wavelengths.

Photoluminescence spectroscopy has not been used extensively to characterize these materials as signals have already been extremely weak or undetectable; the 10 μm PL of Fig. 20 was only obtained at 5 K using a sensitive Fourier Transform technique with a 400 mW Ar$^+$ laser pump. However, improved growth techniques have recently led to observations of intense band-to-band luminescence in isolated 200 Å thick superlattices with 5 and 10 μm gaps [53]. While maximum PL intensity was observed at a temperature of approximately 80 K, appreciable luminescence was still apparent from a 5 μm structure at 300 K. In addition to providing new spectroscopic opportunities, these results illustrate a dramatic improvement in epilayer quality, likely to translate directly into improved IR detectors and light emitters. This promise was recently realized through the demonstration of a 3.2 μm superlattice laser [53]. Experiments to quantify improvements in lifetime and to fabricate and analyze detectors from similarly grown material are in progress but have not yet been completed; all further results cited here come from superlattices of comparatively low optical efficiency.

The dependence of superlattice band gap on layer thicknesses and compositions has been studied in greater detail by examining spectral cutoffs in rudimentary photoconductive detectors [56]. These devices have also been used to determine carrier lifetimes, as will be described later. Superlattices of InSb fraction $x = 0.0-0.35$ have been studied by this method, with InAs layers as thick as 45 Å yielding cutoff wavelengths as long as 17 μm. (A semimetallic band structure has been demonstrated by Hall analysis for somewhat thicker InAs layers [57]. Measured gaps are found to be in good agreement with theory, although different groups use InAs/GaInSb valence band offsets of 510-560 meV [1, 23, 58, 59]. Discrepancies are likely attributable, in part, to uncertainties in appropriate deformation potentials for these materials.

Carrier Lifetimes

As was described in Sec. 1, promise of operating temperatures or detectivities higher than those of HgCdTe stem largely from predictions of significantly longer intrinsic lifetimes in GaInSb/InAs superlattices. Experimental data illustrated in Fig. 21 confirm these predictions at carrier concentrations greater than 10^{17}cm^{-3} [60]. While the

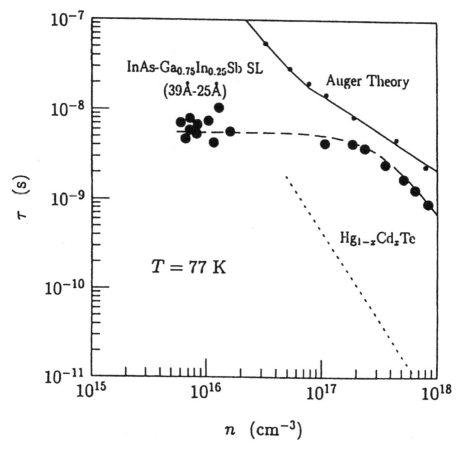

Fig. 21 Comparison of recombination lifetime in a 9 μm cutoff GaInSb/InAs superla-tice and a HgCdTe alloy of the same energy gap. The superlattice lifetime is determined primarily by Shockley-Read-Hall processes at moderate carrier concentrations ($< 10^{17}$ cm^{-3}) and by Auger processes at high electron-hole densities. Auger lifetimes for the superlattice appreciably exceed those of HgCdTe, in good agreement with theory. From Ref. [60].

superlattice examined here has a Shockley-Read limited carrier life-time less than 10 ns, optical excitation permits the carrier lifetime to be probed at densities for which Auger processes dominate. In par-ticular, the superlattice data shown in Fig. 21 are well fit by the dashed line, of the form $\tau(n) \approx (\tau_{SR}^{-1} + \gamma_3 n^2)^{-1}$, for a Shockley-Read lifetime $\tau_{SR} \approx 6$ ns and an Auger coefficient $\gamma_3 \approx 1.3 \times 10^{-27}$ cm^6/s (according to the approximation $\tau_A \approx 1/\gamma_3 n^2$). Data are seen to be in good agreement with Auger recombination rates calculated using a model [25] which employs no free parameters. The concurrence of experiment and theory lends credence to the predicted advantages of ideal superlattices described in Sec. 1.3, which are based on these and more straightforward calculations. Qualitative advantages with respect to HgCdTe are immediately apparent from the figure; Auger lifetimes more than two orders of magnitude greater than those of the II–VI alloy are observed for this superlattice, even though it is not tailored to the suppression of such processes.

Although comparatively long lifetimes have been observed at high carrier concentrations, the density of extrinsic recombination centers in superlattices examined to date has precluded demonstrating the μs minority carrier lifetimes necessary for high performance detectors at 10^{15} cm^{-3} doping levels. There is one report of parallel recombination channels characterized by 2 ns and 2 μs time scales [61]. However, comparison of transient and steady-state photocurrents led to the conclusion that the slow decay process corresponded to recombination of majority carriers with a small percentage of photogenerated holes trapped on sites with low electron capture cross-sections. While the recent observation of strong band-to-band PL in some superlattices suggests that minority carrier lifetimes now exceed those previously measured, this has not yet been confirmed experimentally. Until benefits of reduced Auger processes can be realized through reductions in Shockley-Read recombination, device performance is unlikely to exceed that of HgCdTe.

Optical Absorption

As predicted by Smith and Mailhiot [1,2], large IR absorption coefficients have been demonstrated in GaInSb/InAs superlattices [53, 61]. This is a consequence of the thin layers typical of these structures. While the "spatially indirect" nature of the optical transitions (i.e., the partial localization of electrons in InAs layers and holes in GaInSb layers) tends to reduce oscillator strengths relative to those of 3-dimensional systems, appreciable absorption coefficients are recovered through an increased density of states. Typical absorption spectra near the fundamental energy gap are shown in Fig. 22. This particular superlattice is seen to shift in cutoff by about 20 meV between room and low temperature.

Unlike materials such as HgCdTe, the superlattice absorption spectrum consists of a series of plateaus, characteristic of 2-dimensional

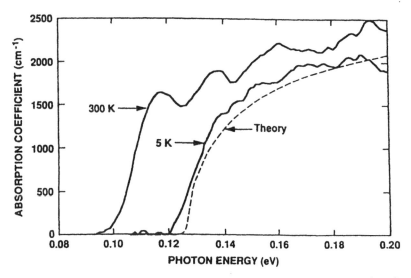

Fig. 22 Room and low temperature absorption coefficients for a 19 Å/35 Å $Ga_{0.75}In_{0.25}$ Sb/InAs superlattice. Although higher energy absorption displays a series of plateaus unlike that of HgCdTe, absorption near the band edge is very similar to that the II–VI alloy. From Ref. [61].

systems. However, more critical to the infrared detector application is the magnitude and abruptness of the absorption coefficient near the energy gap. In this respect the superlattice is similar to the II–VI, typically rising quickly to a coefficient of 1500–3000 cm^{-1}, depending upon the particular thicknesses and compositions of the constituent layers [28]. Despite early concerns about absorption coefficients in a type-II system[4], GaInSb/InAs superlattices are essentially equivalent to HgCdTe in this property.

It has been predicted that the optical properties of LWIR GaInSb/InAs superlattices should be sensitive to interfacial chemistry [23]. As was addressed in Sec. 2, two distinct interfaces are possible: "InSb-like" interfaces, in which planes of atom are stacked in the growth direction as follows:

···Ga Sb Ga S̲b̲ I̲n̲ As In As In As I̲n̲ S̲b̲ Ga Sb Ga···

---

[4]Concerns stemmed from low absorption coefficients measured in other type-II superlattices, for which layers were necessarily thick to achieve LWIR cutoffs (e.g., GaSb/InAs [15] and InSb/InAsSb [17]).

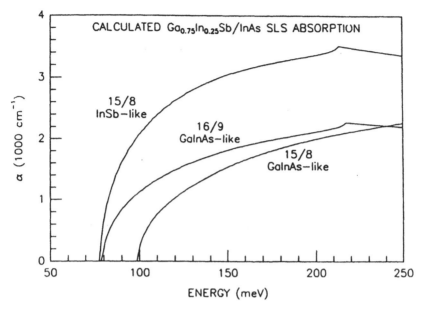

Fig. 23 Calculated absorption coefficients for superlattices with InSb-like and $Ga_{0.75}In_{0.25}$ As-like interfaces, illustrating the shift in energy gap and quantitatively different absorption coefficients obtained in the two cases. Experimental work has focussed on the InSb interface. From Ref. [23].

or "GaInAs-like" interfaces:

$$\cdots Ga\ Sb\ \underline{Ga\ As}\ In\ As\ In\ As\ In\ As\ In\ \underline{As\ Ga}\ Sb\ Ga\cdots.$$

Experiments have qualitatively confirmed the dependence of cutoff wavelength on interfacial chemistry [40,62]; the energy gaps of selected structures with InSb-like interfaces have been measured to be 25–50 meV smaller than those with GaInAs-like interfaces. As illustrated in Fig. 23, interfacial chemistry has also been calculated to have a significant effect on the magnitude and abruptness of the absorption edge of a particular superlattice. While the dependence has not yet been studied experimentally, samples are typically grown with InSb-rich interfaces, for which superior optical absorption properties are calculated. Measured absorption spectra for these superlattices [54,61] are in excellent agreement with theory. We are not aware of experimental absorption data for samples with GaInAs-like interfaces.

Differences in calculated optical absorption coefficients for the two types of interfaces derive from the increased electron-hole overlap at

InSb-like interfaces, relative to that at GaInAs-like interfaces. In principle, cross-sectional scanning tunneling microscopy (STM) provides a measure of the Fermi level in the superlattices and hence can probe wavefunctions for the two cases.[5] Such measurements have been performed on superlattices likely possessing InSb-like interfaces atop GaInSb layers and interfaces of mixed character atop InAs layers [43]. The results indeed seem to suggest significantly greater hole wavefunction spillover at the InSb-like interfaces [47]. This interpretation of the STM data is strengthened by consideration of perpendicular lattice spacings, which appears both to confirm the interfacial stoichiometry of the superlattices and to identify the positions of the interfaces [63]. However, it has also been suggested that observed asymmetries reflect segregation of Sb at the growth front, effectively smearing-out the group-V interface atop the antimonide layers [46]. Such an effect has been observed in isolated heterostructures [51], and would account for the observed data. It is our opinion that both effects are likely present in GaInSb/InAs superlattices studied to date.

3.2. Electrical Properties

Background Doping

Background doping is one of several factors closely correlated with detector performance as it influences minority carrier lifetimes, concentrations, and diffusion lengths. Superlattices with background doping densities of $3 \times 10^{15} \, cm^{-3}$ n-type can now be routinely grown by the procedures described in Sec. 2 [64]. Typical carrier concentrations and mobilities are shown in Fig. 24, as a function of temperature. While fabrication of superlattices with this level of unintentional doping is reproducible and not overly sensitive to the choice of specific constituent layers (suggesting that samples are not greatly compensated), background doping is found to vary somewhat with many parameters, including interfacial stoichiometry, layer thicknesses, and structural quality as inferred from RHEED patterns and x-ray diffraction. As the last two factors are difficult to deconvolve (superlattices employing thick GaInSb layers, especially, seem to be of slightly poorer structural quality) and interfacial stoichiometry is hard to probe directly, it has not yet been possible to definitively identify the

---

[5]Such a test is at present only qualitative, as calculations of STM image contrast require integration over many additional states beyond those at the zone center modeled to date.

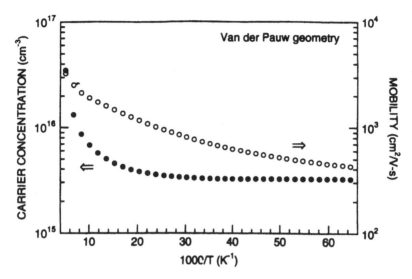

Fig. 24 Typical residual (unintentional) n-type background doping and mobility in a 39 Å/25 Å $Ga_{0.75}In_{0.25}Sb/InAs$ superlattice, derived from a van der Pauw geometry Hall measurement.

source of background doping. Further complications arise from the extensive use of semi-insulating GaAs substrates for in-plane electrical studies. While the inevitable $10^{16} - 10^{17} \, cm^{-3}$ doping of bulk grown GaSb precludes most such measurements on these lattice-matched substrates, growth on GaAs results in high dislocation densities which may alter the electrical characteristics of epilayers, as they do the optical properties [28].

Superlattices grown with GaInAs-like interfaces, rather than those employing 5 sec Sb soaks to achieve InSb-like interfaces, have been found to display higher levels of background doping. Unfortunately, interpretation of this observation is not singular, as superlattices grown with these interfaces typically exhibit a slight degradation of structural quality, which has been found to raise doping densities irrespective of other factors[6]. The n-type background doping characteristic of the superlattices suggests that GaInSb layers are not the dominant source of extrinsic carriers, as antimonides are usually

[6]This is not necessarily to suggest that poorer structural characteristics are a cause of increased doping; it is possible that both are symptoms of changes in growth temperatures or beam fluxes.

accompained by a variety of native p-type structural defects. While doping in GaSb is known to be in the range of 10^{15}–10^{16} cm^{-3} (p-type) at typical superlattice growth temperatures, it has not been possible to grow thick $Ga_{0.75}In_{0.25}Sb$ epilayers of good morphology at this temperature. InAs epilayers typically display n-type background doping in the 10^{15}–10^{16} cm^{-3} range. Associating background doping with interface states, native defects in constituent layers, or impurities, is the subject of ongoing research.

Finally, background carrier concentrations have been observed to drop during the first ten to twenty growths after closing and baking an MBE machine, presumably due to continued outgassing of residual contaminants in the sources and/or machine after bakeout. Such an observation is common to many MBE grown device materials [65].

Carrier Mobilities

Studies of in-plane carrier transport in GaInSb/InAs superlattices have yielded mobilities slightly lower than those of HgCdTe, but wholly adequate for IR detectors. To our knowledge, vertical mobilities have not yet been measured, although experimental data allow upper and lower bounds to be identified.

Results of two sets of experiments suggest that perpendicular carrier mobilities are less than approximately 830 cm^2/Vs at 4.2 K [66], and higher than 40 cm^2/Vs in the case of electrons at room temperature [28]. These results are derived from superlattices with parameters close to those of a 25 Å / 39 Å $Ga_{0.75}In_{0.25}Sb/InAs$ 10 μm "baseline" structure studied extensively. The upper bound is determined from cyclotron resonance experiments, which failed to yield peaks that could be resolved for electron and hole orbits out of the plane of the superlattice layers (i.e., Voigt, rather than Faraday, geometry). Assuming that the criterion for resolution of the two peaks is $\mu B \geqslant 1$, an upper bound of $\mu_\perp \leqslant 830$ cm^2/Vs (4.2 K) is derived, since a 12T field was used in the measurement. The lower limit $\mu_{e,\perp} > 40$ cm^2/Vs (300 K) is inferred from the dominance of lateral resistivities over vertical in small mesa devices. As vertical mobility influences the quantum efficiency of a mesa photodiode, more precise measurement of this parameter is currently being undertaken by several groups. Minority carrier mobilities of several hundred cm^2/Vs are typical of HgCdTe detector material, but somewhat lower values would yield equivalent diffusion lengths in the superlattice if higher minority carrier lifetimes were realized.

In-plane electron mobilities in these superlattices have been found to be significantly higher than those in other thin-layer heterostructures. Dependence of this mobility on InAs layer thickness is illustrated in Fig. 25 while in-plane transport in GaAs/AlGaAs [67], HgTe/CdTe [68], and InAs/AlSb [69] quantum well structures drops precipitously for thin wells (roughly as $\mu \propto d^6$), it is apparent from the figure that electron mobilities of almost 10^4 cm^2/Vs have been observed in GaInSb/InAs superlattices with layers less than 40 Å thick [57]. This is due to differences in the effectiveness of interface roughness scattering as a mobility limiting mechanism in each of the narrow layer structures. Although dominant in each case, the scattering potential is weak in GaInSb/InAs superlattices due to poor confinement of electrons in the InAs layers.

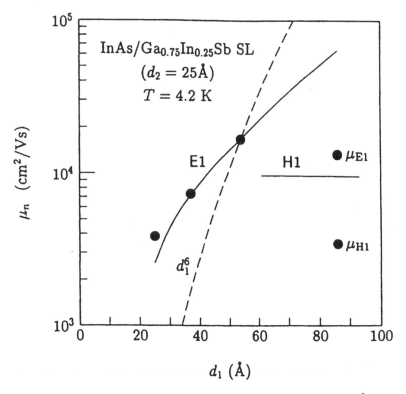

Fig. 25 In-plane electron mobilities for superlattices consisting of 25 Å thick Ga$_{0.75}$In$_{0.25}$Sb layers and InAs layers of variable thickness (d_1). The comparatively high mobilities are well described by a weak interface roughness scattering mechanism, except in the semimetallic sample ($d_1 = 86$Å). From Ref. [64].

Results of detailed band structure calculations (indicated by the solid line in the figure) reveal a much weaker dependence on layer thickness than the classical d^6 result, in reasonable agreement with experiment. In addition to showing the correct functional form, calculations agree quantitatively with the data when interfacial scattering sites are assumed to be approximately 100 Å in lateral extent. Fourier transforming real space STM images of GaSb/InAs superlattices reveals a high frequency component to interfacial roughness, and confirms the presence of a low-frequency peak corresponding to islands roughly 100 Å wide [70].

Last, while perhaps not critical to operation of detectors, it should be mentioned that there is evidence for localized electronic states and for anisotropic in-plane transport in some GaInSb/InAs superlattices. Localization is inferred from large oscillatory components of the Hall conductivity, which exceed Shubnikov-de Haas oscillations in magnitude for some structures with thin InAs layers [59]. Pronounced transport anisotropies have been observed at low temperatures, in superlattices with GaInSb layers thicker than are typical for LWIR detectors [28].

Effective Masses

Unusually high perpendicular effective masses and a qualitatively different band structure from that of HgCdTe are the basis of predicted reductions in tunneling currents in GaInSb/InAs superlattice detectors, as described in Sec. 1.3. Perpendicular effective masses have not yet been measured in the superlattices, but observed in-plane masses are in good agreement with theory. Cyclotron resonance measurements have yielded an in-plane electron effective mass $m_{e,\parallel}^* \approx 0.03\, m_e$ in a 24 Å/38 Å $Ga_{0.71}In_{0.29}Sb$/InAs superlattice with a gap of 10 μm [71,72], in agreement with results of an 8-band $\mathbf{k}\cdot\mathbf{p}$ calculation incorporating strain [28]. Band-to-band magnetoabsorption measurements on similar superlattices yield reduced effective masses $m_{r,\parallel}^* \approx 0.013 - 0.015\, m_e$, from which in-plane hole masses are inferred to be equal to those of electrons, within experimental error [71]. Further experiments have demonstrated an in-plane electron effective mass $m_{e,\parallel}^* \approx 0.017\, m_e$ in a 25 Å/54 Å $Ga_{0.75}In_{0.25}Sb$/InAs superlattice with a cutoff of approximately 24 μm [73]. This, too, is in agreement with calculation. The in-plane hole mass has not been probed in this structure.

As described in the previous section, failure to observe cyclotron resonance absorption lines yielding perpendicular masses is an indication of low growth-direction mobilities in the handful of superlattices examined to date [66]. As poor perpendicular transport of electrons likely reflects an effective scattering process or fluctuation-induced localized states, it is not possible to make inferences about the nature of the extended Bloch states based on results of these measurements. More specifically, there is at present no reason to believe that calculations of suppressed tunneling derived from increased perpendicular effective masses are in error. However, while the marriage of theory and experiment for the in-plane case increases confidence in calculated vertical masses, further measurements are needed to address this issue.

4. DETECTORS BASED ON GaInSb/InAs SUPERLATTICES

In this section we summarize GaInSb/InAs processing techniques employed to date and data from initial devices. Detector performance is examined in the light of measured materials parameters, and improved device designs are proposed. As explained in Sec. 1, suppression of Auger and tunneling processes should lead to GaInSb/InAs superlattice detectors with operating temperatures and/or detectivities appreciably exceeding those of HgCdTe. Such benefits have not yet been realized. Indeed, although the performance of initial devices exceeds that of most QWIPs, superlattice detectors competitive with conventional devices have not yet been demonstrated. However, analysis of available data reveals that recent breakthroughs in material quality and better heterojunction device designs make feasible significantly improved detectors. Experiments to demonstrate these gains are currently in progress.

4.1. Device Fabrication

Several means of fabricating mesa and lateral photovoltaic and photoconductive devices have been explored to date. Both wet and dry processes have been investigated, and ion implantation has been briefly examined for the purpose of pixel delineation in planar devices and as an alternate doping scheme. A number of mesa fabrication techniques currently yield devices of equivalent (material-limited) performance. Here we briefly summarize the available data and the apparent strengths and pitfalls of each approach.

Dry processing has been found to yield mesas with excellent vertical sidewalls, but devices fabricated in this manner display zero bias resistances slightly lower than those prepared by wet etching [74].

However, it is possible that this approach may prove fruitful in the future if followed by a light wet etch. It has been found that a CH_4/H_2 plasma etch, at a ratio of 0.5, a total pressure of 50 mTorr, and a total RF power of 150W is nearly optimal for achieving planar etched surfaces with vertical sidewalls. A two layer aluminum on silicon nitride mask has been found to be effective for avoiding both charging of the sample surface during etching and metal-semiconductor alloying.

Wet etching has more routinely been used to define mesas. Two etches have been utilized, typically employing Al or Au contacts as masks: $Br_2:HBr:H_2O$ (0.5:100:100) and citric acid $(1M):H_2O_2:H_3PO_4:H_2O$ (110:10:2:440). The first etch leaves a rough surface and results in trenching around mesas. For these reasons we have found the latter to be preferable, although etch rates are sometimes found to vary appreciably from one sample to another.

Photojunctions with silicon nitride and silicon dioxide passivating layers have been fabricated, but the utility of these layers is unclear at present. Specifically, areal scaling of the zero bias resistances of each set of devices examined so far suggests that even unpassivated devices are not limited by surface leakage. This is illustrated in Fig. 26, which shows measured $I-V$ characteristics for some of the first unpassivated diodes fabricated, varying in diameter between 5 and 30 μm. Should passivation prove an issue for this class of detectors, it is clear that there are many more candidate insulator and semiconductor passivating layers to be examined.

Doping of GaInSb/InAs superlattices through ion implantation of Si has been examined briefly [74]. Experiments on a 2500 Å thick superlattice show almost ideal n-type activation for a 4×10^{12} cm^{-2} implant after a 30s rapid thermal anneal at 450°C, but an excess of donors at 500°C. However, data at higher implantation doses reveal p-type doping after a 450°C heat treatment, becoming n-type only after a 500°C anneal. These studies have not been carried further. While implant doping of these superlattices with Be or Te may ultimately prove feasible, the amphoteric nature of Si in GaInSb of appreciable In composition makes this an uncertain choice of dopant. Further, it seems possible that implantation of any sort may be complicated by the propensity of the antimonides to form electrically active defects.

4.2. Detector Characteristics

Little of the work on GaInSb/InAs superlattices has been devoted to devices. This is primarily because extrinsic nonradiative recombination

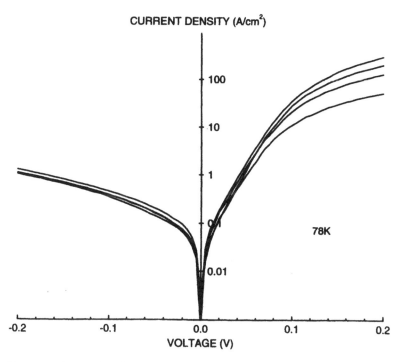

Fig. 26 Scaling with device area of the $I-V$ characteristics of a GaInSb/InAs p-i-n diode. Devices with diameters of 5, 10, 15, and 30 μm were tested. The results indicate that zero bias resistance is dictated by bulk junction properties, rather than surface shunting.

mechanisms have until recently been observed to greatly reduce photoexcited carrier lifetimes. So long as this condition persisted, there was little hope of achieving devices competitive with HgCdTe based detectors. This situation has recently changed, with the demonstration of radiatively dominated recombination in some structures [53]. However, devices fabricated using these improved growth techniques have not yet been tested. In light of this, we choose only to summarize detector data gathered to date, coming from superlattices known to be of inferior optical quality, and emphasize instead performance limiting mechanisms and prospects for improvement. This analysis is largely a straightforward extension of the work of Kinch and Borrello, which those authors applied to HgCdTe [4].

Demonstrated Device Performance

Unpassivated lateral photoconductors have been fabricated by several groups to determine the dependence of superlattice energy gaps on layer thicknesses and compositions [43,61]. However, detectivity data have been obtained only for photovoltaic devices. Both heterojunction [75] and homojunction [76] devices have been fabricated and analyzed. While values of R_0A as high as 75–400 Ω-cm^2 at 78 K have been reported in heterojunction detectors with 8.5 μm cutoff wavelengths, it is unlikely that these values are meaningful detector figures of merit in these structures, due to the presence of transport blocking heterojunctions between the narrow gap superlattices and wider gap contact layers in the devices examined to date. In particular, *n*-type superlattices were surrounded by *n*-type GaSb layers and *n*-type InAs layers. Consideration of the band alignments in this system reveals that barriers to the transport of electrons likely exist at the superlattice/InAs interfaces, in addition to the GaSb/superlattice *p-n* junctions.

R_0A values in homojunction devices were initially reported to be approximately 0.1 Ω-cm^2 at 78 K for 10–12 μm cutoff wavelengths. More recent results indicate small improvements in detectivities coincident with reduction in background doping densities and slight increases in photoluminescent efficiencies. In particular, values of R_0A of approximately 1 Ω-cm^2 are now more typical for cutoff wavelengths in excess of 10 μm. However, detectivities of devices with shorter cutoff wavelengths show little improvement, remaining significantly lower than those of HgCdTe. Analysis of a 5×5 miniarray with a cutoff wavelength of 7.5 μm showed an average R_0A of 5 Ω-cm^2 at 78 K. While a bimodal distribution of impedances suggested that not all pixels had been etched beyond the junction, the impedances of even the best devices ($\approx 25\Omega$-cm^2) were sufficiently low that this point was not investigated further. Fig. 27 illustrates $I-V$ and $R-V$ characteristics of one of these 50 μm × 50 μm pixels. A quantum efficiency of approximately 10% has been estimated for these devices, based on measurements made by illuminating through a Si fanout. This was the intended efficiency of the test device, based on a 2000 cm^{-1} absorption coefficient and an active region 0.5 μm thick. (While the ease with which thin devices are fabricated and processed makes them desirable for test purposes, much thicker active regions are readily grown by MBE, making possible devices with near unity LWIR absorption.)

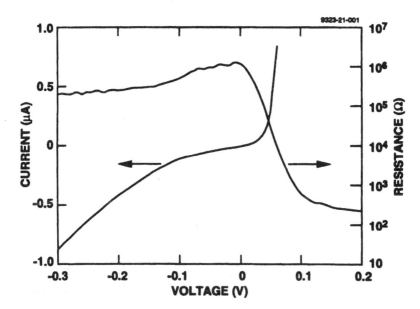

9323-21-001

Fig. 27 Typical *I–V* and *R–V* characteristics of a p-i-n homojunction superlattice detector. Data are from a 50 μm × 59 μm mesa device, at a temperature of 78 K. From Ref. [76].

Translating the above zero bias resistances into thermally limited detectivities

$$D^* = \frac{q\eta}{2h\nu}\sqrt{\frac{R_0 A}{kT}}$$

yields 78 K detectivities D^* in excess of 1×10^{10} cm$\sqrt{\text{Hz}}$/W at wavelengths beyond 10 μm, assuming the previously quoted values of quantum efficiency η to be correct. We note in passing that no direct measurement of detectivity has yet been made for a GaInSb/InAs diode, to our knowledge. However, in view of the simplicity of the homojunction devices and the observed independence of $R_0 A$ on mesa size and contact area, it is reasonable to assume that measured impedances are indeed those of the relevant photojunctions.

In consideration of the magnitude of homojunction zero bias resistances measured to date and the short lifetimes characteristic of these superlattices, it is likely that the performance of past devices has been generation-recombination limited. Per Kinch and Borrello [41], the zero bias resistance in this regime is given by[7]

$$R_0 A = \frac{\tau_0 E_g}{q^2 n_i W},$$

where τ_0 reflects an average carrier capture time, E_g is the energy gap, n_i is the intrinsic carrier concentration, and the depletion width $W = \sqrt{2\varepsilon\varepsilon_0 V_b / n_0 q}$. Considering the case of the well studied 10 μm superlattice 25 Å/39 Å $Ga_{0.75}In_{0.25}Sb/InAs$, for which $\varepsilon \approx 12.25$ and $n_i \approx 3 \times 10^{12} cm^{-3}$ at 77 K,

$$R_0 A = 200 \tau_0 \sqrt{n_0}.$$

For a carrier concentration of $3 \times 10^{15} cm^{-3}$ and an observed $R_0 A \approx 1\Omega\text{-cm}^2$, this yields a capture lifetime τ_0 of approximately 90 ps. Similar analyses of the 7.5 μm and 12 μm junctions examined to date (for which $n_i \approx 2 \times 10^{11} cm^{-3}$ and $2 \times 10^{13} cm^{-3}$, respectively, at 77 K) yield lifetimes of 80 ps and 50 ps.

Unfortunately, no direct determinations of lifetime have been made in superlattices grown under the conditions used for these detector structures, in part because the lifetimes have been found to be too short to be measured by conventional means. It is likely from the absence of photoluminescence from these superlattices that $\tau \ll 1$ ns, since band-to-band luminescence was fairly readily detected from a 10 μm sample with a 5 ns lifetime. While it is not possible to be more quantitative at present, the 50–90 ps values quoted here are entirely plausible. We note that these lifetimes would yield a diffusion length of the order of 2500 Å at 77 K in the 10 μm superlattice. This value is not inconsistent with the measured quantum efficiency of 10% and absorption coefficient of about 2000 cm^{-1}, in view of the approximate nature of these numbers and the inequivalence, strictly speaking, of the relevant diffusion and depletion region lifetimes.

It is less likely that the performance of past devices has been tunneling or diffusion limited. In the latter case [4],

$$R_0 A = \frac{kT \tau_p}{q^2 p_n L_p},$$

where τ_p, p_n, and L_p represent minority carrier lifetimes, concentrations, and diffusion lengths, respectively ($L_p = \sqrt{(kT/q)\mu_h \tau_p}$). Even in

---

[7] Note that a factor of q was ommited from the denominator of Eq. (43) in Ref. 4.

the worst plausible case (e.g., $\tau_p = 10$ps, $\mu_2 = 100$ cm^2/V-s, $\mu_e = 1000$ cm^2/V-s), the zero bias resistances in such a regime appreciably exceed measured values. Likewise, we consider tunneling unlikely to dominate at these wavelengths owing to the appreciable depletion length at the doping levels employed here and the significant reduction in tunneling relative to HgCdTe afforded by the experimentally confirmed superlattice band structure.

Design Considerations

It has not yet been necessary to address the issue of surface passivation in GaInSb /InAs based detectors. While successful approaches have been developed for InSb, these and other schemes remain untested in this system as junction impedances have been low for reasons related to materials. In the following analysis we ignore passivation as there seems little that can be contributed theoretically and as there is as yet no relevant data on the subject. However, we mention this issue as discussion of these detectors would be incomplete without acknowledging the considerable uncertainty it introduces.

Assuming that surface shunting does not limit diode performance, detectivity can be presumed to rise rapidly with improved material lifetime. In particular, $R_0 A$ scales as τ_0 in the generation-recombination limited regime, then as $\tau^{1/2}$ in a diffusion limited regime (assuming the sample thickness to be greater than the diffusion length). As mentioned in Sec. 1.3, detectors based on this class of superlattices should benefit appreciably from suppressed Auger recombination. However, since an intrinsic upper limit on the superlattice lifetime is imposed by radiative processes, the Auger advantage is realized in practice by raising the majority carrier concentration to fairly high levels without reducing the lifetime. The advantage in this comes from reducing the background of minority carriers against which photoexcited carriers must be discriminated. Simple scaling laws strictly applicable only for nondegenerate statistics reveal that maximum detectivity is achieved when the carrier concentration is raised to a level for which the two lifetimes are equal, $\tau_r \approx \tau_A$.

Both recombination and transport mechanisms favor use of p-type superlattice active layers. Particularly in the LWIR and VLWIR spectral regions, suppression of Auger-1 processes at low temperatures is small for n-type materials in which the absorption coefficient is appreciable. By contrast, minigaps are readily introduced in the growth direction dispersion relation dominant in p-type material. Further, the poor growth direction mobility of holes [77] would severely limit the quantum efficiency of many n-type devices, but serves to suppress leakage currents from the highly doped side of the juction in detectors with p-type active layers. The limit this leakage current places on the detectivity can be inferred from measured parameters. Specifically, τ has been measured to be approximately 1ns in a 10 μm superlattice with 10^{18} cm^{-3} free carriers. Making the worst case assumptions that this corresponds to the Auger lifetime for 10^{18} cm^{-3} doped n-type material and that the hole mobility is 100 cm^2/V-s we derive an upper bound of $R_0 A \leqslant 1.7 \times 10^5 \Omega$-cm^{-2} for a 10 μm photovoltaic operating at 77 K.

Detectivities for diffusion limited diodes have already been shown in Fig. 4. To illustrate explicitly the arguments outlined above, we consider again the case of the 10 μm cutoff superlattice, for which the diffusion limit yields

$$R_0 A = 5 \times 10^6 n_0 \sqrt{\tau_n}.$$

Considering the case of 10^{17} cm^{-3} p-type material, for which the Auger lifetime is still well above the radiative lifetime of 4μs, we arrive at $R_0 A = 3.4 \times 10^5 \Omega$-cm^2. We note that in this case the impedance might be dominated by the more highly doped n-type side. Should such limits ever be achieved in practice, use of a suitable heterojunction such as an InAs/AlSb superlattice on the highly doped side of the junction would in principle eliminate leakage from this side of the device.

Finally, while surface shunting may limit the performance of devices, we consider it unlikely that tunneling will constrain doping levels and performance to those practical for HgCdTe. The magnitude of the imaginary wavevector in the superlattice gap can be expected to greatly reduce both direct band-to-band tunneling and tunneling through intermediate states.

5. CONCLUSIONS

Development of detectors based on $Ga_{1-x}In_xSb/InAb$ superlattices is at a pivotal point. Many of the intrinsic properties predicted to favor these structures have now been confirmed experimentally. Despite this, 78 K detectivities of superlattice-based LWIR photovoltaic devices examined to date are only slightly in excess of 10^{10} cm $\sqrt{Hz}$/W. While this value exceeds the detectivities of 8–10 μm QWIP detectors

[78] operating near their theoretical limits [3], it is not yet competitive with HgCdTe devices. It is possible that producibility arguments and niche applications will favor use of the III–V superlattices in some cases, even at the current level of detectivity, as is being argued for the MQW detectors. However, at this time there is real hope for a period of rapid improvement in superlattice performance, owing to the great improvement in optical quality of the latest structures. With maturation of the materials it is probable that the emphasis will shift to development of surface passivation techniques, which has not yet proven necessary. Progress in this area will likely determine whether detectors surpassing HgCdTe can be demonstrated in the near future. Barring problems with passivation it seems probably that GaInSb /InAs-based detectors could fill many needs spanning tactical MWIR to low background VLWIR applications.

Acknowledgments

The authors are grateful to numerous colleagues for stimulating discussions and insights. Special thanks are due T.C. McGill, H. Ehrenreich, C.H. Grein, J.R. Meyer, and D.L. Smith. Parts of this work were supported under DARPA/ONR Contracts Nos. N00014-89-C-0203 and N00014-92-C-0228.

References

1. D.L. Smith and C. Mailhiot, *J. Appl. Phys.*, **62**, 2545 (1987).
2. C. Mailhiot and D.L. Smith *J. Vac. Sci. Technol., A*, **7**, 445 (1989).
3. M.A. Kinch and A. Yariv, *Appl. Phys. Lett.*, **55**, 2093 (1989) and **56**, 2354 (1990).
4. M.A. Kinch and S.R. Borrello, *Infrared Physics*, **15**, 111 (1975).
5. P.M. Young, C.H. Grein, H. Ehrenreich, and R.H. Miles, *J. Appl. Phys.*, **74**, 4774 (1993).
6. R.L. Whitney, K.F. Cuff, and F.W. Adams, in *Semiconductor Quantum Wells and Superlattices for Long-Wavelength Infrared Detectors*, edited by M.O. Manasreh (Artech House, Boston, 1993) pp. 55–108.
7. O.K. Wu, private communication.
8. G.A. Sai-Halasz, R. Tsu, and L. Esaki, *Appl. Phys. Lett.*, **30**, 651 (1977).
9. J.N. Schulman and T.C. McGill, *Appl. Phys. Lett.*, **34**, 663 (1979).
10. D.L. Smith, T.C. McGill, and J.N. Schulman, *Appl. Phys. Lett.*, **43**, 160 (1983).
11. J.P. Baukus, A.T. Hunter, J.N. Schulman, T.C. McGill, and J.P. Faurie, *J. Appl. Phys.*, **64**, 283 (1988).
12. S. Jost, presented at the *JPL Innovative Long Wavelength Infrared Detector Workshop*, Pasadena, April 1992 (unpublished).
13. L.L. Chang and L. Esaki, *Surf. Sci.*, **98**, 70 (1980).
14. L.L. Chang, N.J. Kawai, G.A. Sai-Halasz, R. Ludeke, and L. Esaki, *Appl. Phys. Lett.*, **35**, 939 (1979).
15. D.K. Arch, G. Wicks, T. Tanoue, and J.-L. Staudenmann, *J. Appl. Phys.*, **58**, 3933 (1985).
16. G.C. Osbourn, *J. Vac. Sci. Technol., A*, **2**, 176 (1984).

17. S.R. Kurtz, G.C. Osbourn, R.M. Biefeld, L.R. Dawson, and H.J. Stein, *Appl. Phys. Lett.*, **52**, 831 (1988).
18. I.T. Ferguson, A.G. Norman, B.A. Joyce, T.-Y. Seong, G.G. Booker, R.H. Thomas, C.C. Phillips, and R.A. Stradling, *Appl. Phys. Lett.*, **59**, 3324 (1991).
19. S.R. Kurtz, L.R. Dawson, R.M. Biefeld, D.M. Follstaedt, and B.L. Doyle, *Phys. Rev.*, *B*, **46**, 1909 (1992).
20. S.R. Kurtz, presented at the *Quantum Structures for Infrared Applications Symposium, Electrochemical Society Meeting*, New Orleans, October 1993 (unpublished).
21. S.R. Kurtz, L.R. Dawson, T.E. Zipperian, and R.D. Whaley, *IEEE Electr. Dev. Lett.*, **11**, 54 (1990).
22. C.H. Grein, H. Cruz, M. Flatte, and H. Ehrenreich, submitted to *Appl. Phys. Lett.*
23. R.H. Miles, J.N. Schulman, D.H. Chow, and T.C. McGill, *Semicond. Sci. Technol.*, **8**, S102 (1993).
24. D.L. Smith, unpublished.
25. C.H. Grein, P.M. Young, and H. Ehrenreich, *Appl. Phys. Lett.*, **61**, 2905 (1992).
26. C.H. Grein, M. Flatte, P.M. Young, and H. Ehrenreich (in preparation).
27. J.A. Bailey, private communication.
28. R.H. Miles, unpublished.
29. L. Esaki, in *The Technology and Physics of Molecular Beam Epitaxy*, edited by E.H.C. Parker (Plenum, New York, 1985), Chapter 6.
30. H.K. Choi, S.J. Eglash, and G.W. Turner, *Appl. Phys. Lett.*, **64**, 2474 (1994).
31. H. Q. Le, G.W. Turner S.J. Eglash, H.K. Choi, and D.A. Coppeta, *Appl. Phys. Lett.*, **64**, 152 (1994).
32. E.R. Brown, J.R. Söderström, C.D. Parker, L.J. Mahoney, K.M. Molvar, and T.C. McGill, *Appl. Phys. Lett.*, **58**, 2291 (1991).
33. K. Yoh, T. Moriuchi, and M. Inoue, *IEEE Electron Device Lett.*, **11**, 526 (1990).
34. J.D. Werking, C.R. Bolognesi, L.D. Chang, C. Nguyen, E.L. Hu, and H. Kroemer, *IEEE Electron Device Lett.*, **13**, 164 (1992).
35. X. Li, K.F. Longenbach, Y. Wang, and W.I. Wang, *IEEE Electron Device Lett.*, **13**, 192 (1992).
36. D.H. Chow, R.H. Miles, C.W. Nieh, and T.C. McGill, *J. Cryst. Growth*, **111**, 683 (1991).
37. B.V. Shanabrook, J.R. Waterman, J.L. Davis, R.J. Wagner, and D.S. Katzer, private communication.
38. M. Yano, K. Yamamoto, T. Utatsu, and M. Inoue, *J. Vac. Sci, Technol.*, *B*, **12**, (1994).
39. B.M. Clemens and J.G. Gay, *Phys. Rev.*, *B*, **35**, 9337 (1987).
40. D.H. Chow, R.H. Miles, and A.T. Hunter, *J. Vac. Sci. Techonl.*, *B*, **10**, 888 (1992).
41. H. Kroemer, T.Y. Liu, and P.M. Petroff, *J. Cryst. Growth*, **95**, 96 (1989).
42. P.M. Petroff, *Semiconductors and Semimetals*, *Vol.22*, *part A* (Academic, Orlando, 1985), Chap. 6, p. 379.
43. R.H. Miles, D.H. Chow, and W.J. Hamilton, *J. Appl. Phys.*, **71**, 211 (1982).
44. D.A. Reich, A.M. Womchak, P.P. Chow, and J.M. Van Hove, submitted to *J. Cryst. Growth*.
45. D.A. Collings, T.C. Fu, T.C. McGill, and D.H. Chow, *J. Vac. Sci., and Technol. B*, **10**, 1779 (1992).
46. R.M. Feenstra, D.A. Collins, D.Z.-Y. Ting, M.W. Wang, and T.C. McGill, submitted to *J. Vac. Sci., Technol., B*.
47. A.Y. Lew, E.T. Yu, D.H. Chow, and R.H. Miles, *Apppl. Phys. Lett.*, **65**, 201 (1994).
48. G. Tuttle, H. Kroemer, and J.H. English, *J Appl. Phys.*, **67**, 3032 (1990).
49. I. Sela, I.H. Campbell, B.K. Laurich, D.L. Smith, L.A. Samoska, C.R. Bolognesi, A.C. Gossard, and H. Kroemer, *J. Appl. Phys.*, **70**, 5608 (1991).
50. J.R. Waterman, B.V. Shanabrook, R.J. Wagner, M.J. Yang, J.L. Davis, and J.P. Omaggio, *J. Vac. Sci, Technol.*, to be published.

51. M.W. Wang, D.A. Collins, T.C. McGill, and R.W. Grant, *J. Vac. Sci. Technol.*, *B*, **11**, 1418 (1993).
52. T.M. Rossi, D.A. Collins, D.H. Chow, and T.C. McGill, *Appl. Phys. Lett.*, **57**, 2256 (1990).
53. R.H. Miles, D.H. Chow, Y.-H. Zhang, and P.D. Brewer, submitted to *Appl. Phys. Lett.*
54. R.H. Miles, D.H. Chow, J.N. Schulman, and T.C. McGill, *Appl. Phys. Lett.*, **57**, 801 (1990).
55. P. Voisin, G. Bastard, C.E.T. Goncalves da Silva, M. Voos. L.L. Chang, and L. Esaki, *Solid State Commun.*, **39**, 79 (1981).
56. See, for example, R.H. Miles, D.H. Chow, and T.C. McGill, *SPIE.* **1285**, 132 (1990).
57. C.A. Hoffman, J.R. Meyer, E.R. Youngdale, F.J. Bartoli, and R.H. Miles, *Appl. Phys. Lett.*, **63**, 2210 (1993).
58. C.H. Grein, P.M. Young, and H. Ehrenreich, *Appl. Phys. Lett.*, **61**, 2905 (1992).
59. C.A. Hoffman, J.R. Meyer, E.R. Youngdale, F.J. Bartoli, and R.H. Miles, *Solid State Electronics*, **37**, 1203 (1994).
60. E.R. Youngdale, J.R. Meyer, C.A. Hoffman, F.J. Bartoli, C.H. Grein, P.M. Young, H. Ehrenreich, R.H. Miles, and D.H. Chow, *Appl. Phys. Lett.*, **64**, 3160 (1994).
61. I.H. Campbell, I. Sela, B.K. Laurich, D.L. Smith, C.R. Bolognesi, L.A. Samoska, A.C. Gossard, and H. Kroemer, *Appl. Phys Lett.*, **59**, 846 (1991).
62. J.R. Waterman, B.V. Shanabrook, R.J. Wagner, M.J. Yang, J.L. Davis, and J.P. Omaggio, *Semicond. Sci. Technol.*, **8**, S106 (1993).
63. A.Y. Lew, E.T. Yu, D.H. Chow, and R.H. Miles, submitted to *Proceedings of the Materials Research Society*.
64. R.H. Miles and D.H. Chow, presented at *JPL Innovative Long Wavelength Infrared Detector Workshop*, Pasadena, 1992.
65. C.T. Foxon, T.S. Cheng, P. Dawson, D.E. Lacklison, J.W. Orton, W. Van der Vleuten, O. H. Hughes, and M. Henini, *J. Vac. Sci. Technol.*, *B*, **12**, 1026 (1994).
66. M.J. Yang, R.J. Wagner, D.H. Chow, and R.H. Miles, unpublished.
67. H. Sakaki, T. Noda, K. Hirakawa, M. Tanaka, and T. Matsusue, *Appl. Phys. Lett.*, **51**, 1934 (1987).
68. J.R. Meyer, D.J. Arnold, C.A. Hoffman, and F.J. Bartoli, *Appl. Phys. Lett.*, **58**, 2523 (1991).
69. C.R. Bolognesi, H. Kroemer, and J.H. English, *Appl. Phys. Lett.*, **61**, 213 (1992).
70. R.M. Feenstra, D.A. Collins, D.Z.-Y. Ting, M.W. Wang, and T.C. McGill, *Phys. Rev. Lett.*, **72**, 2749 (1994).
71. J.P. Omaggio, J.R. Meyer, R.J. Wagner, C.A. Hoffman, M.J. Yang, D.H. Chow, and R.H. Miles, *Appl. Phys. Lett.*, **61**, 207 (1992).
72. J.P. Omaggio, R.J. Wagner, J.R. Meyer, C.A. Hoffman, M.J. Yang, D.H. Chow, and R.H. Miles, *Semicond. Sci. Technol.*, **8**, S112 (1993).
73. J.R. Meyer, E.R. Youngdale, C.A. Hoffman, F.J. Bartoli, R.H. Miles, and L.R. Ram-Mohan, *J. Electronic Materials*, to be published.
74. A.T. Hunter, D.H. Chow, and R.H. Miles, unpublished.
75. M.D. Jack, B. Baumgratz, G.R. Chapman, S.M. Johnson, K. Kosai, R.H. Miles, D.H. Chow, J. Johnson, L. Samoska, A.C. Gossard, and J.L. Merz, *Proc. IRIS Device Specialty Group* (1993).
76. R.H. Miles, D.H. Chow, M.H. Young, B.A. Baumgratz, G.R. Chapman, and M.D. Jack, *Proc. IRIS Materials Specialty Group* (1993).
77. E. Runge and H. Ehrenreich, private communication.
78. See, for example, L.J. Kozlowski, G.M. Williams, G.J. Sullivan, C.W. Farley, R.J. Anderson, J. Chen, D.T. Cheung, W.E. Tennant, and R.E. DeWames, *IEEE Trans. on Elector. Dev.*, **38**, 1124 (1991).

CHAPTER 8

Novel InTlSb Infrared Detectors

MANIJEH RAZEGHI

Center for Quantum Devices, Department of Electrical Engineering and Computer Science, Northwestern University, Evanston, Illinois 60208 USA

1. INTRODUCTION

Infrared detectors and imaging arrays operating at wavelengths in the 8–12 μm atmospheric windows have attracted considerable interest because of their wide range of applications in areas such as infrared missile seeker-tracer systems, space surveillance, and medical imaging. HgCdTe has been the dominant material system for such applications. HgCdTe offers the freedom of tailoring its bandgap all the way from the HgTe ($-0.30\,eV$) value to the CdTe value ($1.60\,eV$) by simply varying the Hg/Cd ratio within the crystal. However, this II–VI compound is notorious for metallurgical problems: the weak Hg-Te bonding restricts the strength of the material, and the high Hg vapor pressure makes it difficult to grow alloys with uniform composition over large substrate areas and is also the reason for poor thermal stability.

A variety of novel material systems and structures based on III–V semiconductors have been proposed as alternatives. Among the III–V binary semiconductors, InSb has the lowest bandgap of 0.17 eV at 300 K. It is currently the preferred material to $Hg_{0.69}Cd_{0.31}Te$ for the detection in the mid-infrared [1] because of its superior mechanical strength. However, the bandgap of InSb is too wide for $8-12\,\mu m$ infrared region and so as an alternative InAsSb has been considered. $InAs_{1-x}Sb_x$ has a maximum cut-off wavelength of about $10\,\mu m$ at 77 K ($x = 0.65$) [2] and to realize detectors that can span the entire $8-12\,\mu m$ range, InAsSb/InSb strained-layer superlattices (SLS) have been proposed [3]. In this concept the infrared response of small bandgap component of the SLS, InAsSb, can be extended to longer wavelength by the tensile strain-induced decrease of its bandgap. Detectors have been demonstrated using these InAsSb/InSb SLS [4] but the strain-induced material problems are expected to limit their further development. GaInSb/InAs strained type II superlattice has also been considered as a promising structure for infrared detection [5]. Type II superlattice is distinguished by its staggered band alignment, in which the conduction band of one of its constituent material is lower in energy than the valence band maximum of the other [6,7]. Because of this alignment, type II superlattices have a bandgap smaller than that of either of the constituent materials. However, also as a result of this energy band alignment electrons and holes are localized in adjacent layers. This leads to the problem of weak electron-hole wave function overlap for thick layer thicknesses in the superlattice. In GaInSb/InAs superlattice, thin InAs and GaInSb layers can produce the required superlattice bandgap and consequently, large optical absorption is possible in this system [5]. Nevertheless, practical GaInSb/InAs detector performance has not been demonstrated due to material problems such as high background doping density and strain induced defects.

Among the proposed III–V photodetectors, the most dramatic progress have been made by AlGaAs/GaAs multiquantum well (MQW) detectors [7,8]. Focal plane arrays and discrete detectors have already been demonstrated using this material system. The detection mechanism is based on the electron intersubband transition between the ground state and the first excited state of the quantum wells. The energy separation between these levels is determined by the quantum well width and barrier height. One major drawback of MQW detectors is the necessity of an oblique incident angle for optical absorption. A number of solutions, including the use of grating [9] and valence intersubband transition [10], have been proposed. However, none of these novel approaches have yet proven to be superior to HgCdTe.

Bulk III–V semiconductors utilizing direct interband optical transition have been investigated as possible infrared materials, capable of providing the flexibility of HgCdTe such as simpler device structures and higher quantum efficiency but with the added metallurgical and processing advantages offered by III–V technology. InSbBi [11,12] and InAsSbBi [13,14] have been considered as possible candidates but they are metastable and have never evolved into practical device technology. InTlSb has also been proposed as another infrared material system. Wood et al. first suggested that alloying InSb with thallium might dilate the crystal lattice and reduce the bandgap [15]. They attempted the growth of InTlSb alloys by molecular beam epitaxy but never succeeded in observing any change in the physical properties of InSb. Schilfgaarde et al. [16] performed band-structure calculations and predicted TlSb to be a semi-metal with a low-temperature bandgap of $-1.5\,eV$. $In_{1-x}Tl_xSb$ is thus expected to reach a bandgap of $100\,meV$ at $x = 0.083$ while exhibiting a similar lattice constant as InSb since the radius of Tl atom is very similar to In. Even though TlSb was expected to favor the CsCl-type structure, InTlSb alloy was estimated to exhibit a zinc-blende structure up to 20% Tl [16].

This chapter presents the physical properties of this new InTlSb alloy and the advancements made towards realizing InTlSb infrared photodetectors. Firstly, the growth and characterization of device quality InSb is reviewed as the preliminary step towards the growth of InTlSb. Next, the first successful growth of InTlSb alloys exhibiting extended infrared response is described. Detailed discussion of the various structural, electrical, and optical characterization results are also given to gain insight into the material properties of InTlSb. Finally, the first InTlSb photodetectors are demonstrated and their detectivities are evaluated to assess the merits of InTlSb as a novel III–V material for infrared detection.

2. GROWTH AND CHARACTERIZATION METHODS

2.1. Growth

Growth Technique

Metalorganic chemical vapor deposition (MOCVD) [17] is a non-equilibrium growth technique that has established itself as a unique and important epitaxial crystal growth technique. It is an ideal growth technique for InTlSb because of its success in growing new and meta-stable materials such as InAsBi and InAsBiSb [14]. In addition, MOCVD is attractive due to its ability to grow uniform layers of low background doping density and sharp interfaces, and for mass production.

The MOCVD reactor used for growing InSb based alloys consists of four major parts: gas cabinet, control panel, growth chamber and waste venting system. The gas cabinet is a network of electropolished stainless steel tubing, valves, pumping system, and gas flow controllers to transport the organometallic and hydride sources to the growth chamber. All the metalorganic sources are contained in stainless steel bubblers, which are held in temperature controlled baths. Mass flow controllers are used to control the flow rate of the Pd-diffused H_2 carrier gas with an accuracy of $\pm 1\%$ of the full scale. The control panel consists of gas flow controlling electronics, which are used to control the gas flow to either the growth chamber or vent. The growth chamber is a horizontal quartz version developed by Bass [18], in which the gas flow is parallel to the substrate surface. The substrate is placed on a SiC-coated graphite susceptor. The growth is carried out at a pressure of 76 Torr and radio frequency induced heating is used. A chromel-alumel thermocouple, placed below the susceptor, provides a feedback to the temperature controller and an infrared pyrometer is used as an independent second temperature monitoring system.

Substrates

For the growths of InSb and InTlSb, both InSb and semi-insulating GaAs substrates were used. An additional GaAs coated Si substrate was used in the study of InSb growth because of the possibility for future integration of infrared detection and signal processing circuits on the same substrate. Superclean GaAs and GaAs/Si substrates were directly loaded into the growth chamber without any preparation. During the heating of the substrate, a pregrowth arsine overpressure was maintained in order to prevent any surface degradation. InSb substrates were degreased using trichloroethylene, acetone and iso-propanol, and etched in lactic acid:nitric acid (10:1) mixture for 7 minutes.

2.2. Characterization Methods

In order to investigate the physical properties of as-grown films, a number of structural, electrical, and optical characterizations were carried out on a routine basis. Structural characterizations were performed on all the samples. However, the study of electrical and optical properties were limited to epilayers grown on semi-insulating GaAs substrate because of the requirements of the characterization measurements. For electrical characterization such as Hall measurements, semi-insulating GaAs does not contribute to electrical conduction while for optical transmission measurements, it provides a medium that is relatively transparent in the $3-12 \mu m$ wavelength region and has significantly different optical properties from those of InSb alloys. With the exception of a few InSb reference samples, the optical characterizations were further restricted to InTlSb alloys since the wavelength of interest is beyond the cutoff wavelength of InSb.

High Resolution X-ray Diffraction Measurement

This structural characterization provides in-depth information about the crystal structure. Full width at half maximum (FWHM) of the x-ray rocking curve is used as a measure of the epitaxial film quality. The x-ray curve peak angular position of any crystal can be used to evaluate its lattice constant a using Bragg's law. Thus the lattice mismatch $\Delta a/a$ of epilayers can also be evaluated from the difference in the x-ray peak angular positions corresponding to the epilayers and substrate. Combining x-ray diffraction measurements with Vegard's law concerning the linear variation of the solid solution composition with lattice constant, the composition of III–V ternary alloys can also be determined, if the lattice constant of its constituent binary alloys are known.

Ball-Polishing Thickness Measurement

Ball polishing measurement is a very quick method for determining the thickness of the epilayers. There are no required sample preparation and the system consists of a sample holder and a stainless ball bearing which is placed in a motor driven (rotated) ball bearing holder. The thickness is determined by milling a hole in the sample using the rotating ball bearing and a fine diamond paste, deep enough to reveal the boundaries of the epilayer of interest. The diameter of the epilayer boundaries can then be measured under an optical microscope and then the thickness of the epilayer can be calculated using

$$e = \frac{d_{ex}^2 - d_{in}^2}{8RG},\qquad(1)$$

where e is the actual epilayer thickness, R is the radius of the ball bearing, and G is the microscope magnification factor. d_{ex} and d_{in} are the measured exterior and interior diameter of the epilayer boundaries, respectively.

Hall Measurement

Hall measurement is the most common characterization method used to determine the electrical properties. For an uncompensated semiconductor the carrier mobility and carrier concentration can be estimated from the measured conductivity (σ) and the Hall coefficient (R_H) by using the relationships $\mu_H = \sigma R_H$ and $p = 1/(qR_H)$ (p-type), $n = -1/(qR_H)$ (n-type). In the case of mixed conduction where both electrons and holes are dominant in the transport, the Hall coefficient becomes sensitive to the magnetic field strength B and to the mobility ratio $b = \mu_n/\mu_p$. In this case the Hall coefficient is expressed as [19]

$$R_H = \frac{(p - b^2 n) + (\mu_n B)^2 (p - n)}{q[(p + bn)^2 + (\mu_n B)^2 (p - n)^2]}\qquad(2)$$

The Hall measurements are performed using the Van der Pauw method on clover-shaped samples. The ohmic contacts are made on the samples by alloying with In-Sn at 350°C under 10% H_2:90%N_2 ambient. The current and magnetic field directions are separately reversed to eliminate any Hall-probe misalignment.

Transmission/Absorption Measurements

Optical transmission measurements have been carried out using a Mattson Fourier Transform Infrared (FTIR) spectrometer operating in the 2.5 μm–25 μm wavelength range. It is equipped with a temperature-controlled cryostat that enables measurements down to 77 K. The transmission ($\mathfrak{T}$) of monochromatic radiation of incident intensity I_o (source) through an absorbing medium is given by

$$\mathfrak{T} = I/I_o = \exp(-\alpha L), \tag{3}$$

where I is the transmitted light intensity, L is the optical path length, and α is the absorption coefficient of the medium. α has units of reciprocal length, making αL a dimensionless quantity.

3. PRELIMINARY PHOTOCONDUCTORS

As preliminary photoconductors, the same clover-shaped samples employed in Hall measurements were used. The active area was measured to be approximately 4.4 mm^2 with 3 mm electrode spacing.

3.1. Photoconductivity Measurement

The photoconductor spectral response was obtained by comparing the photoconductor under test to the spectrometer internal pyroelectric detector, and by correcting for the slow frequency response of the pyroelectric detector. This correction is necessary since the photoconductors such as the ones to be characterized exhibit a flat frequency response but exhibit strong spectral variations as opposed to a flat spectral response and slow speed of response of the internal pyroelectric detector.

The absolute responsivity was determined as a function of bias by comparing the photoconductor under test to a calibrated HgCdTe photoconductor. Various bias voltages were applied using a circuit board that acts as both a dc-current source and an ac-preamplifier delivering an output signal compatible with the Fourier Transform input of the spectrometer. Such circuit board is the standard that is used in FTIR spectrometers equipped with HgCdTe photoconductors. When the bias is varied, the overall shape of the photoresponse spectrum does not change, but its magnitude changes.

3.2. Detectivity Estimation

The responsivity $\mathfrak{R}$ of a photoconductor is expressed by

$$\mathfrak{R} = \eta \frac{e}{h\nu} G, \tag{4}$$

where η is the absorption quantum efficiency, e the electron charge, $h\nu$ is the incident photon energy and G is the photoconductive gain. At low bias, the responsivity increases linearly with bias, because the photoconductive gain is a linear function of the bias as shown by the following relation

$$G = \frac{\mu\tau V_a}{d^2}, \tag{5}$$

where $\mu\tau$ is the carrier mobility-lifetime product, V_a is the applied bias ·voltage, and d is the spacing between the contacts. At higher bias, the responsivity saturates and eventually decreases because of sample heating. Similar behavior has been observed for HgCdTe photoconductors [20].

Neglecting Fabry-Perot oscillations for a first approximation, the quantum efficiency η is expressed as:

$$\eta = (1 - R)(1 - e^{-\alpha L}), \tag{6}$$

where R is the reflection coefficient at the front surface, α is the absorption coefficient, and L is the epilayer thickness. The reflection coefficient R can be calculated through

$$R = \frac{(n-1)^2}{(n+1)^2}, \tag{7}$$

where n is the refractive index of the epilayer. In the case of InTlSb, $n \approx 4$ which implies $R \approx 0.36$, assuming the refractive index of InTlSb is comparable to that of InSb. The absorption coefficient α values are collected from transmission measurements. Combining these values, the quantum efficiency can be estimated using Eq. (6).

From the slopes of the linear sections of the responsivity versus bias data and from the calculated quantum efficiencies, the gain per applied bias voltage (G/V_a) can be determined. With this information, the noise current i_n can be calculated. The two contributions to the detector noise are generation-recombination noise and Johnson noise

[21], so the total noise current is expressed by:

$$i_n = \sqrt{4eGi\Delta f + \frac{4kT}{r_d}\Delta f}, \qquad (8)$$

where Δf is the frequency interval over which the noise is measured, r_d is the photoconductor resistance, i is the average current flowing through the photoconductor ($i = V_a/r_d$), and kT is the thermal energy. The specific detectivity D^* is then given by:

$$D^* = \frac{\Re}{i_n}\sqrt{A}\sqrt{\Delta f}, \qquad (9)$$

where A is the photoconductor area.

4. GROWTH AND PROPERTIES OF HIGH QUALITY InSb EPITAXIAL FILMS

This section discusses the growth and properties of InSb epitaxial films. These are key steps for establishing the proper initial growth conditions for InTlSb as well as understanding the effects of thallium on the physical characteristics of InSb. Reproducible growth of high quality InSb films is also prerequisite for fabricating InSb/InTlSb/InSb p-i-n photodetectors.

4.1. Growth

Epitaxial techniques such as liquid phase epitaxy (LPE) [22], molecular beam epitaxy (MBE) [23–25], and metalorganic chemical vapor deposition (MOCVD) [26–28] have been previously used to grow InSb films. Compared to homoepitaxial InSb growth, heteroepitaxial growth on GaAs and Si has added appeal because of the possibility for integration of infrared detection and signal processing circuits on the same substrates. However, InSb has a large lattice mismatch with GaAs (14.5%) and Si (19%) and as a result, only limited amount of work has been carried out in this subject. To date, there has been no report of InSb growth on Si by MOCVD.

Recently, high quality InSb films have been grown on InSb, GaAs, and Si substrates using low-pressure MOCVD [29]. Trimethylindium (TMI) and trimethylantimony (TMSb) were used as sources for indium and antimony, respectively. Their respective bubbler temperatures were kept at 18°C and 0°C. In order to find the optimum InSb growth conditions, growth temperature and V/III ratio were varied; the growth temperature was varied from 425°C to 475°C and V/III ratio was varied from 5 tc 15. Based upon the surface morphology of InSb films, V/III ratio of 10 was found to be optimum at a growth temperature of 455°C, as listed in Table 1. Further decrease in V/III ratio degraded the morphology, and indium droplets formed on the surface; the latter phenomenon was confirmed by dissolution of these droplets in hydrochloric acid, which preferentially etches indium [30].

4.2. Structural Properties

Under the optimum conditions, mirror-like InSb epilayers were grown on InSb, GaAs and Si substrates. However, since these films are lattice matched to the InSb substrate, they are expected to yield better overall film quality on InSb than the ones grown on GaAs and Si. This was confirmed from the comparison of x-ray FWHM of approximately 3 μm-thick InSb films grown on InSb, GaAs, and GaAs/Si substrates as shown in Fig. 1. The film grown on InSb had x-ray FWHM of 14 arcsec while films grown on GaAs and GaAs/Si exhibited a FWHM of 171 and 361 arcsec, respectively. As expected, broader FWHM values on GaAs and Si substrates were obtained due to higher dislocation density. Nevertheless, these FWHM values are one of the best reported for InSb films of comparable thickness on the respective substrates independent of the growth technology. An estimate of the dislocation density of the films can be made using the relation [31]:

$$D = \beta^2/9b^2 \tag{10}$$

where D is the actual dislocation density, β is the FWHM in radians, and b is the length of the Burgers vector of the dislocations. This relation is valid when the dominant contribution to the FWHM is due

TABLE 1
Optimum growth conditions for InSb.

Growth temperature	455°C
Growth pressure	76 Torr
TMI flow rate	50 cc/min
TMSb flow rate	20 cc/min
V/III ratio	10
Total H_2 flow rate	1.5 l/min

to dislocations, as in this case. Using the value $b = \sqrt{2} \cdot a$ for 60° dislocations, where $a = 6.48$ Å is the lattice parameter of InSb, $D = 9.1 \times 10^6$ cm^{-2} was obtained for InSb film with FWHM of 171 arcsec. Similarly, for InSb film on Si with FWHM of 361 arcsec, the estimated dislocation density was $D = 4.1 \times 10^7$ cm^{-2}.

X-ray diffraction measurements were also performed on unintentionally doped InSb films grown on GaAs substrate, with film thickness varying from 0.95 µm to 4.85 µm. The results are presented in Fig. 2. It is seen that the x-ray FWHM decreases with increasing film thickness, indicating improved crystalline quality away from the highly mismatched InSb/GaAs interface. This is consistent with the behaviors reported by Biefeld et al. [27] and McConville et al. [32], who have shown the reduction of dislocation density with increasing layer thickness through transmission electron microscopy.

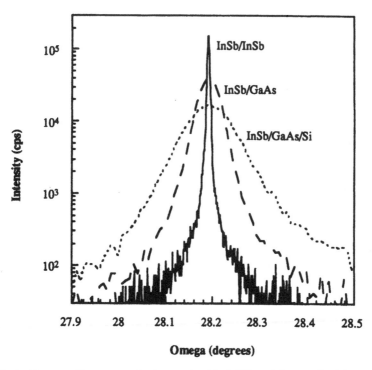

Fig. 1 X-ray rocking curves for homoepitaxial (InSb) and heteroepitaxial (GaAs, GaAs/Si) InSb films.

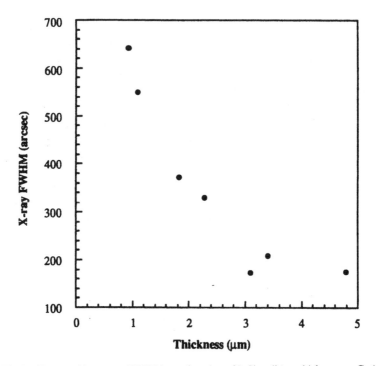

Fig. 2   X-ray rocking curve FWHM as a function of InSb epilayer thickness on GaAs.

4.3. Electrical Properties

Room temperature Hall mobilities for different InSb film thicknesses grown on GaAs substrate also improved with increasing thickness as shown in Fig. 3. These trends again reflect the decrease of dislocation density away from the interface. The Hall coefficient R_H was negative and the carrier concentration of the InSb films was generally in the range of 1×10^{16} cm^{-3} to 3×10^{16} cm^{-3} at 300 K, which is in good agreement with the intrinsic value determined through [33]

$$n_i = 5.63 \times 10^{14} T^{3/2} \exp(-0.127/kT) \qquad (11)$$

where k is the Boltzmann constant and T is the temperature.

As a mean of determining the effect of dislocation on the electrical properties, Hall effect measurements were performed at low temperatures where lattice scattering is reduced and impurity and defect scattering is enhanced. InSb/GaAs samples of different thicknesses were

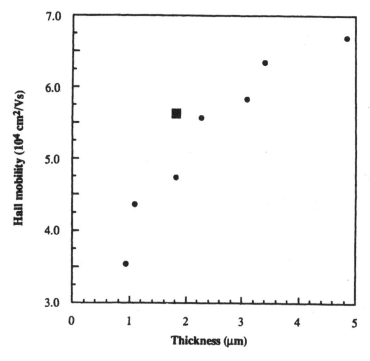

Fig. 3 Room temperature Hall mobility as a function of InSb epilayer thickness on GaAs.

used and the temperature was varied from room temperature down to 4 K. Typical spectra of temperature dependent Hall coefficient and Hall mobility of a InSb/GaAs are given in Fig. 4. The measured Hall data displayed a behavior that could not be explained just by the combined effects of various scattering mechanisms; the Hall coefficient remained negative over entire temperature range (indicating n-type), but the Hall mobility decreased dramatically far below expected values for n-type InSb. In order to explain this behavior, the Hall mobility and Hall coefficient have been simulated using a three layer model [34]. This model is an extension of Petritz's layer model [35] for a three-layer system with a surface electron accumulation layer as postulated by Soderstrom et al. [36], an interface layer with high density of dislocations [37], and a bulk like layer with a highly reduced defect density [32, 38].

This theoretical analysis has shown that the InSb films are p-type and the behavior is attributed to donor-like dislocations caused by the

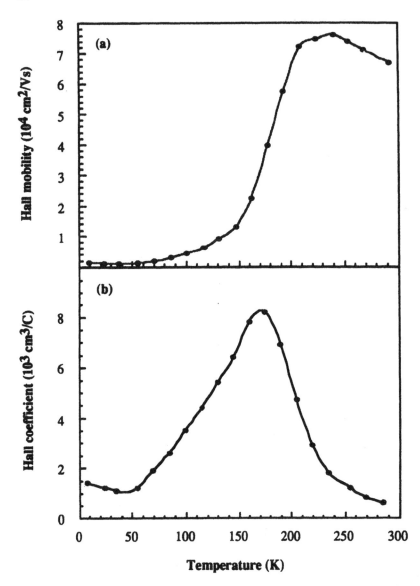

Fig. 4 Temperature dependence of the (a) Hall mobility and (b) Hall coefficient. The sign of the Hall coefficient remained negative over the entire temperature range.

large mismatch and to a surface layer which dominates the transport in the material at low temperatures. At higher temperatures the conductivities of the interface and bulk-like layer rapidly increase due to the ionization of deep-level donors. The effects of the surface layer becomes significant especially in thin layers with high surface conductivity. This keeps the sign of the Hall coefficient negative at low temperatures, and this can lead to a misinterpretation of the measured Hall data unless a multilayer model is considered. The theoretical results of this model provides good agreement with the experimental observations as well as with other studies reporting the growth of p-type InSb on GaAs substrates [32, 36, 39].

In spite of these results, independent studies on these InSb samples transport properties have suggested it might be n-type [40]. This study was based on magnetic field dependence of Hall mobility and magnetoresistance. The magnitude of the observed changes as a function of magnetic field strength was too large to be explained using a conventional two carrier model. By considering the material to be n-type and the existence of a surface conduction layer, the calculated results were consistent with the measured values. These investigations suggest a complex electrical conduction scheme in InSb that needs to be addressed further not only to get a better understand of InSb electrical properties but also to help analyze the electrical properties of its alloys.

4.4. Doping

Doping of InSb is a required step towards realization of photovoltaic detectors, which are preferred over photoconductive devices for focal plane arrays because of the high power usage of photoconductors. One of the most widely used n-type dopant source in III–V semiconductor industry is silane (SiH_4). However, silane and hydrogen selenide have been reported to result in poor surface morphologies and reduced growth rate [41]. These results are attributed to a reaction between the dopant and the metalorganics in the vapor phase. The mechanism of the interaction is unclear and other effects such as impurities in the source could also be the cause of the observed results. Disilane (Si_2H_6) which has a higher pyrolysis efficiency than silane at the growth temperature [42] has also been used in order to investigate its doping efficiency in InSb. No noticeable change in surface morphology or growth rate was observed but there were no conclusive evidence of significant doping levels. Successful n-type doping have been reported using dimethyltellurium (DMTe) [41]. In experiments, in which hydrogen was passed directly through the DMTe bubbler resulted in very high carrier concentration around 10^{19} cm^{-3} and there was no systematic dependence of the concentration on the flow. Through the use of mixtures of DMTe diluted to 10 ppm in hydrogen, doping levels down to 8×10^{16} cm^{-3} have been achieved. However, for reproducible doping results thorough cleaning of the reaction chamber and susceptor after each growth is necessary because of the strong memory effect associated with Te. Another alternative is tetraethyltin (TESn) which has been used to achieve doping levels in the range of 6×10^{15} to 4×10^{18} cm^{-3}. Both Hall and secondary ion mass spectrometry measurements on Sn doped samples indicated that Sn is incorporated as a n-type dopant with very little diffusion, no memory effect and no self-compensation [43]. These qualities of TESn makes Sn a much better doping source than Te or Se.

As a p-type dopant, dimethylzinc (DMZ) was used with its bubbler held at a constant temperature at $-25°C$. DMZ flow, which varied from 2 cc/min and 10 cc/min, were diluted after passing through the bubbler. Dilution flow was kept at 1200 cc/min. The lowest concentration of 5×10^{18} cm^{-3} was obtained with a flow of 2 cc/min. Diethylzinc (DEZ) was also reported to result in very high doping concentrations [41]. Biefeld et al. have reported Cd as an effective p-type dopant in InSb, having a lower incorporation efficiency than Zn in InSb [41]. A similar trend was found for the incorporation efficiency of Zn and Cd in doping experiments in InP growth [44]. Using a mixture of dimethylcadmium (DMCd) diluted to 25 ppm in hydrogen, doping range from 5×10^{15} cm^{-3} to 9×10^{19} cm^{-3} has been realized [41].

5. GROWTH AND PROPERTIES OF InTlSb EPITAXIAL FILMS

5.1 Band and Crystal Structure

The band structure of InTlSb have been investigated by Schilfgaarde et al. [16,45] based on various material parameters of its binary constituents, InSb and TlSb. However, very limited experimental data is available on TlSb. One of the earliest studies was done by Williams

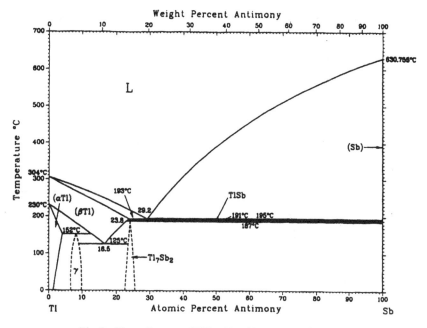

Fig. 5 Phase diagram of TlSb (After Sharma et al. [51]).

in 1906 [46], who first determined the fundamental Tl-Sb binary phase diagram. Since then, other intermediate Tl-Sb phases have been reported [47–51] and they are included in the phase diagram of Fig. 5. As the figure shows, TlSb is found to be stable only in the narrow temperature range from 191°C to 195°C [50] but other TlSb parameters such as the crystal structure are not known.

The material characteristics of TlSb utilized by Schilfgaarde et al. were thus calculated theoretically, from the full-potential linear muffin-tin orbital method (FP-LMTO) within the local density approximation (LDA). The results of the investigation showed that TlSb slightly favors the CsCl over the zinc blende structure at zero temperature and the zinc blende solid solution of InTlSb is stable at low thallium compositions, as illustrated in Fig. 6. Moreover, InTlSb is predicted to be more robust than HgCdTe because of higher cohesive energies of InSb and TlSb compared to those of CdTe and HgTe. In the 8–12 μm wavelength range, InTlSb has stronger bonding InSb as its majority component.

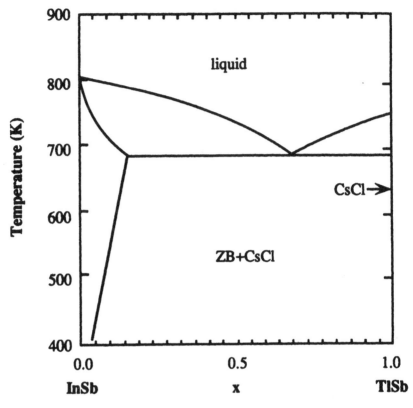

Fig. 6   Phase diagram of $In_{1-x}Tl_xSb$ alloy (After Chen et al. [45]).

The bandgap of TlSb was determined through LDA, which is known to underestimate the semiconductor bandgaps as much as 50% [52,53]. This uncertainty was accounted for by exploiting the systematic nature of the LDA errors; assuming that TlSb has a dielectric constant close to InSb and HgTe, the average of the underestimates of these two were used to obtain the corrected TlSb bandgap of $-1.5\,eV$ [45]. Based on a scaled-virtual crystal approximation, and the bandgaps of TlSb and InSb, the bandgap for $In_{1-x}Tl_xSb$ was calculated at zero temperature as a function of the alloy concentration x. The estimated x value for $0.1\,eV$ bandgap at zero temperature is $x = 0.083$ [45].

TABLE 2
Growth conditions for InTlSb.

Growth temperature	455 °C
Growth pressure	76 Torr
TMI flow rate	50 cc/min
TMSb flow rate	20 cc/min
CPTl flow rate	3–20 cc/min
CPTl bubbler temperature	0–80 °C

5.2. Growth

In addition to the TMI and TMSb sources, cyclopentadienylthallium (CPTl) was used as the thallium source for the growth of InTlSb. CPTl is a solid with low vapor pressure at room temperature and sublimes at 120°C in vacuum (~ 0.1 torr). It has been previously used for growing thallium(III) oxide, a compound that has drawn interest with the discovery of high-temperature superconducting Tl-Ba-Ca-Cu oxide [54]. The thallium flow was introduced into the growth chamber without perturbing other growth parameter settings. This has lead to the first successful growth of InTlSb, exhibiting an extended absorption in the infrared [55].

The flux of thallium was then varied in order to tailor the thallium incorporation and to further extend the absorption edge of the material. As summarized in Table 2, this was achieved by varying the thallium flow rate and bubbler temperature. In this study, a nominal structure was adopted: it consisted in a 1 hour growth of InSb buffer layer followed by a 1 hour growth of InTlSb. All the layers were nominally undoped. A few samples having thicker InTlSb epilayers were grown in order to improve the external quantum efficiency of the photoconductors.

5.3. Optical Properties

Optical transmission measurement was used to determine a shift in the absorption edge of the InTlSb [55, 56]. As shown in Fig. 7, the absorption edge clearly extends beyond that of InSb ($\lambda^{-1} = 1818$ cm^{-1}). However, an accurate assessment of the absorption edge by this method is difficult due to thin epilayers and Fabry-Perot oscillations. These oscillations are produced by multiple reflections at the air-epilayer and substrate-epilayer interfaces, and from their period other interesting optical parameters such as refractive index of the epilayers have been extracted [56].

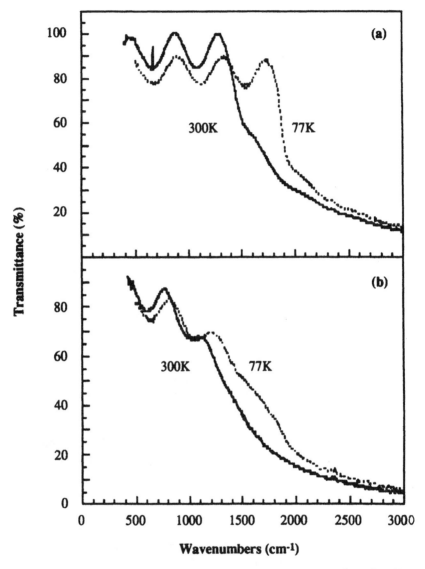

Fig. 7 Optical transmission of the epitaxial layers versus wavenumber (10^4/wavelength) at 77 K and 300 K for (a) InSb layer and (b) InTlSb/InSb layer.

The period of the oscillations yields the phase shift for light passing through the epilayer, from which the product **n·d** can be extracted, where **n** is the refractive index of the epilayer, and **d** is its thickness. Assuming tabulated values for the refractive index of InSb, **n** = 3.9 ± 0.1

within the wavelength range of interest [57], this provides an independent estimate of the InSb epilayer thickness, $\mathbf{d} = 2.9 \pm 0.1\mu m$ which is about the same as the value determined from ball-polishing measurement. Looking at the InTlSb/InSb spectrum, the product $\mathbf{n \cdot d}$ (actually, the sum of the individual products $\mathbf{n \cdot d}$ for each layer) appears slightly larger than for the InSb sample. No refractive index data is available for InTlSb, but since it has a smaller bandgap than InSb its refractive index should be higher. However, this difference is expected to be only slight compared to the large refractive index difference between InSb and GaAs. This is confirmed by the fact that no additional interference feature that might occur from the InSb/InTlSb interface is observed. In brief, the slightly higher value of the product $\mathbf{n \cdot d}$ can originate from a slight increase in the refractive index as well as from a slight difference in total epilayer thickness that would lie within the ball-polishing measurement uncertainty.

A careful observation of Fig. 7 reveals that the product $\mathbf{n \cdot d}$ increases with temperature because the fringe spacing is slightly reduced. By recording precisely the position of the Fabry-Perot extrema and their shift with temperature, the average value of the temperature coefficient of the product $\mathbf{n \cdot d}$ in the transparent wavelength range can be experimentally determined: $1/\mathbf{n \cdot d} \, d(\mathbf{n \cdot d})/dT = 7.7 \times 10^{-5} K^{-1}$ for the InSb sample, and $2.8 \times 10^{-4} K^{-1}$ for the InTlSb/InSb sample. It is apparent from these values that the Fabry-Perot fringes undergo a much larger change with temperature in the case of the InTlSb/InSb sample. Comparing these experimental values with tabulated values for InSb [57]:

$$\frac{1}{\mathbf{n}} \frac{d\mathbf{n}}{dT} = 4.5 \times 10^{-5} K^{-1} \text{ for the temperature coefficient of the refractive}$$

index,

$$\frac{1}{\mathbf{d}} \frac{d\mathbf{d}}{dT} = 0.5 \times 10^{-5} K^{-1} \text{ for the thermal expansion coefficient,}$$

from which we deduce:

$$\frac{1}{\mathbf{n \cdot d}} \frac{d(\mathbf{n \cdot d})}{dT} = \frac{1}{\mathbf{n}} \frac{d\mathbf{n}}{dT} + \frac{1}{\mathbf{d}} \frac{d\mathbf{d}}{dT} = 5.0 \times 10^{-5} K^{-1}.$$

Note that the change in the product $\mathbf{n \cdot d}$ occurs mainly through the variation of the refractive index. The experimental value is in reasonable agreement with the tabulated one in the case of InSb. No such tabulated data is available for InTlSb. One of the possible reasons for the much larger temperature dependence in the case of InTlSb is the enhanced refractive index dispersion near the absorption edge which results in an enhanced dispersion for its temperature coefficient as well. Since fewer fringes are visible in the case of the InTlSb/InSb sample, the experimental value has been averaged over a narrower wavelength region close to the InTlSb absorption edge.

In order to determine the absorption coefficient, the epilayer has been modeled as a conventional Fabry-Perot etalon filled with an absorbing medium [56]. The reflection and transmission coefficient of each interface were determined using tabulated values for the refractive index of GaAs and InSb, and the refractive index of InTlSb was assumed to be the same as that of InSb. The phase shift through the epilayer was determined from the position of the experimental Fabry-Perot extrema. The resulting absorption spectra at 77 K are given on Fig. 8: the InSb sample exhibits a sharp absorption edge at 5.5 μm. The InTlSb/InSb sample exhibits overall higher absorption, and an absorption tail extending beyond InSb absorption edge is observed. It should be noted that the computed absorption coefficient for the InTlSb/InSb bilayer given in Fig. 8 corresponds to the average absorption of the two layers weighted by their respective thicknesses:

$$\alpha = \frac{\alpha_{InTlSb} d_{InTlSb} + \alpha_{InSb} d_{InSb}}{d_{InTlSb} + d_{InSb}}, \tag{12}$$

where α_{InSb} and α_{InTlSb} designate the absorption coefficients of each individual layer. d_{InSb} and d_{InTlSb} correspond to their respective thicknesses. This tends to attenuate any feature of the InTlSb absorption spectrum, and the observed shift should be even more apparent for a thick InTlSb epilayer.

An accurate assessment of the cutoff wavelength of the InTlSb has been obtained through photoresponse measurements [56, 58–60] at 77 K, at which temperature the signal-to-noise ratio is much higher than at room temperature. A characteristic set of normalized photoresponse spectra of various InTlSb alloys is given in Fig. 9. The cut-off wavelength ranges from 5.5 μm to 9 μm for the alloys grown so far. The narrow feature around 4.2 μm wavelength is due to absorption by carbon dioxide in the sample compartment and can be eliminated by proper nitrogen purge. The broader feature around 3 μm wavelength is attributed to electrical resonance within the experimental set-up.

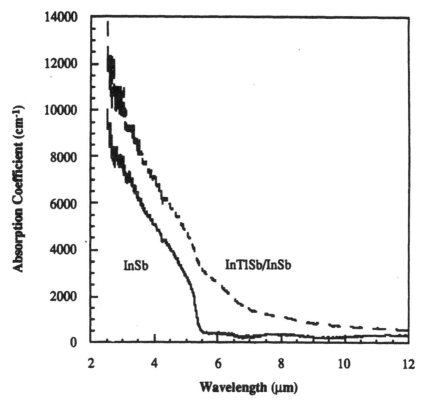

Fig. 8 Optical absorption spectra at 77 K extracted from transmission measurements for both InSb (solid spectrum) and InTlSb/InSb (dashed spectrum).

5.4. Structural Properties

Auger electron spectroscopy was employed as a structural characterization tool to confirm the presence of thallium. From the measured spectra (Fig. 10), a clear peak corresponding to thallium is observed for the InTlSb/InSb sample, confirming the presence of thallium. The absolute composition of the alloy has not been determined since a InTlSb with known Tl concentration is not available for calibration purpose. However, a rough estimate of the alloy stoichiometry can be obtained assuming a low temperature bandgap of -1.5 eV for TlSb and a linear dependence of the bandgap on thallium content [16]. A cutoff wavelength shift from 5.5µm for InSb up to 9µm for $In_{1-x}Tl_xSb$

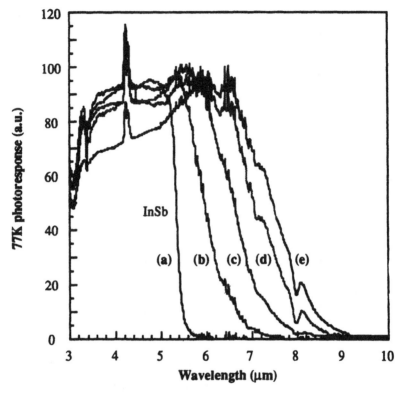

Fig. 9 Normalized photoconductor spectral response of various $In_{1-x}Tl_xSb$ alloys at 77K. The estimated Tl contents are approximately (a) $x = 0$, (b) $x = 0.028$, (c) $x = 0.041$ (d) $x = 0.046$, and (e) $x = 0.050$.

corresponds to an approximately 88meV bandgap reduction and to an estimated $x = 0.050$ thallium content (see Fig. 9).

Morphology of the samples were poorer than our best InSb results and this is attributed to the non-optimum growth conditions of InTlSb films. This was reflected in the FWHM of the InTlSb x-ray peak at (800) orientation, where InSb and InTlSb peaks are clearly resolved. Figure 11 shows a typical (800) x-ray profile. As the thallium flow is increased, the InTlSb peak broadens and shifts away from the InSb peak. Increase in dislocation density, as a result of the increase in mismatch with respect to InSb, is believed to be the cause of the InTlSb peak broadening. This·assertion is supported by the plot of the InTlSb x-ray FWHM as a function of lattice-mismatch between InTlSb and InSb (determined from the x-ray profile), shown in Fig. 12.

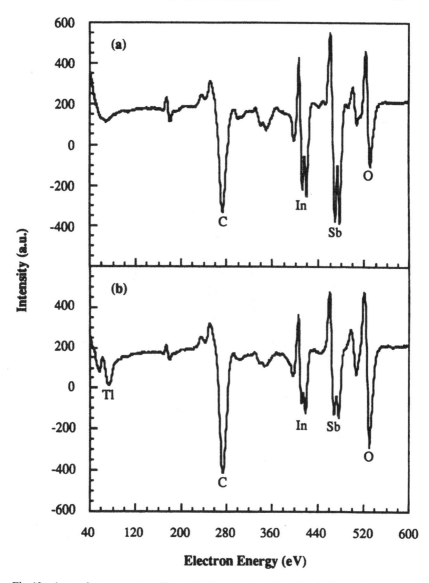

Fig. 10 Auger electron spectra of the (a) InSb and of the (b) InTlSb/InSb sample.

The reduction in lattice constant is attributed to increasing thallium content. Surprisingly, a lattice contraction is observed instead of the predicted dilation [60]. Figure 13 shows the quasi-linear relationship between the lattice mismatch and the cutoff wavelength (determined

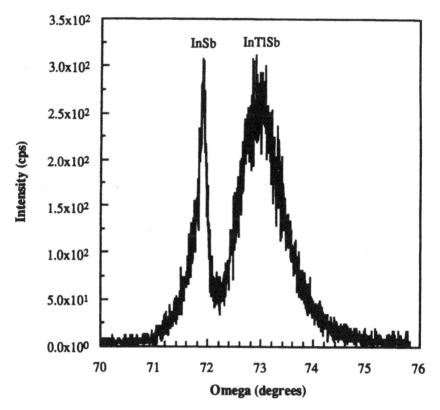

Fig. 11 X-ray diffraction rocking curve of an InTlSb sample. This was measured at (800) diffraction orientation. In addition to the InSb buffer layer peak, a peak corresponding to InTlSb is resolved.

as the wavelength for which photoresponse falls at 10% of its maximum value). This suggests that the photoresponse cutoff wavelength can be increased further by increasing thallium content. This is in marked contrast with InAsSb which exhibits a minimum bandgap as a function of As content. Furthermore, a 9 μm cutoff wavelength corresponds to a lattice mismatch of -1.3% which is significantly smaller than that for InAsSb alloy having a similar cutoff wavelength. This is important for the development of device-quality material with reduced dislocation.

Effects of thermal annealing have also been carried out to assess the thermal stability of InTlSb. A number of InTlSb/InSb samples were annealed at 350°C under N_2 ambient for 15 minutes, 30 minutes, and

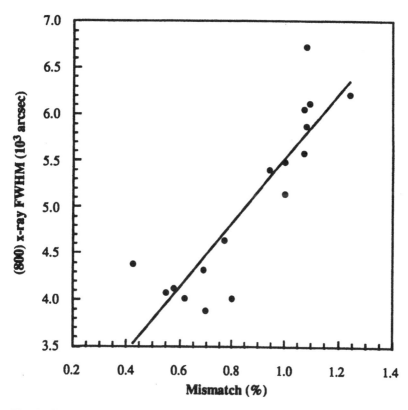

Fig. 12 Relationship between $In_{1-x}Tl_xSb$ ($0 \leqslant x \leqslant 0.05$) x-ray FWHM and its lattice mismatch with InSb.

1 hour. All of the InTlSb samples tested were found to be thermally stable showing no signs of degradation in terms of surface morphology, x-ray diffraction rocking curve, and photoresponse. This is encouraging in light of the doubts that had arisen about the feasibility of InTlSb alloys, based on phase-diagram considerations, which indicate that TlSb was highly unstable [50]. This is an added advantage over HgCdTe, in which high Hg diffusion coefficient decreases the stability of its devices.

5.5. Electrical Properties

A number of trends have been observed in the electrical properties of InTlSb. Hall measurements indicated a room temperature Hall

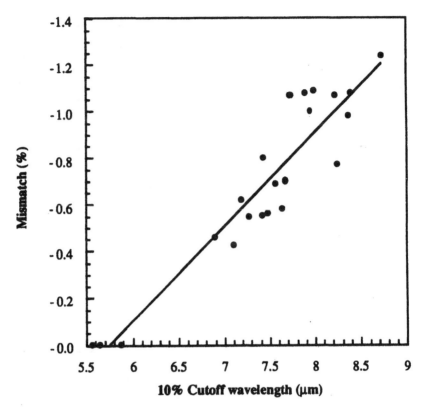

Fig. 13 Lattice mismatch between $In_{1-x}Tl_xSb$ $(0 \leqslant x \leqslant 0.05)$ and InSb versus 77 K photoresponse cutoff wavelength (defined as the wavelength for which the photoresponse falls at 10% of its maximum value).

mobility ranging from 50,000 cm²/Vs to 20,000 cm²/Vs, and an electron concentration ranging from 1×10^{16} cm⁻³ to 5×10^{16} cm⁻³ for the nominal InTlSb samples, as shown in Fig. 14. The Hall mobility decreased monotonically with increasing thallium flow, while the electron concentration simultaneously increased. The increase in electron concentration is typical of an intrinsic semiconductor with decreasing bandgap. The decrease in mobility is attributed to an increase in alloy scattering. Measurements at 77 K indicate an opposite trend for the mobility and scattered data for the electron concentration (Fig. 15). This behavior is likely to occur because of the dominant contribution of parallel conduction layers at low temperature, as evidenced in InSb epilayers through temperature-dependent Hall measurements.

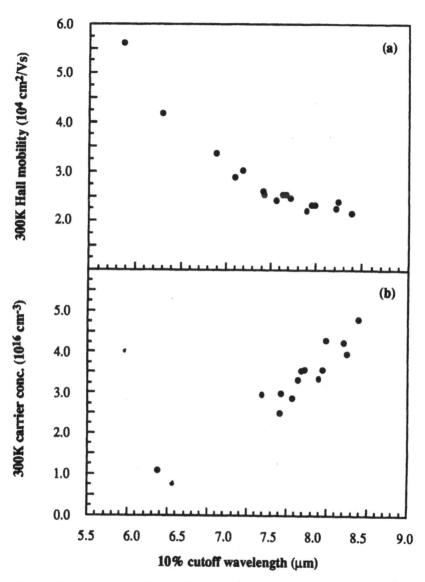

Fig. 14 Room temperature Hall mobility (a) and carrier concentration (b) as a function of 77 K photoresponse cutoff wavelength.

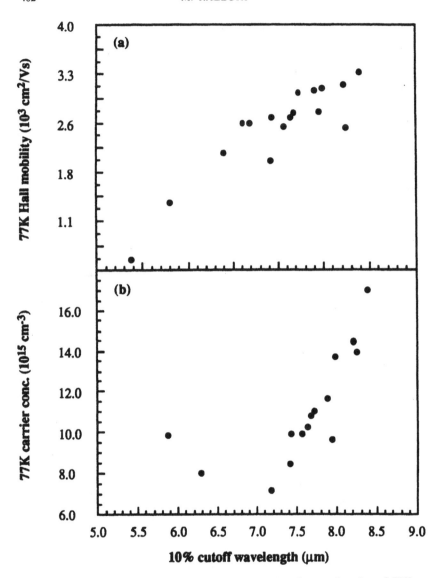

Fig. 15 77 K Hall mobility (a) and carrier concentration (b) as a function of 77 K photoresponse cutoff wavelength.

5.6. Performance of InTlSb Photoconductors

For the InTlSb/InSb having the standard nominal structure, photo-conductivity measurements were carried out using the same experimental set-up and calculations described previously. The detectivity was determined at 77 K, at which temperature the generation-recombination noise predominates over Johnson noise [21]. Hence, the specific detectivity can be expressed as

$$D^* = \frac{\Re}{\sqrt{4eGi\Delta f}} \sqrt{A} \sqrt{\Delta f}. \tag{13}$$

As the thallium flow was varied, the specific detectivity D^* at 7μm increased from 10^8 cmHz$^{1/2}$W^{-1} to 3×10^8 cmHz$^{1/2}$W^{-1} [58–60]. This results mainly from the increase in quantum efficiency. This quantum efficiency is improved in the case of the thick InSb/InTlSb/InSb sample grown under the condition that yielded a 9μm cutoff wavelength. X-ray diffraction pattern and photoresponse spectrum were quite similar to those of the thin structure, thus indicating good composition control. The room temperature Hall mobility increased from 2×10^4 cm^2/Vs for the thin structure to 3×10^4 cm^2/Vs for the thicker one. This improvement in material quality was confirmed by an increase in the photoconductive gain over bias ratio G/V. The thicker sample presents both a higher absorption quantum efficiency and a higher photoconductive gain which results in a significantly larger specific detectivity $D^* = 1 \times 10^9$ cmHz$^{1/2}$W^{-1} at $\lambda = 7$ μm.

6. CONCLUSION

High quality InSb films and novel InTlSb alloys of various compositions have been successfully grown by low pressure metalorganic chemical vapor deposition as preliminary steps towards fabrication of InTlSb infrared detectors. The films have been studied using various structural, electrical, and optical characterization methods. The results of these characterizations of homoepitaxial and heteroepitaxial InSb films on InSb, GaAs, and Si substrates have proven to be one of the best reported so far for InSb films.

The photoresponse cutoff wavelength of the InTlSb alloys were tailored from 5.5 μm up to 9 μm. The lattice mismatch between InTlSb and InSb for the latter cutoff wavelength is only − 1.3%. Experimental observations indicate that the bandgap can be further decreased by increasing the thallium content. A thick InTlSb epilayer was grown and a photoconductor was fabricated out of it. Owing to both an increase in absorption quantum efficiency and an overall increase in material quality, the specific D* increased significantly; at $\lambda = 7\mu m$, $D^* = 10^9$ cmHz$^{1/2}$W^{-1} for a photoconductor having a 9 μm cutoff wavelength.

Additional research is being currently pursued to increase the thallium incorporation and hence extend the cutoff wavelength. Doping of InTlSb is also being investigated to realize photovoltaic detectors. Nevertheless, the presented results outline the promise of InTlSb alloy as an infrared detection material to challenge the industry standard HgCdTe.

Acknowledgments

This work has been supported by the Office of Naval Research under contract number N00014-92-J-1951 and Strategic Defense Initiative Organization under contract number N00014-93-0409. The author thanks Dr. G. Wright and Dr. Y.S. Park of the Office of Naval Research and Dr. R. Balcerak of the Advanced Research Projects Agency for their support and encouragement. Assistance of Y.H. Choi and Dr. E. Bigan in this work is also acknowledged.

References

1. R. Balcerak, *Semicond. Sci. Technol.*, **6**, C1 (1991).
2. C.G. Bethea, B.F. Levine, M.Y. Yen, and A.Y. Cho, *Appl. Phys. Lett.*, **53**, 291 (1988).
3. G.C. Osbourn, *J. Vac. Sci. Technol.*, **B2**, 176 (1984).
4. S.R. Kurtz, L.R. Dawson, R.M. Biefeld, I.J. Fritz, and T.E. Zipperian, *IEEE Elec. Dev. Lett.*, **10**, 150 (1989).
5. D.L. Smith and C. Mailhiot, *J. Appl. Phys.*, **62**, 2545 (1987).
6. G.A. Sai-Halasz, R. Tsu, and L. Esaki, *Appl. Phys. Lett.*, **30**, 651 (1977).
7. J.M. Kuo, S.S. Pei, S. Hui, S.D. Gunapala, and B.F. Levine, *J. Vac. Sci. Technol.*, **B10**, 995 (1992).
8. B.F. Levine, C.G. Bethea, G. Hasnain, J. Walker, and R.J. Malik, *Appl. Phys. Lett.*, **53**, 296 (1988).
9. K. Goossen, S.A. Lyon, and K. Alavi, *Appl. Phys. Lett.*, **53**, 1027 (1988)
10. B.F. Levine, S.D. Gunapala, J.M. Kuo, S.S. Pei, and S. Hui, *Appl. Phys. Lett.*, **59**, 1864 (1991).
11. A.M. Jean-Louis and C. Hamon, *Phys. Status Solid*, **34**, 329 (1969).
12. A.J. Noreika, W.J. Takei, M.H. Francombe, and C.E.C. Wood, *J. Appl. Phys.*, **53**, 4932 (1982).
13. T.P. Humphreys, P.K. Chiang, S.M. Bedair, and N.R. Parikh, *Appl. Phys. Lett.*, **53**, 142 (1988).
14. K.Y. Ma, Z.M. Fang, D.H. Jaw, R.M. Cohen, G.B. Stringfellow, W.P. Kosar, and D.W. Brown, *Appl. Phys. Lett.*, **55**, 2420 (1989).

15. C.E.C. Wood, A. Noreika, and M. Francombe, *J. Appl. Phys.*, **59**, 3610 (1986).
16. M. van Schilfgaarde, A. Sher, and A.B. Chen, *Appl. Phys. Lett.*, **62**, 1857 (1993).
17. M. Razeghi, "The MOCVD Challenge," Adam Hilger, London (1989).
18. S.J. Bass, *J. Crystal Growth*, **31**, 172 (1975).
19. R.A. Smith, "Semiconductors," Cambridge University Press, Cambridge, (1959).
20. R.M. Broudy and V.J. Mazurczyk, "(HgCd)Te photoconductive detectors," Semiconductors and Semimetals, vol. **18**, 157, edited by R.K. Wilardson and A.C. Beer, Academic Press, (1981).
21. A. Rose, "Concepts in photoconductivity and allied problems", Wiley, New York, (1963).
22. D.E. Holmes and G.S. Kamath, *J. Electron. Mater.*, **9**, 95 (1980).
23. P.E. Thompson, J.L. Davis, J. Waterman, R.J. Wagner, D. Gammon, D.K. Gaskill, and R. Stahlbush, *J. Appl. Phys.*, **69**, 7166 (1991).
24. J.-I. Chyi, D. Biswas, S.V. Iyer, N.S. Kumar, H. Morkoc, R. Bean, K. Zanio, H.-Y. Lee, and H. Chen, *Appl. Phys. Lett.*, **54**, 1016 (1989).
25. J.L. Davis and P.E. Thompson, *Appl. Phys. Lett.*, **54**, 2235 (1989).
26. P.K. Chiang and S.M. Bedair, *J. Electrochem. Soc.*, **131**, 2422 (1984).
27. R.M. Biefeld and G.A. Hebner, *J. Crystal Growth* **109**, 272 (1991).
28. D.K. Gaskill, G.T. Stauf, and N. Bottka, *Appl. Phys. Lett.*, **58**, 1905 (1991).
29. Y.H. Choi, R. Sudharsanan, C. Besikci, E. Bigan, and M. Razeghi, *Mat. Res. Soc. Symp. Proc.*, **281**, 375 (1993).
30. R.M. Biefeld, *J. Crystal Growth*, **75**, 255 (1986).
31. N. Chand, J. Allam, J.M. Gibson, F. Capasso, F. Beltram, A.T. Macrander, A.L. Hutchinson, L.C. Hopkins, C.G. Bethea, B.F. Leviine, and A.Y. Cho, *J. Vac. Sci. Technol.*, **B5**, 822 (1987).
32. C.F. McConville, C.R. Whitehouse, G.M. Williams, A.G. Cullis, T. Ashley, M.S. Skolnick, G.T. Brown and S.J. Courtney, *J. Crystal Growth*, **95**, 228 (1989).
33. Y.J. Jung, M.K. Park, S.I. Tae, K.H. Lee, and H.J. Lee, *J. Appl. Phys.*, **69**, 3109 (1991).
34. C. Besikci, Y.H. Choi, R. Sudharsanan, and M. Razeghi, *J. Appl. Phys.*, **73**, 5009 (1993).
35. R.L. Petritz, *Phys. Rev.*, **110**, 1254 (1958).
36. J.R. Soderstrom, M.M. Cumming, J-Y. Yao, and T.G. Andersson, *Semicond. Sci. Technol.*, **7**, 337 (1992).
37. M. Yano, T. Takase, and M. Kimata, *Phys. Status Solidi* **A54**, 707 (1979).
38. A.J. Noreika, J. Greggi, Jr., W.J. Takei, and M.H. Francombe, *J. Vac. Sci. Technol.* **A1**, 558 (1983).
39. G.M. Williams, C.R. Whitehouse, C.F. McConville, A.G. Cullis, T. Ashley, S.J. Courtney, and C.T. Elliot, *Appl. Phys. Lett.*, **53**, 1189 (1988).
40. S.N. Song, J.B. Ketterson, Y.H. Choi, R. Sudharsanan, and M. Razeghi, *Appl. Phys. Lett.*, **63**, 964 (1993).
41. R.M. Biefeld, S.R. Kurtz, and I.J. Fritz, *J. of Electron. Mat.*, **18**, 775 (1989).
42. G.B. Stringfellow, "Organometallic Vapor-Phase Epitaxy: Theory and Practice," Academic Press (1989)
43. R.M. Biefeld, J.R. Wendt, and S.R. Kurtz, *J. of Crystal Growth*, **107**, 836 (1991).
44. A.W. Nelson and L.D. Westbrook, *J. Crystal Growth*, **68**, 102 (1984).
45. A.-B. Chen, M. van Schilfgaarde, and A. Sher, *J. of Electron. Mater.*, **22**, 843 (1993).
46. R.S. Williams, *Z. Anorg. Allg. Chem.*, **50**, 127–132 (1906).
47. T. Barth, *Z. Phys. Chem.*, **127**, 113 (1927).
48. E. Persson and A. Westgren, *Z. Phys. Chem.*, **136**, 208 (1928).
49. R.W.G. Wyckoff, "Crystal Structures," *Vol.* **1**, 2nd ed. (1963).
50. B. Predel and W. Schwermann, *Z. Naturforsch.* **A25**, 877–886 (1970).
51. R.C. Sharma and Y.A. Chang, "Binary Alloy Phase Diagrams,' 2nd edition, edited by T.B. Massalski, 3312–3314 (1990).
52. L.J. Sham and M. Schluter, *Phys. Rev. Lett.*, **51**, 1888 (1983).

53. M.T. Yin and M.L. Cohen, *Phys. Rev.*, **B26**, 5668 (1982).
54. A.D. Berry, R.T. Holm, R.L. Mowery, N.H. Turner, and M. Fatemi, *Chem. Mater.*, **3**, 72 (1991).
55. Y.H. Choi, C. Besikci, R. Sudharsanan, and M. Razeghi, *Appl. Phys. Lett.*, **63**, 361 (1993).
56. Y.H. Choi, P.T. Staveteig, E. Bigan, and M. Razeghi, *J. Appl. Phys.*, Vol. 75, 3196, (1994).
57. Landolt-Bornstein, "Numerical Data and Functional Relationships in Science and Technology," Vol. **17**, 310, Springer-Verlag, New York (1982).
58. P.T. Staveteig, Y.H. Choi, G. Labeyrie, E. Bigan, and M. Razeghi, *Appl. Phys. Lett.* Vol. 64, 460 (1994).
59. M. Razeghi, Y.H. Choi, P.T. Staveteig, and E. Bigan, Proceedings of the 184th Meeting of the Electrochemical Society, vol. 94, 147 (1993).
60. E. Bigan, Y.H. Choi, G. Labeyrie, and M. Razeghi, Proceedings of SPIE, Vol. 2145, 2 (1994).

Index

487

For Product Safety Concerns and Information please contact our EU
representative GPSR@taylorandfrancis.com Taylor & Francis Verlag GmbH,
Kaufingerstraße 24, 80331 München, Germany

Printed and bound by CPI Group (UK) Ltd, Croydon, CR0 4YY
01/05/2025
01858556-0004